AF255291

ALGEBRAIC GEOMETRY

GEORGE BELL AND SONS
LONDON: PORTUGAL ST., LINCOLN'S INN
CAMBRIDGE: DEIGHTON, BELL & CO.
NEW YORK: THE MACMILLAN CO.
BOMBAY: A. H. WHEELER & CO.

ALGEBRAIC GEOMETRY

A NEW TREATISE ON
ANALYTICAL CONIC SECTIONS

BY

W. M. BAKER, M.A.

HEADMASTER OF THE MILITARY AND CIVIL DEPARTMENT
AT CHELTENHAM COLLEGE

LONDON
GEORGE BELL & SONS
1906

GLASGOW: PRINTED AT THE UNIVERSITY PRESS
BY ROBERT MACLEHOSE AND CO. LTD.

PREFACE

THE recent reforms in the teaching of Mathematics will, it is hoped, be considered sufficient justification for the appearance of a new treatise on Algebraic Geometry, and attention may be called to the following points which differentiate the present volume from its predecessors on the same subject:

As it is written for beginners, and for the average boy, the straight line and the circle are very fully treated. The elementary ideas of the Calculus have been utilised, and full use made of one of the best of recent reforms, viz., the abolition of the water-tight compartment between Geometry and Algebra.

The examples are numerous and varied, and a number of new types have been introduced, such as have been set in recent Examinations. Revision questions which may be taken orally, and Revision papers, are also given at various stages.

The use of squared paper is encouraged, and another special feature of the work is the very free use of diagrams.

All the most important properties of the Conic Sections are proved, either by Algebra or by Geometry, and the book claims to be complete as far as, and including the use of Cartesian and Polar co-ordinates.

The author's thanks are due to Mr. J. M. Dyer, of Eton, for permission to use examples from his collection of Mathematical Examples, to the Controller of His Majesty's Stationery Office for permission to use examples set in recent examinations held by the Civil Service Commissioners, and to Mr. T. Hyett, of Cheltenham College, for valuable suggestions and help.

W. M. B.

May, 1906.

CONTENTS

CHAPTER   PAGE

I. RECTANGULAR CO-ORDINATES -   -   -   -   -   -   -   1

II. THE STRAIGHT LINE   -   -   -   -   -   -   -   15

III. PAIRS OF STRAIGHT LINES, ETC. -   -   -   -   -   49

IV. OBLIQUE AXES. POLAR CO-ORDINATES -   -   -   -   60

V. THE CIRCLE. (RECTANGULAR AXES.)   -   -   -   -   74

VI. REVISION PAPERS -   -   -   -   -   -   -   100

VII. ORTHOGONAL CIRCLES. RADICAL AXIS. THE CIRCLE
(OBLIQUE AXES). POLAR EQUATION OF A CIRCLE   -   104

VIII. THE PARABOLA   -   -   -   -   -   -   -   119

IX. REVISION PAPERS -   -   -   -   -   -   -   165

X. THE ELLIPSE -   -   -   -   -   -   -   -   170

XI. REVISION PAPERS -   -   -   -   -   -   -   230

XII. THE HYPERBOLA   -   -   -   -   -   -   234

XIII. POLAR EQUATION OF A CONIC SECTION -   -   -   282

XIV. CURVE TRACING   -   -   -   -   -   -   294

XV. PLANE SECTIONS OF A CONE -   -   -   -   -   313

XVI. REVISION PAPERS -   -   -   -   -   -   -   318

ALGEBRAIC GEOMETRY FOR BEGINNERS.

CHAPTER I.

IN Algebraic, or Analytical, Geometry, we enlist the services of Algebra in aid of Geometry.

By means of algebraic symbols we shall show how the position of a point in a plane may be determined, and how equations may be taken to represent straight lines or curves. From equations we shall determine the geometrical properties of the curves which they represent, and how a curve may be drawn to scale when its equation is known.

It is hoped that these ideas will not be altogether unfamiliar to the student. He has probably learnt the geometrical interpretation of certain algebraic formulae, such as

$$(a + b)(a - b) = a^2 - b^2,$$

and how to draw the *graphs* of simple and quadratic algebraic expressions.

In his study of this subject, above all things let the beginner remember that he is dealing with Algebra in combination with Geometry. Let it therefore become a constant habit with him to illustrate his algebraic work with geometrical figures. By so doing he will remove many difficulties from his path. Moreover, the results, as seen from a well-drawn geometrical diagram, will often serve to corroborate the algebraic work.

RECTANGULAR CO-ORDINATES.

1. Take two straight lines xOx', yOy' at right angles to one another.

Let P be any point in their plane, and in the quadrant xOy, and draw PN, PM perpendicular to Ox and Oy respectively.

$$\text{Let } PM = x \text{ and } PN = y.$$

PM and PN are called the co-ordinates of P.

These values x and y define the position of the point P in the quadrant xOy; so that if we are given the values of x and y, we can find or place the point P.

For instance, if it is required to find the point indicated by the values $x = 3$ and $y = 4$, along Ox measure $ON = 3$, and along Oy measure $OM = 4$ units of length.

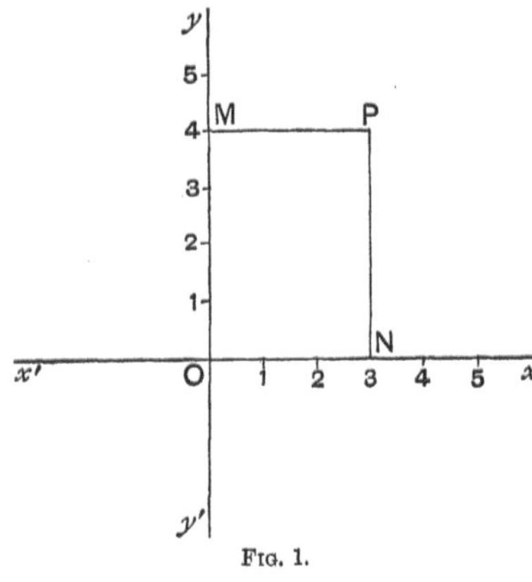

Fig. 1.

Draw MP parallel to Ox, and NP parallel to Oy. P being the point at which they intersect.

Then $PM = ON = 3$, and $PN = OM = 4$, and therefore P is the point required.

x and y are the **co-ordinates** of the point P.

xOx', yOy' are called the **axes of co-ordinates**, or more shortly the **axes**, O the **origin**.

P is usually described as the point (x, y).

x is called the **abscissa**, and y the **ordinate** of the point P.

2. If lines drawn in one direction are taken as positive, then lines drawn in the opposite direction must be taken as negative.

Lines drawn in the directions Ox, Oy are considered positive, and therefore lines drawn in the directions Ox', Oy' are taken as negative.

For example, in the accompanying figure, at Q the abscissa is negative, and the ordinate positive. At R, the abscissa is negative, and also the ordinate. At S the abscissa is positive, and the ordinate negative.

The student will see that this **convention of signs** is the same as in Trigonometry.

In practice, it is simplest to draw the point $(3, 4)$ in the following way :

Along Ox measure $ON = 3$; and at N draw NP perpendicular to ON in the direction Oy, the positive direction, and make

NP $= 4$. We then have the same point as in the paragraph above.

Note that the *signs* of the co-ordinates determine the quadrant in which a point lies.

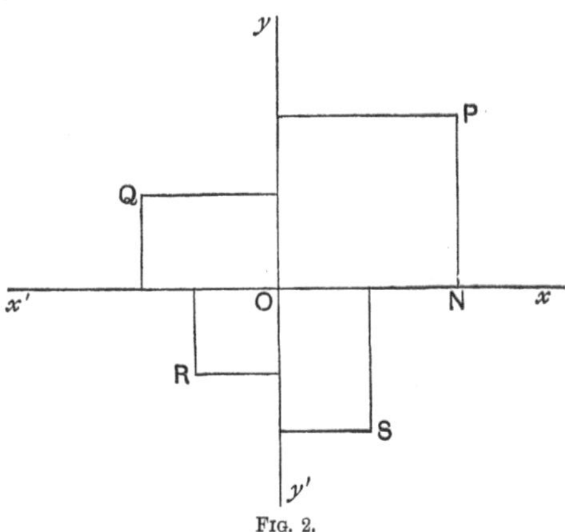

Fig. 2.

Thus the point $(-2, 3)$ lies in the second quadrant $y\mathrm{O}x'$, for its abscissa is negative and its ordinate positive.

The point $(-3, -2)$ lies in the third quadrant $x'\mathrm{O}y'$.

Thus when the co-ordinates of a point are given with their proper signs, we can determine the exact position of the point in the plane of $x\mathrm{O}x'$ and $y\mathrm{O}y'$.

3. *The distance of the point (x, y) from the origin is equal to*

$$\sqrt{\mathbf{x}^2 + \mathbf{y}^2}.$$

Let P be the point. Draw the ordinate PN, so that

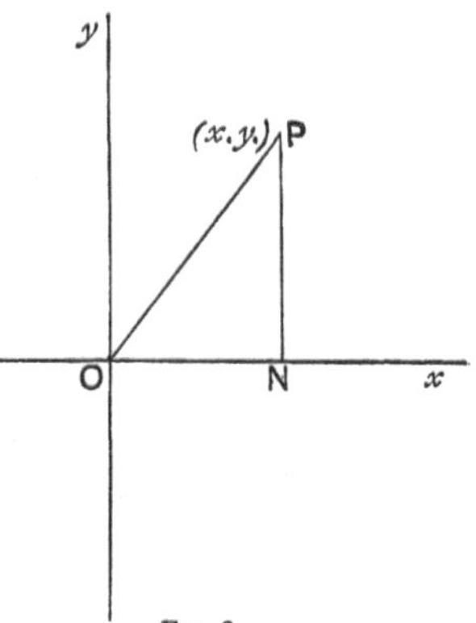

$$\mathrm{ON} = x, \quad \mathrm{PN} = y,$$

$$\mathrm{OP}^2 = \mathrm{ON}^2 + \mathrm{PN}^2 = x^2 + y^2 \; ;$$

$$\therefore \; \mathrm{OP} = \sqrt{x^2 + y^2}.$$

Fig. 3.

4. *The distance between the points (x_1, y_1), (x_2, y_2) is equal to*

$$\sqrt{(x_1 - x_2)^2 + (y_1 - y_2)^2}.$$

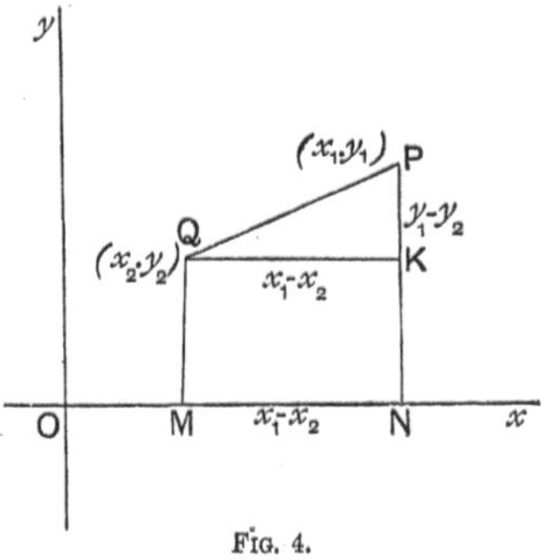

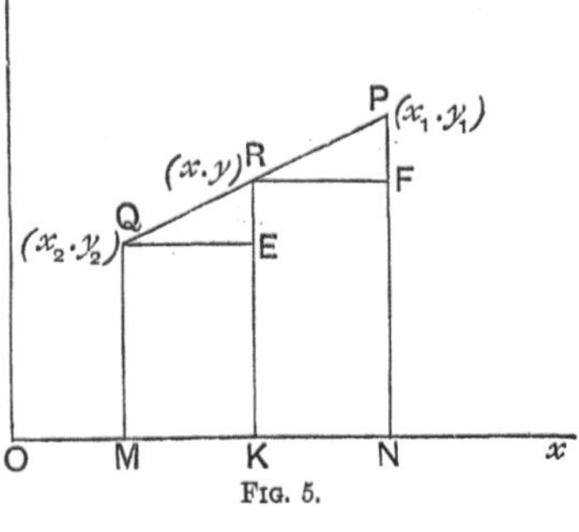

FIG. 4.

Let P be the point (x_1, y_1), Q the point (x_2, y_2).

Draw the ordinates PN, QM, and draw QK perpendicular to PN.

$$QK = MN = ON - OM = x_1 - x_2,$$

$$PK = PN - KN = PN - QM$$

$$= y_1 - y_2 ;$$

$$\therefore \ PQ^2 = QK^2 + PK^2$$

$$= (x_1 - x_2)^2 + (y_1 - y_2)^2$$

and $\ PQ = \sqrt{(x_1 - x_2)^2 + (y_1 - y_2)^2}.$

5. *The co-ordinates of the middle point of the line joining the points* (x_1, y_1), (x_2, y_2) *are*

$$\frac{x_1 + x_2}{2}, \ \frac{y_1 + y_2}{2}.$$

Let P, Q be the points (x_1, y_1), (x_2, y_2), R the middle point of PQ. Draw the ordinates PN, QM, RK. Also draw QE perpendicular to RK, and RF perpendicular to PN.

Let (x, y) be the co-ordinates of R.

From the similar $\triangle^s$ PFR, REQ,

$$\frac{PF}{PR} = \frac{RE}{QR}.$$

But $\ PR = QR. \ \therefore \ PF = RE.$

Also $\ PF = PN - FN$

$$= PN - RK = y_1 - y.$$

And $\ RE = RK - EK$

$$= RK - QM = y - y_2.$$

$$\therefore \ y_1 - y = y - y_2,$$

$$2y = y_1 + y_2,$$

$$y = \frac{y_1 + y_2}{2}.$$

FIG. 5.

In the same way $QE = FR,$

$$i.e. \quad OK - OM = ON - OK;$$
$$\therefore \quad x - x_2 = x_1 - x,$$
$$2x = x_1 + x_2,$$
$$x = \frac{x_1 + x_2}{2}.$$

Note that R is the Centre of Gravity of equal particles at P and Q.

6. *To find the co-ordinates of the point dividing in a given ratio the line joining two given points.*

Let $P(x_1, y_1)$, $Q(x_2, y_2)$ be the given points, and let the point R divide PQ so that $\dfrac{PR}{QR} = \dfrac{a_1}{a_2}$.

Let (x, y) be the co-ordinates of R.

It is required to find x and y in terms of the co-ordinates of P and Q and a_1, a_2.

Draw the ordinates PN, QM, RK. Also draw QE, RF perpendicular to RK and PN.

From the similar $\triangle^s$ PFR, REQ,

$$\frac{PF}{PR} = \frac{RE}{RQ} \quad i.e. \quad \frac{PF}{a_1} = \frac{RE}{a_2}.$$

But $\quad PF = PN - FN$
$$= PN - RK = y_1 - y,$$

and $\quad RE = RK - EK$
$$= RK - QM = y - y_2,$$

$$\therefore \quad \frac{y_1 - y}{a_1} = \frac{y - y_2}{a_2},$$

$$a_2 y_1 - a_2 y = a_1 y - a_1 y_2$$

and $\quad y = \dfrac{a_2 y_1 + a_1 y_2}{a_1 + a_2}.$

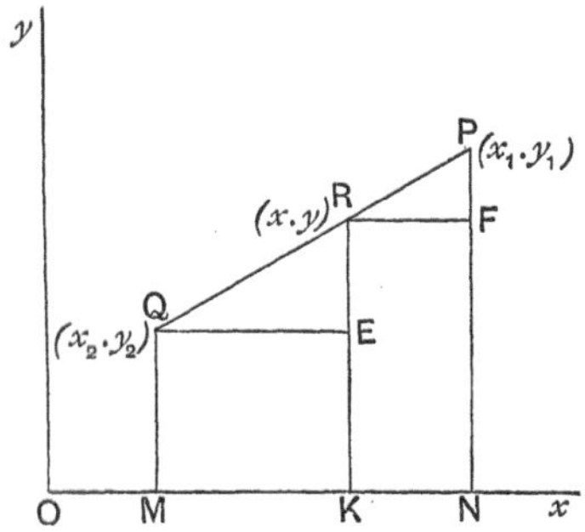

In the same way,

$$\frac{QE}{a_2} = \frac{RF}{a_1} \quad i.e. \quad \frac{x - x_2}{a_2} = \frac{x_1 - x}{a_1}$$

and $\quad x = \dfrac{a_2 x_1 + a_1 x_2}{a_1 + a_2}.$

Fig. 6.

Note that R is the Centre of Gravity of particles of masses a_2, a_1, placed at P and Q respectively; or R is the point of application of the resultant of *like* parallel forces of magnitudes a_2, a_1 at P and Q respectively.

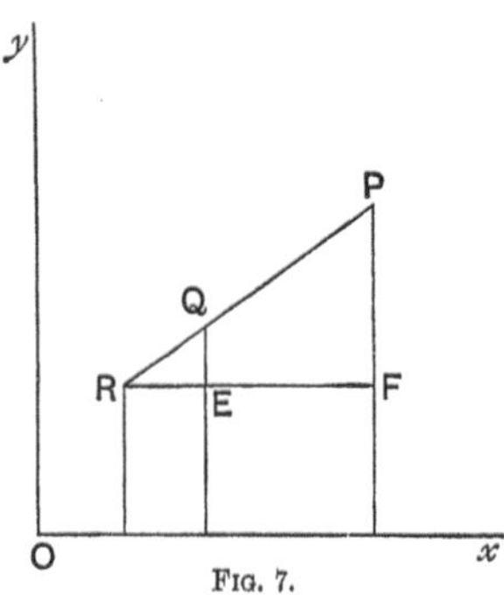

If R divides PQ *externally*, so that $\dfrac{PR}{QR}=\dfrac{a_1}{a_2}$ as in the figure, we shall have

$$\frac{PF}{PR}=\frac{QE}{QR},\quad \frac{y_1-y}{a_1}=\frac{y_2-y}{a_2}$$

and $\quad y=\dfrac{a_1y_2-a_2y_1}{a_1-a_2}.$

Similarly

$$x=\frac{a_1x_2-a_2x_1}{a_1-a_2}.$$

Fig. 7.

Note that R is the point of application of the resultant of unlike parallel forces of magnitudes a_2 and a_1 at P and Q respectively.

Examples I. a.

1. Read off the co-ordinates of the points P_1, P_2 ... shown in the diagram on the next page (Fig. 8).

2. In what line does a point lie (1) if its abscissa is zero, (2) if its ordinate is zero?

3. Find the co-ordinates of the middle point of the line joining the points (3, 4), (5, 2). Verify your result by means of a diagram on squared paper.

4. What are the co-ordinates of the middle point of the line joining the points (3, 5), (-3, -5)? Verify on squared paper.

5. Construct the triangle whose vertices are the points (2, 2), (-4, 2), (0, -4), using a half-inch unit on squared paper.

6. Find the co-ordinates of the middle points of the sides of the triangle in the previous question.

7. Find the length of the line joining the points (6, 5), (3, 1). Verify on squared paper.

8. Find the length of the line joining the points (13, 2), (1, -3). Verify on squared paper.

Find the lengths of the straight lines joining the following pairs of points :

9. (o, o), (a, b). **10.** (a, b), $(-a, -b)$.

11. (a, b), $(a, -b)$. **12.** $(a\cos\theta, a\sin\theta)$, $(b\cos\theta, b\sin\theta)$.

13. $\left(\dfrac{a}{m_1{}^2},\ \dfrac{2a}{m_1}\right),\ \left(\dfrac{a}{m_2{}^2},\ \dfrac{2a}{m_2}\right).$ **14.** $(a\cos\theta,\ b),\ (a,\ b\cos\theta).$

15. $(6, 3),\ (2, 1)$ correct to two decimal places.

16. $(-3,\ -4),\ (-1,\ 5)$,,   ,,   ,,

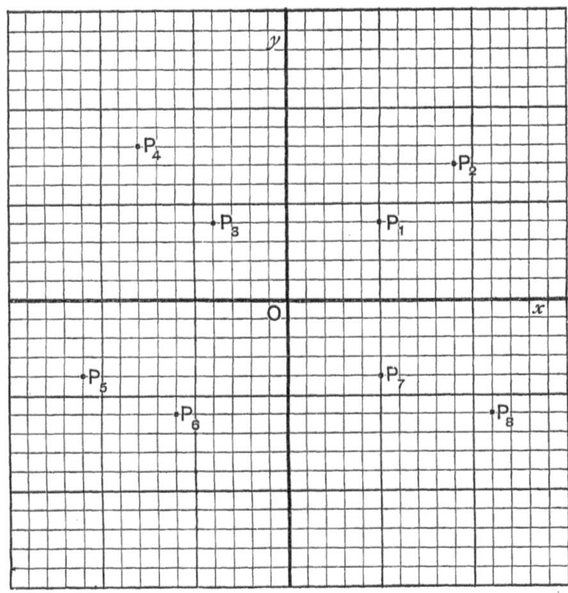

Fig. 8.

17. Find the co-ordinates of the points of trisection of the line joining the points $(3, 2),\ (6, 8)$.

18. Find the co-ordinates of that point of trisection of the line joining $(7, 5),\ (-2,\ -1)$, which is nearer to the point $(7, 5)$.

AREA OF A TRIANGLE.

7. *To find the area of a triangle formed by joining the origin to the points* $(x_1,\ y_1),\ (x_2,\ y_2)$.

Let P, Q be the points $(x_1,\ y_1),\ (x_2,\ y_2)$ respectively.

Draw the ordinates PN, QM.

The area of $\triangle$ OPQ $= \triangle$ OQM $+$ trapezium QMNP $- \triangle$ ONP

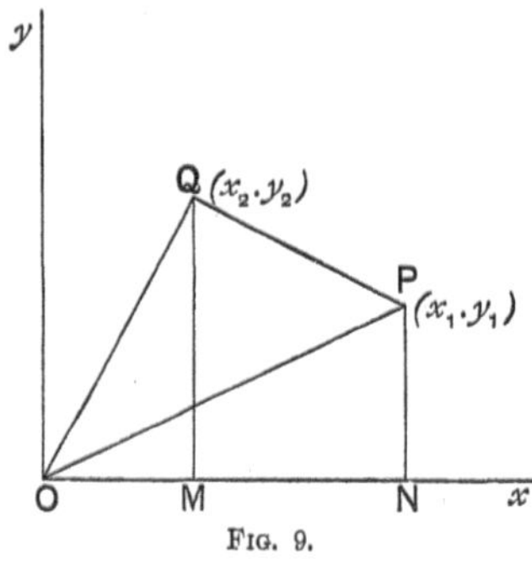

FIG. 9.

[The area of a trapezium
$= \frac{1}{2}$(sum of parallel sides)
$\times$ altitude.]

$= \frac{1}{2}$[OM . QM $+$ (QM $+$ PN) MN
$\quad -$ ON . PN]

$= \frac{1}{2}[x_2 y_2 + (y_2 + y_1)(x_1 - x_2)$
$\quad - x_1 y_1]$

$= \frac{1}{2}(\mathbf{x_1 y_2 - x_2 y_1}).$

8. We may use the formula of the preceding article to find the area of any triangle.

Find the area of the triangle formed by joining the points $(-2, 1)$, $(3, 2)$, $(2, 4)$.

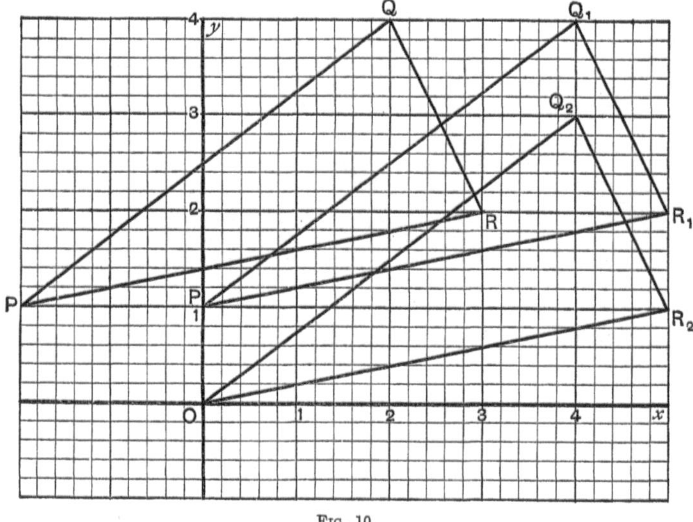

FIG. 10.

Add 2 to each abscissa, and subtract 1 from each ordinate. This, as seen by the figure, is the same as moving the triangle through 2 units parallel to Ox, and then through 1 unit in a negative direction parallel to Oy.

The original $\triangle PQR$ first moves into the position $P_1Q_1R_1$, and then into the position OQ_2R_2, the lengths of its sides remaining the same.

The co-ordinates of O, Q_2, R_2 are $(0, 0)$, $(4, 3)$, $(5, 1)$ respectively.

$$\therefore \quad \triangle PQR = \triangle OQ_2R_2 = \tfrac{1}{2}[5 \times 3 - 4 \times 1] \quad [\tfrac{1}{2}(x_1y_2 - x_2y_2)]$$
$$= \tfrac{1}{2}(15 - 4) = 5\tfrac{1}{2}.$$

9. *The area of the triangle formed by joining the points* (x_1, y_1), (x_2, y_2), (x_3, y_3) *is equal to* $\tfrac{1}{2}[x_1(y_2 - y_3) + x_2(y_3 - y_1) + x_3(y_1 - y_2)]$.

Let ABC be the three points. Draw the ordinates AL, BM, CN.

$\triangle ABC =$ trapezium $ALNC +$ trapezium $CNMB -$ trapezium $ALMB$.

The area of a trapezium
$= \tfrac{1}{2}$(sum of parallel sides) $\times$ altitude ;

$\therefore$ trapezium $ALNC$
$= \tfrac{1}{2}(AL + CN)LN$
$= \tfrac{1}{2}(y_3 + y_1)(x_3 - x_1).$

Trapezium $CNMB$
$= \tfrac{1}{2}(CN + BM)MN$
$= \tfrac{1}{2}(y_2 + y_3)(x_2 - x_3).$

Trapezium $ALMB$
$= \tfrac{1}{2}(AL + BM)LM$
$= \tfrac{1}{2}(y_2 + y_1)(x_2 - x_1).$

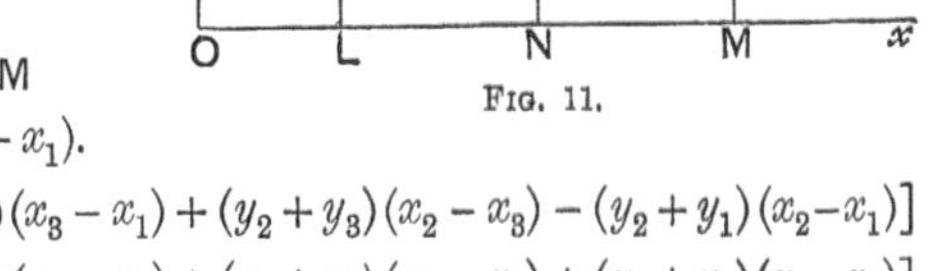

Fig. 11.

$$\therefore \triangle ABC = \tfrac{1}{2}[(y_3 + y_1)(x_3 - x_1) + (y_2 + y_3)(x_2 - x_3) - (y_2 + y_1)(x_2 - x_1)]$$
$$= \tfrac{1}{2}[(y_3 + y_1)(x_3 - x_1) + (y_1 + y_2)(x_1 - x_2) + (y_2 + y_3)(x_2 - x_3)].$$

Or simplifying,

$$\triangle ABC = \tfrac{1}{2}[x_1(y_2 - y_3) + x_2(y_3 - y_1) + x_3(y_1 - y_2)].$$

An area is necessarily positive. In order to ensure that the above expression is positive, it will be found necessary to arrange the co-ordinates in the formula in the order that they occur when the triangle is circumscribed in an anti-clockwise direction.

E.g. With the following figure the above expression would be negative:

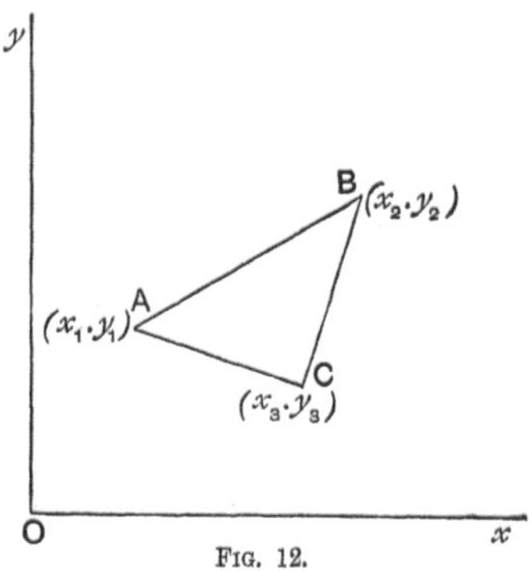

FIG. 12.

10. *Find the area of the triangle joining the points* (3, 1), (− 1, 3), (− 3, − 2).

We may use the formula of Art. 9, or the following method :

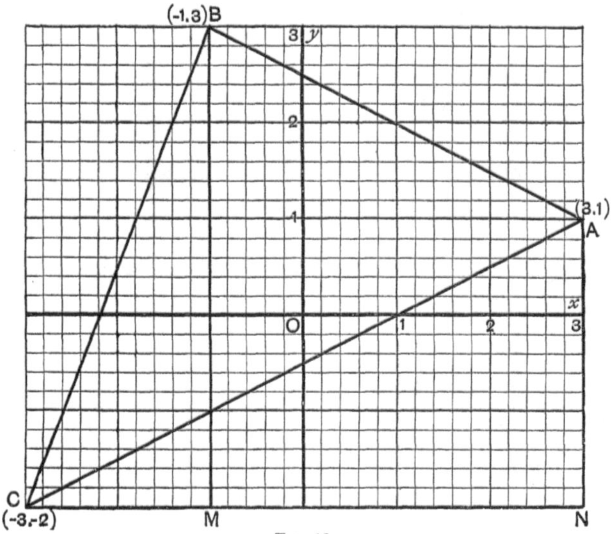

FIG. 13.

Plotting the triangle ABC on squared paper as in the diagram, and drawing CN parallel to Ox, and BM, AN parallel to Oy as shown,

$$\triangle ABC = \triangle CBM + \text{trapezium BMNA} - \triangle ACN$$
$$= \tfrac{1}{2}CM \cdot BM + \tfrac{1}{2}(BM + AN)MN - \tfrac{1}{2}AN \cdot CN$$
$$= \tfrac{1}{2} \cdot 2 \times 5 + \tfrac{1}{2}(5 + 3)4 - \tfrac{1}{2} \cdot 3 \times 6 = 12.$$

The area of a quadrilateral or polygon, when the co-ordinates of its angular points are given, may be found by dividing the area into triangles.

Examples I. b.

Find, without using a formula, the area of the figure having for vertices the points:

1. $(4, 2)$, $(4, 5)$, $(-2, 2)$.   2. $(0, 0)$, $(3, 3)$, $(3, -5)$.
3. $(0, 0)$, $(2, 6)$, $(4, 4)$.   4. $(0, 5)$, $(4, 5)$, $(4, 8)$.
5. $(-4, 3)$, $(-1, -2)$, $(3, -2)$.   6. $(0, 0)$, (x_1, y_1), $(x_1, -y_1)$.
7. $(0, 0)$, (x_1, y_1), $(-x_2, y_2)$.   8. $(0, 0)$, $(0, c)$, (h, k).

In this example the result is independent of k. What geometrical theorem does this prove?

9. (x_1, y_1), $(x_1, -y_1)$, $(0, -y_1)$.   10. $(1, -2)$, $(3, 3)$, $(-3, 2)$.
11. $(3, 1)$, $(3, -1)$, $(-1, 3)$.

Find the areas of the figures formed by joining the following points, using the formula for the area of a triangle *when advisable*:

12. (h, k), (k, h), (a, a).   13. $(0, 0)$, $(0, 3)$, $(4, 0)$, $(4, 5)$.
14. $(0, 0)$, $(6, 0)$, $(4, 3)$, $(0, 7)$.   15. $(1, -3)$, $(3, 1)$, $(1, 4)$, $(-5, -2)$.
16. $(5, 0)$, $(3, 4)$, $(-4, 1)$, $(2, -3)$.

EQUATION OF A LOCUS.

11. Def. *If a point moves but is subject to certain restrictions, the path it traces out is called its* **locus.**

The student will be familiar with certain well-known loci.

E.g. The locus of a point which moves at a constant distance from a fixed point is a circle, whose centre is at the fixed point.

The locus of a point equidistant from two fixed points is a straight line bisecting at right angles the straight line joining the two fixed points.

The locus of a point which is equidistant from two given intersecting straight lines is the two bisectors of the angles between the straight lines.

12. Equation of a locus.

Plot a number of points all at equal distances, say 2 units, from the axis of x.

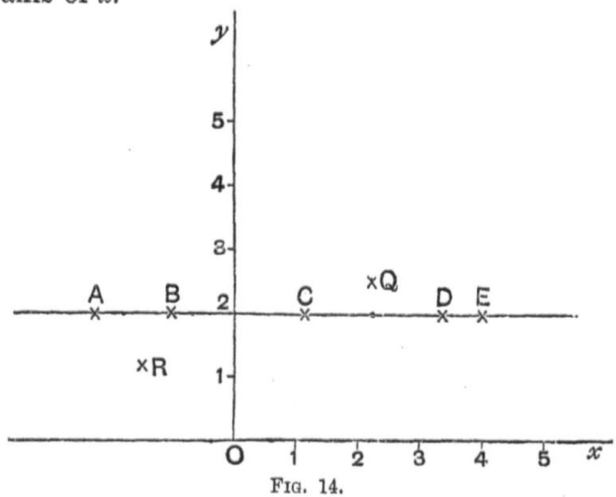

FIG. 14.

If **A, B, C, D,** ... are the points, we see that the locus is a straight line **ABC**, parallel to **O**x and at a distance 2 from it.

If (x, y) are the co-ordinates of *any* point on the locus, we see that $y = 2$.

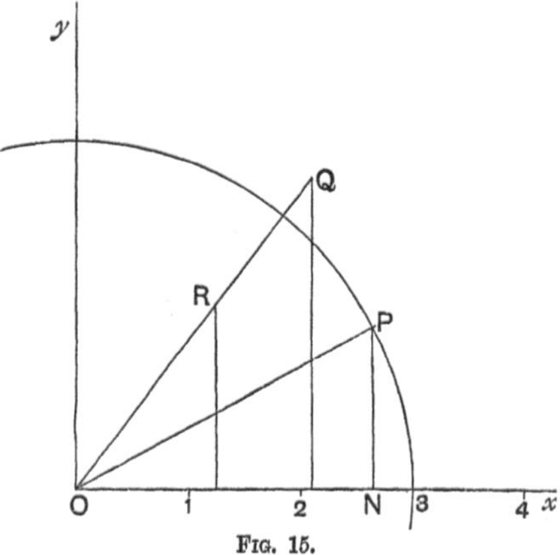

FIG. 15.

Also y = 2 for every point on the straight line, and for no other point.

E.g. At **Q** $y > 2$, at **R** $y < 2$;

$\therefore$ $y = 2$ is said to be the *equation of the locus.*

Take a circle of radius 3 units, and whose centre is at the origin. If (x, y) are the co-ordinates of any point **P** on its circumference, we see from the figure that

$$x^2 + y^2 = \mathrm{ON}^2 + \mathrm{PN}^2 = \mathrm{OP}^2 = 9 ;$$

$\therefore$ for any point on the circle $x^2 + y^2 = 9$.

Moreover this equation is not true unless the point (x, y) lies on the circumference.

At Q $x^2 + y^2 = \mathsf{OQ}^2 > \mathsf{OP}^2 > 9$.

At R $x^2 + y^2 = \mathsf{OR}^2 < \mathsf{OP}^2 < 9$.

$\therefore\ x^2 + y^2 = 9$ is said to be the equation of the locus, in this case a circle.

It will be seen, as we proceed with the subject, that if a point (x, y) moves in any given manner, we shall always be able to find an equation, connecting x, y and given quantities, which is true for all points on the locus of the point and for no other point.

Example i. *Find the equation of the locus of a point which is always equidistant from the points $(a, 0)$, $(-a, 0)$.*

Let (x, y) be any point on the locus.

The distance of (x, y) from the point $(a, 0) = \sqrt{(x-a)^2 + y^2}$ (Art. 4.)

,,   ,,   ,, $(-a, 0) = \sqrt{(x+a)^2 + y^2}$ (Art. 4.)

$\therefore$ by hypothesis, $\sqrt{(x-a)^2 + y^2} = \sqrt{(x+a)^2 + y^2}$.

Squaring, $x^2 - 2ax + a^2 + y^2 = x^2 + 2ax + a^2 + y^2$,

$$4ax = 0,$$

and $x = 0$ is the reqd. equation.

We see from a figure that the locus itself is a straight line bisecting the line joining the given points at right angles.

Example ii. *Find the equation of the locus of a point whose distance from the point (a, β) is constant and equal to c.*

Let (x, y) be any point on the locus.

The distance of (x, y) from the point $(a, \beta) = \sqrt{(x-a)^2 + (y-\beta)^2}$. (Art. 4.)

$$\therefore\ \sqrt{(x-a)^2 + (y-\beta)^2} = c,$$

and $(x-a)^2 + (y-\beta)^2 = c^2$ is the reqd. equation.

It is at once seen from a figure that the locus itself is a circle whose centre is at the point (a, β) and whose radius is c.

Examples I. c.

[When using squared paper indicate clearly, on the diagram, the unit employed.]

1. Plot six or seven points whose abscissae are 6. Thus determine the locus of a point which moves so that its abscissa is 6. What is its equation?

2. Plot a number of points whose ordinates are 4. What is the locus, and what is its equation?

3. Plot a number of points whose abscissae are -3.   What is the locus, and what is its equation?

4. Plot a number of points whose ordinates are -5.   What is the locus, and what is its equation?

5. Find the equation of the locus of a point equidistant from the points $(3, 0)$, $(-4, 0)$.

6. Find the equation of the locus of a point equidistant from the points $(3, 4)$, $(-2, 1)$.

7. What is the equation of the locus of a point which moves so that its perpendicular distance from the axis of x is three times its perpendicular distance from the axis of y?

8. A point moves so that its distance from the origin is always equal to 5: find the equation of its locus.

9. A point moves so that its distance from the point $(0, 0)$ is three times its distance from the point $(2, -4)$.   Find the equation of its locus.

10. The distance of a point from the axis of x is always equal to its distance from the point $(-2, 5)$: find the equation of its locus.

11. Find the equation of the locus of a point whose distance from the point $(-2, 4)$ is always equal to 5.

12. On squared paper plot a number of points whose distances from the axis of y are always twice their distances from the axis of x.   What can you see about the locus of such points?
Find the equation of the locus.

13. On squared paper plot a number of points such that in each case the ordinate is greater than the abscissa by unity.   What do you see about the locus of such points?
Find the equation of the locus.

14. Plot the points $(3, 1)$, $(6, 2)$ and join them by a straight line.
Prove that if (x, y) is *any* point on this line, then $x = 3y$.

15. Plot the points given in the following table:

When

$x =$	-2	-1	0	1	2
$y =$	1	2	3	4	5

Note that the ordinate always exceeds the abscissa by 3; and deduce the equation of the locus of such points.

16. When

$x =$	-5	-4	-3	-2	-1	0
$y =$	12	11	10	9	8	7

Plot the points given in the above table, and determine the equation of the locus of such points.

CHAPTER II.

THE STRAIGHT LINE.

13. *To find the equation of a straight line paralell to the axis of x.*

Let **PB** be the straight line, cutting the axis of y at **B**; and let $OB = b$.

Let the co-ordinates of **P**, *any* point on the line, be (x, y).

Draw the ordinate **PN**.

Then $\qquad\qquad$ **PN** = **OB** or $y = b$.

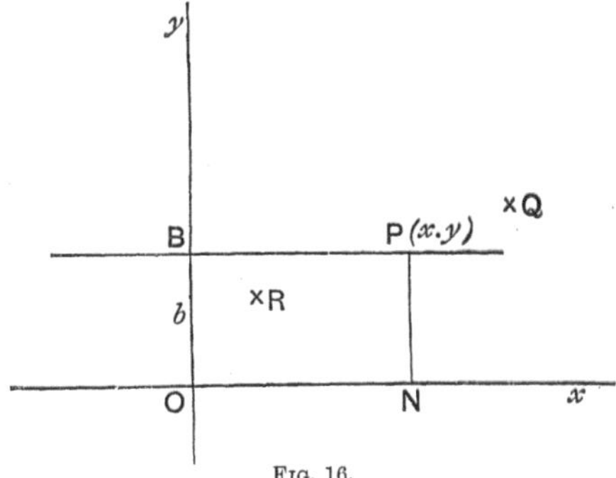

Fig. 16.

Now we see that this equation is true for *any* point on the line **PB**, and that it is true for no other points.

$\qquad$ *E.g.* At **Q**, $y > b$; at **R**, $y < b$.

$\qquad \therefore \; y = b$ is the equation of the line.

14. *To find the equation of a straight line parallel to the axis of y.*

As in the previous article, if a is the intercept of the line on the axis of x, $x = a$ is the equation of the line.

Note. The equation of the axis of y is $x = 0$.

$\qquad\qquad$,, $\qquad\qquad$,, $\qquad$ x is $y = 0$.

When (x, y) is *any* point on a locus, x and y are called **Current Co-ordinates**.

15. *To find the equation of a straight line through the origin and drawn in a given direction.*

Let OP be a straight line, the $\angle$ POx (θ) being given.

Let (x, y) be the co-ordinates of *any* point P on the line. Draw the ordinate PN. We have now to find the equation connecting x, y, and the given $\angle \theta$.

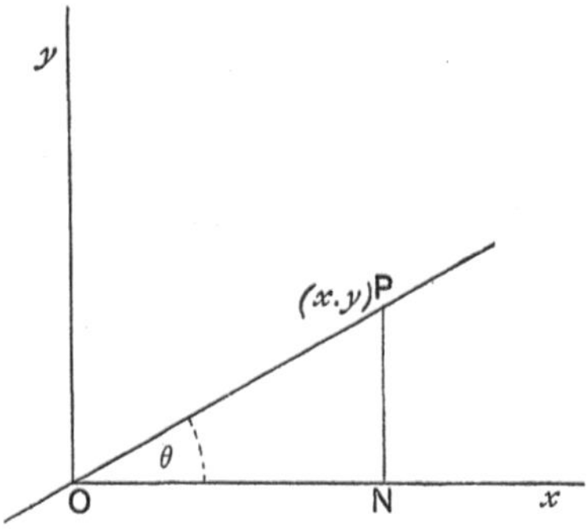

From the figure

$$\frac{PN}{ON} = \tan \theta,$$

$$\text{or } \frac{y}{x} = \tan \theta,$$

and $\qquad y = mx,$

when $\tan \theta = m$.

$$y = mx$$

is the reqd. equation.

Fig. 17.

16. Slope or Gradient. The letter m is generally used to denote the tangent of the angle which a straight line makes with the axis of x, and is called the **slope** or **gradient** of the line. It is measured in the following manner :

If the straight line cuts the axis of x at A, let Ax revolve about the point A *in a positive direction* until it coincides with the straight line. The tangent of the angle which Ax has turned through is the **slope** of the line. Thus in the figure, the slopes of the different lines are

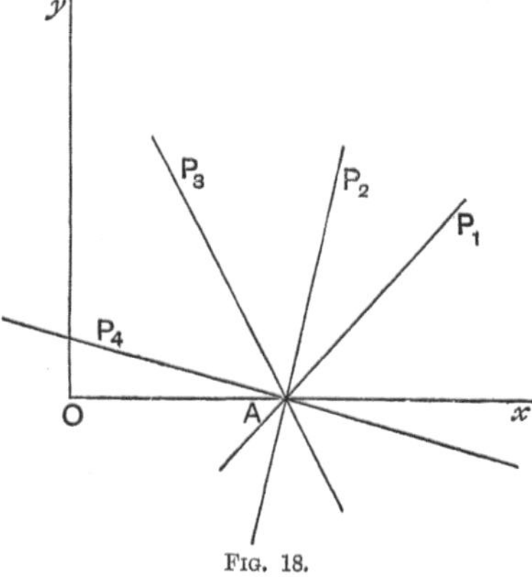

Fig. 18.

$$\tan P_1 A x, \quad \tan P_2 A x, \quad \tan P_3 A x, \quad \tan P_4 A x.$$

17. *To find the equation of a straight line passing through the origin, and also through a given point* (x_1, y_1).

Let OP be the straight line passing through the given point Q (x_1, y_1), and also through the origin O.

Let (x, y) be *any* point P on the line.

It is required to find the equation connecting x, y, x_1, and y_1.

Draw the ordinates PN, QM.

From the similar $\triangle$'s PNO, QMO,

$$\frac{PN}{QM} = \frac{ON}{OM},$$

or

$$\frac{y}{y_1} = \frac{x}{x_1}.$$

This is the reqd. equation.

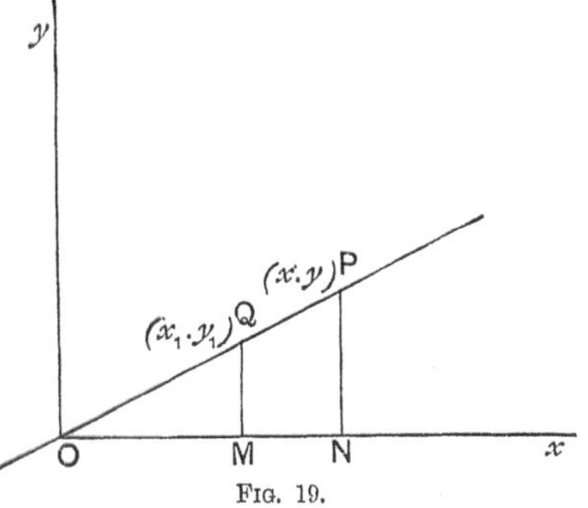

FIG. 19.

18. *Given the* **slope** *of a straight line, and its intercept on the axis of* y, *to find its equation.*

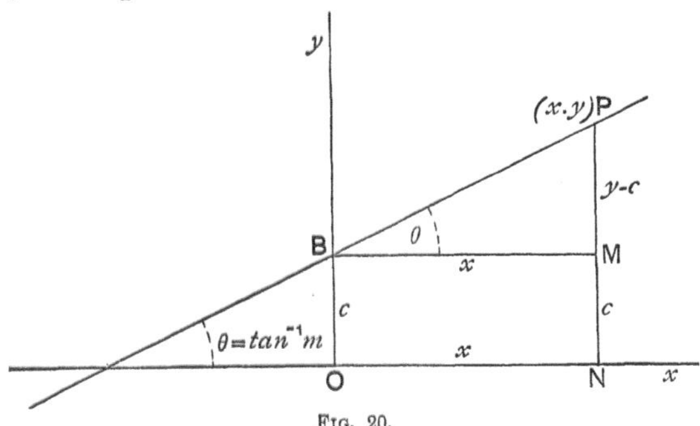

FIG. 20.

Let BP be the straight line, making an angle θ with the axis of x, so that $\tan \theta = m$, the given *slope*. Also let $c = $ OB, the intercept on the axis of y.

If P (x, y) is *any* point on the line, it is required to find the equation connecting x, y, m, and c.

B.A.G.   B

Draw PN perpendicular to the axis of x, and BM perpendicular to PN.

From the $\triangle$ PMB, $\dfrac{PM}{BM} = \tan \theta = m$.

But PM = PN − MN = $y - c$, and BM = ON = x ;

$$\therefore \frac{y-c}{x} = m,$$

or **$y = mx + c$**, the reqd. equation.

Note that if the point (x_1, y_1) is on the line $y = mx + c$, x_1 and y_1 must satisfy the equation $y = mx + c$, *i.e.* $y_1 = mx_1 + c$.

Example.   The point $(2, 9)$ is on the straight line $y = 2x + 5$, for
$$9 = 2 \times 2 + 5.$$

The straight line Ax + By = 0 passes through the origin $(0, 0)$, for these co-ordinates satisfy the equation.

If a straight line passes through the origin its equation has no constant term, for otherwise the co-ordinates $(0, 0)$ of the origin would not satisfy the equation.

19. *Given the intercepts of a straight line on the axes of co-ordinates, to find its equation.*

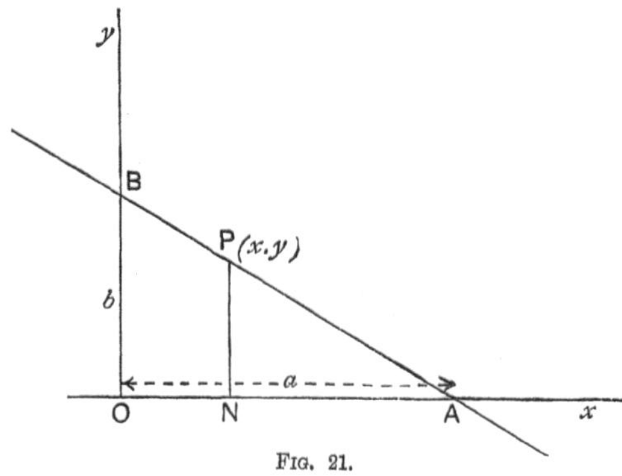

Fig. 21.

Let a be the intercept OA on the axis of x,
and b „   „   OB   „   „ y.
Let P (x, y) be *any* point on the line.
It is required to find the equation connecting x, y, a, and b.

Draw the ordinate PN.
From the similar $\triangle$^s PNA, BOA,

$$\frac{PN}{OB} = \frac{AN}{OA}.$$

But AN $=$ OA $-$ ON $= a - x$;

$$\therefore \frac{y}{b} = \frac{a-x}{a} = 1 - \frac{x}{a},$$

or $\dfrac{\mathbf{x}}{\mathbf{a}} + \dfrac{\mathbf{y}}{\mathbf{b}} = 1$, the reqd. equation.

Note that the straight line $\dfrac{x}{a} + \dfrac{y}{b} = 1$ passes through the point $(0, b)$, for when $x = 0$ and $y = b$ the equation is satisfied.

This straight line also passes through the point $(a, 0)$.

20. *Given that p is the length of the perpendicular from the origin upon a straight line, and that* α *is the angle the perpendicular makes with the axis of x, to find the equation of the straight line.*

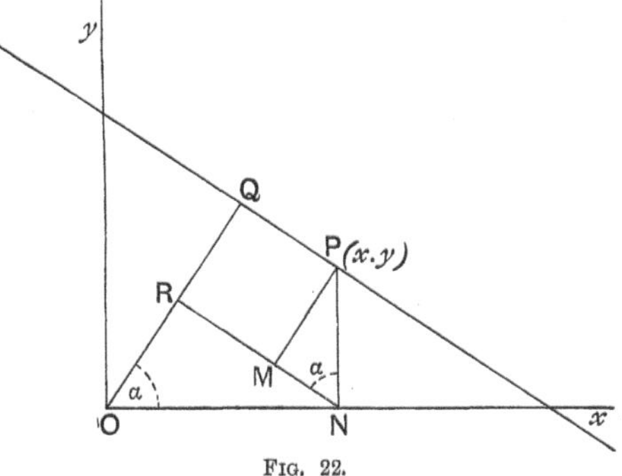

Fɪɢ. 22.

Let PQ be the straight line, OQ$(=p)$ the perpendicular drawn to it from the origin O, and let $\angle$ QOX $= α$.

If P(x, y) is *any* point on the straight line, it is required to find the equation connecting x, y, p, and α.

Draw the ordinate PN. Also draw NR perpendicular to OQ, and PM perpendicular to RN.

$$\angle \text{PNM} = 90° - \angle \text{RNO} = α.$$

$$\therefore\ p = OQ = OR + RQ = OR + PM$$
$$= ON \cos a + PN \sin a,$$

or $p = x \cos a + y \sin a.$

$\therefore$ **x cos** $a +$ **y sin** $a =$ **p** is the required equation.

Note carefully that if any values of x and y satisfy the equation of a locus, the point (x, y) lies on that locus.

Also conversely, if the point (x, y) is on a locus, these values of x and y must satisfy the equation of the locus.

Thus the point $(3, 4)$ lies on the locus whose equation is

$$x^2 + y^2 = 25,\ \text{for}\ 3^2 + 4^2 = 25.$$

The point $(6, 2)$ lies on the straight line whose equation is

$$y = x - 4,\ \text{for}\ 2 = 6 - 4.$$

21. Thus far we observe that in each case the equation of the straight line is **of the first degree** in x and y. We will now prove that such must be the case.

To prove that every equation of the first degree in x and y represents a straight line.

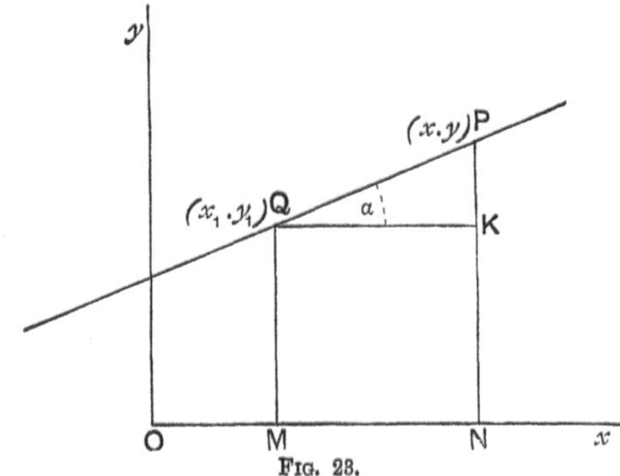

Fig. 23.

The most general equation of the first degree is $Ax + By + C = 0$.

Take Q a fixed point on the locus, and let its co-ordinates be (x_1, y_1).

Let P(x, y) be any other point on the locus.

Join PQ and draw the ordinates PN, QM.

Also draw QK parallel to Ox to meet PN at K. Let $\angle$ PQK $= a.$

P is on the locus, $\therefore$ $Ax + By + C = 0$.

Q　　,,　　,,　　$\therefore$ $Ax_1 + By_1 + C = 0$.

$\therefore$ by subtraction, $A(x - x_1) + B(y - y_1) = 0$.

$$\therefore \frac{y - y_1}{x - x_1} = -\frac{A}{B}, \text{ which is constant ;}$$

i.e. from the figure $\dfrac{PK}{QK}$ is constant.

$\therefore$ $\tan \alpha$ is constant, or α is a fixed angle for all positions of P, for QK is drawn in a fixed direction.

$\therefore$ the locus of P is a straight line through Q, making an angle $\tan^{-1}\left(-\dfrac{A}{B}\right)$ with Ox.

22. The general equation $Ax + By + C = 0$ contains three constants, A, B, C. These, however, are not independent, for we can divide through by any one of them. Thus $Ax + By + C = 0$ is the same equation as

$$\frac{Ax}{C} + \frac{By}{C} + 1 = 0, \quad \text{or} \quad \frac{x}{-\dfrac{C}{A}} + \frac{y}{-\dfrac{C}{B}} = 1.$$

Hence if the values of $-\dfrac{C}{A}$ and $-\dfrac{C}{B}$ are known, the position of the straight line is determined by means of the formula $\dfrac{x}{a} + \dfrac{y}{b} = 1$, and the straight line can be drawn. We see therefore that the constants A, B, C are equivalent to two *independent* constants.

It will always be found that two conditions, sufficient to find two constants, are enough to determine the position of a straight line.

In the equation $y = mx + c$, m and c are the constants, and if their values are known, the straight line can be drawn.

So in the equation $x \cos a + y \sin a = p$, the values of a and p determine the straight line.

23. *To find the co-ordinates of the point of intersection of two given straight lines.*

Let $a_1 x + b_1 y + c_1 = 0$, $a_2 x + b_2 y + c_2 = 0$ be the equations of the lines.

The co-ordinates of the common point of the straight lines satisfy both equations.

$\therefore$ we have to find values of x and y which satisfy both equations.

In other words, we have to solve the two equations, looking upon them as simultaneous.

The given equations may be solved in the usual way by eliminating x, and substituting the value of y, when found, but the student should be able to **write down** the solution, thus:

$$\frac{x}{b_1 c_2 - b_2 c_1} = -\frac{y}{a_1 c_2 - a_2 c_1} = \frac{1}{a_1 b_2 - a_2 b_1}.$$

It is easily remembered in the following manner:

Write down the coefficients, omitting the x's and y's, thus:

$$a_1, \ b_1, \ c_1,$$

$$a_2, \ b_2, \ c_2.$$

To obtain the denominator of x, imagine the a column erased, and take the products of the b's and c's, crossways as indicated, putting the minus sign between them, thus obtaining $b_1 c_2 - b_2 c_1$.

To obtain the denominator of y, imagine the b column erased, take the products of the a's and c's as before, with the minus sign between them, and **place the minus sign before y.**

To obtain the third denominator, imagine the c column erased, and take the products of the a's and b's, as for x.

Example. *Find the point of intersection of the straight lines.*

$$4x - 6y - 24 = 0,$$
$$3x + 7y + 5 = 0.$$

Solving these equations by the above method,

$$\frac{x}{-6 \times 5 + 7 \times 24} = \frac{-y}{4 \times 5 + 3 \times 24} = \frac{1}{4 \times 7 + 3 \times 6},$$

$$\frac{x}{6(-5 + 28)} = \frac{-y}{4(5 + 18)} = \frac{1}{2(14 + 9)}.$$

$$\therefore \ x = 3, \ y = -2.$$

$\therefore$ (3, -2) is the common point of the two lines.

24. If three straight lines are concurrent, *i.e.* meet at a point, the co-ordinates of the common point of any two will satisfy the equation of the third.

The equation of a straight line may be written in any of the forms obtained in Articles 18, 19, 20.

Example i.　*Write the equation $3x - 4y = 7$ in the form $y = mx + c$.*
The equation may be written　$4y = 3x - 7$,

$$\text{or }\ y = \frac{3x}{4} - \frac{7}{4};$$

which is in the form　$y = mx + c$.

The given line makes an angle $\tan^{-1}\frac{3}{4}$ with the axis of x, and passes through the point on the axis of y, whose ordinates are $\left(0,\ -\frac{7}{4}\right)$.

Example ii.　*Write the equation $3x - 4y = 7$ in the form $\frac{x}{a} + \frac{y}{b} = 1$.*

$$\text{It may be written }\ \frac{3x}{7} - \frac{4y}{7} = 1,$$

$$\text{or }\ \frac{x}{2\frac{1}{3}} - \frac{y}{1\frac{3}{4}} = 1,$$

which is of the form $\frac{x}{a} + \frac{y}{b} = 1$.

The given line cuts from the axes lengths equal to $2\frac{1}{3}$ and $-1\frac{3}{4}$ respectively.

Example iii.　*Write the equation $3x - 4y = 7$ in the form $x\cos a + y\sin a = p$.*
The point to be observed here is that $\sin^2 a + \cos^2 a = 1$.
Dividing through by $\sqrt{3^2 + 4^2}$, *i.e.* by 5, the equation becomes

$$\frac{3x}{5} - \frac{4y}{5} = \frac{7}{5}, \text{ which is in the form}$$

$$x\cos a + y\sin a = p.$$

$$\left[\cos a = \frac{3}{5},\ \ \sin a = -\frac{4}{5},\ \text{ and }\ \sin^2 a + \cos^2 a = \frac{3^2 + 4^2}{5^2} = 1.\right]$$

The perpendicular on the given line from the origin makes an angle $\tan^{-1}\left(-\frac{4}{3}\right)$ with the axis of x, and the length of this perpendicular is $\frac{7}{5}$ units.

Example iv.　*Write the equation $Ax + By + C = 0$ in the form*

$$x\cos a + y\sin a - p = 0.$$

Using the method of Example iii., the required equation is

$$-\frac{Ax}{\sqrt{A^2 + B^2}} - \frac{By}{\sqrt{A^2 + B^2}} - \frac{C}{\sqrt{A^2 + B^2}} = 0.$$

25. *Given the equation of a straight line to draw this straight line to scale.*

[It is generally advisable to use squared paper.]
This can be done by finding (by trial) the co-ordinates of two points on the straight line, and joining them.

Example 1. *Draw the straight line $3x - 4y = 12$.*
In this equation, when $y = 0$, $x = 4$;
$$\therefore \text{ the point } (4, 0) \text{ is on the line.}$$

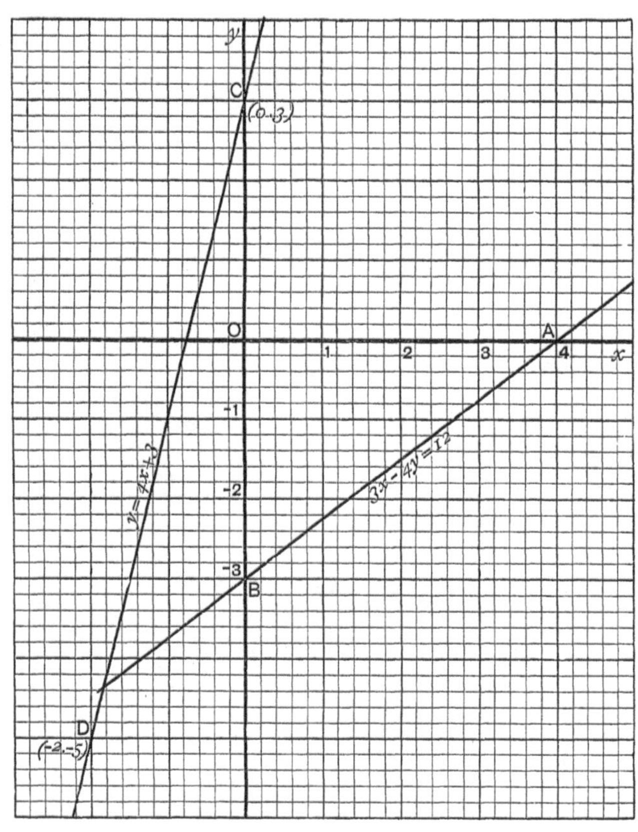

Fig. 24.

Plot this point, A in the figure.
When
$$x = 0, \; y = -3 ;$$
$$\therefore \text{ the point } (0, -3) \text{ is on the line.}$$
Plot this point, B in the figure.   AB is the straight line.

Example ii. *Draw the straight line $y = 4x + 3$.*

When $x = 0$, $y = 3$;

$$\therefore \text{ the point } (0, 3) \text{ is on the line.}$$

Plot this point, C in the figure.

When $y = 0$, $x = -\frac{3}{4}$.

The point $(-\frac{3}{4}, 0)$ is not very convenient to plot; we therefore find another.

When $\qquad\qquad x = -2, \quad y = -8 + 3 = -5$;

$$\therefore \text{ the point } (-2, -5) \text{ is on the line.}$$

Plot this point, D in the figure.

$$CD \text{ is the line.}$$

Examples II. a.

[The following Examples may be taken orally.]

Write down, or read off, in each case quoting the formula used, the equations of the following straight lines :

1. Parallel to, and at a distance 8 from the axis of x, and on the positive side of it.

2. Parallel to, and at a distance 5 from the axis of x, and on the negative side of it.

3. Parallel to, and at a distance 4 from the axis of y, and on the positive side of it.

4. Parallel to, and at a distance 3 from the axis of y, and on the negative side of it.

5. Through the origin and having a slope equal to 4.

6. Through the origin and the point $(3, 2)$.

7. Through the origin and the point $(-4, 5)$.

8. Through the origin and the point $(-5, 5)$.

9. Through the origin and having a slope -3.

10. Having a slope equal to 4 and making an intercept 3 on the axis of y.

11. Having a slope equal to -3 and making an intercept -2 on the axis of y.

12. Making equal intercepts on the axes, each equal to 3.

13. Making intercepts 3 and -4 on the axes of x and y respectively.

14. What are the slopes of the following straight lines :

(i) $y = 2x - 5$? (ii) $y - 3x = 4$? (iii) $5y + x = 6$?

(iv) $3x + 4y = 7$? (v) $4x - 5y + 8 = 0$? (vi) $Ax + By + C = 0$?

(vii) $\dfrac{x}{a} + \dfrac{y}{b} = 1$? (viii) $x \cos a - y \sin a = p$? (ix) $x \sin a + y \cos a = p$?

15. Write down, or read off, the intercepts of the following straight lines on the axis of y:

(i) $y = 3x - 5$. (ii) $3y = 4x + 6$. (iii) $3x - 6y = 9$.

(iv) $3x + 4y + 12 = 0$. (v) $Ax + By + C = 0$. (vi) $x \cos a + y \sin a - p = 0$.

16. Write down, or read off, the intercepts of the following straight lines on the axis of x:

(i) $x + y = 4$. (ii) $y + 2x = 6$. (iii) $y - x = 5$.

(iv) $3x - 4y - 18 = 0$. (v) $Ax + By + C = 0$. (vi) $x \cos a + y \sin a = p$.

17. What is the condition that the point (x_1, y_1) should be on the straight line $xx_1 + yy_1 = a^2$?

18. What is the condition that the point $(0,1)$ should be on the straight line $ax + by = c$?

19. What is the condition that the straight line $ax - by = c$ should be equally inclined to the axes?

20. What is the condition that the straight line $ax + by + c = 0$ should pass through the origin?

21. Where do the straight lines $x = a$, $y = b$ meet?

22. What are the co-ordinates of the points of intersection of $x - y = 2$, $x + y = 2$?

23. The straight lines $3x + 4y = 5$, $3x + 4y = 8$ do not meet at all. How is this?

In drawing to scale indicate clearly, on the sheet containing the diagram, the unit employed.

Draw all diagrams in pencil, letter them in ink, and study neatness.

Let the diagram illustrate the problem as much as possible.

E.g. **Points, whose co-ordinates are known, should have those co-ordinates written against them.**

Keep squared paper for diagrams, giving written explanations, etc., on another sheet.

Examples II. b.

[Work questions 1-9 without assuming any formula.]

1. Find the equation of the straight line passing through the origin and the point (4, 3).

2. Find the equation of the straight line making an intercept -3 on the axis of y, and having a slope equal to unity.

3. Find the equation of the straight line whose intercepts on the axes of x and y are 4 and -4.

4. A straight line makes equal intercepts both of positive sign on the axes, and the perpendicular on it from the origin is equal to 4. Find its equation.

5. A straight line has a slope equal to unity and passes through the point $(5, -5)$. Find its equation.

Find the intercepts of the following straight lines on the axes of co-ordinates :

6. $2x + 3y = 1$.　　　**7.** $3x + 4y = 5$.　　　**8.** $\dfrac{x}{2} - \dfrac{y}{3} = 7$.　　　**9.** $ax + by = c$.

10. In the same diagram, on squared paper, draw the following straight lines, distinguishing each by writing its equation alongside it :

$$\text{(i) } x = 4, \quad \text{(ii) } x = -3, \quad \text{(iii) } y = 5, \quad \text{(iv) } y = -1.$$

11. Do the same with the following straight lines :

$$\text{(i) } y = x, \quad \text{(ii) } y = -x, \quad \text{(iii) } y = 4x, \quad \text{(iv) } y = -2x.$$

12. In the same diagram, on squared paper, draw the straight lines $3x + 4y = 12$ and $4x - 3y = 12$. Explain your method.

Draw the following straight lines to scale on squared paper, explaining your method in each case :

13. $x - 3y = 6$.　　　　**14.** $x - 2y = 0$.　　　　**15.** $5x + 4y - 20 = 0$.

16. $\dfrac{x}{\sqrt{2}} + \dfrac{y}{\sqrt{2}} = 3$.　　　　**17.** $3x - 4y = 7$.　　　　**18.** $13x - 15y = 0$.

19. $5x - 3y = 8$.

Find the points of intersection of the following pairs of straight lines :

20. $x + y = 4$, $x - y = 2$.　　　　**21.** $x - y = 2$, $2x + y = 7$.

22. $3x - 4y = 7$, $3x + 4y = 17$.　　　　**23.** $3x = 4y - 1$, $5y - 6x = 8$.

24. $7x - 6y + 59 = 0$, $3x - 8y + 47 = 0$.

25. Discuss the equation $Ax + By + C = 0$,

　　(i) when $A = 0$,　　　　(ii) when $B = 0$,　　　　(iii) when $C = 0$,

　(iv) when $A = \infty$,　　　(v) when $A = C = 0$,　　　(vi) when $B = \infty$.

26. Find the equation of the straight line through the origin and the middle point of the line joining $(2, 3)$, $(4, -9)$.

27. Find the equation of the straight line through the origin and the middle point of the line joining (x_1, y_1) and (x_2, y_2).

28. What is the value of m if the straight line $y = mx + 4$ passes through the point of intersection of $x - y = 5$ and $x + y = 7$.

29. Prove that the following three straight lines are concurrent :

$$x - y = 6, \quad 4y - 3x + 22 = 0, \quad 6x + 5y + 8 = 0.$$

26. *Given a point (x_1, y_1) on a straight line, and the slope (m) of the line, to find its equation.*

Let PQ be the line, Q being the given point (x_1, y_1). Also let the line make an angle $\tan^{-1} m$ with Ox.

If $P(x, y)$ is *any* point on the line, it is required to find the equation connecting x, y, x_1, y_1, and m.

Draw the ordinates PN, QM ; and draw QK perpendicular to PN.

$$PK = PN - KN = PN - QM = y - y_1,$$
$$QK = MN = ON - OM = x - x_1 ;$$

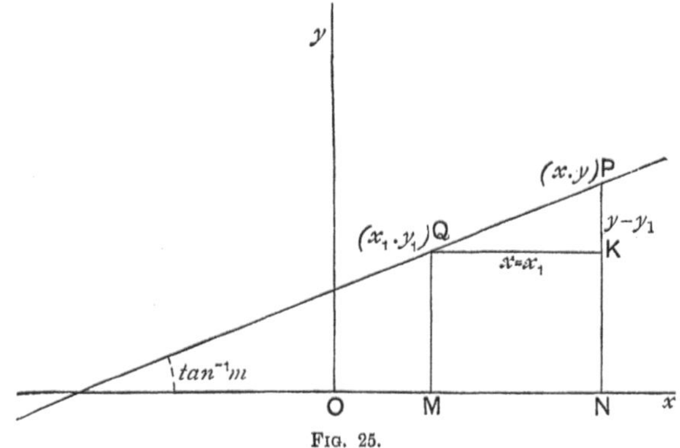

Fig. 25.

$$\therefore \text{ from the } \triangle PKQ, \ \frac{PK}{QK} = \tan PQK = m,$$

$$\text{or } \frac{y - y_1}{x - x_1} = m ;$$

$$\therefore \ \mathbf{y - y_1 = m(x - x_1)} \text{ is the reqd. equation.}$$

27. The following analytical method might be used.

Let $y = mx + c$ be the equation required, *where c is unknown.*
By hypothesis, the point (x_1, y_1) lies on the line.

$$\therefore \ x_1, \ y_1 \text{ satisfy the equation } y = mx + c ;$$
$$\therefore \ y_1 = mx_1 + c.$$

Also $\qquad\qquad\qquad y = mx + c ;$

$\therefore$ by subtraction, $y - y_1 = m(x - x_1)$ is the reqd. equation.

28. *To find the equation of the straight line passing through the two given points* $(x_1, y_1), (x_2, y_2).$

Let A and B be the given points, $(x_1, y_1), (x_2, y_2)$, and P(x, y) *any point on the line.*

It is required to find the equation connecting $x, y, x_1, y_1, x_2,$ and $y_2.$

Draw the ordinates AH, BL, PN, and draw AKM perpendicular to BL and PN.

From the similar triangles PMA, BKA,

$$\frac{PM}{BK} = \frac{AM}{AK}. \quad \dots\dots\dots\dots\dots\dots\dots(1)$$

But

$$PM = PN - MN = PN - AH = y - y_1,$$
$$BK = BL - KL = BL - AH = y_2 - y_1,$$
$$AM = HN = ON - OH = x - x_1,$$
$$AK = HL = OL - OH = x_2 - x_1;$$

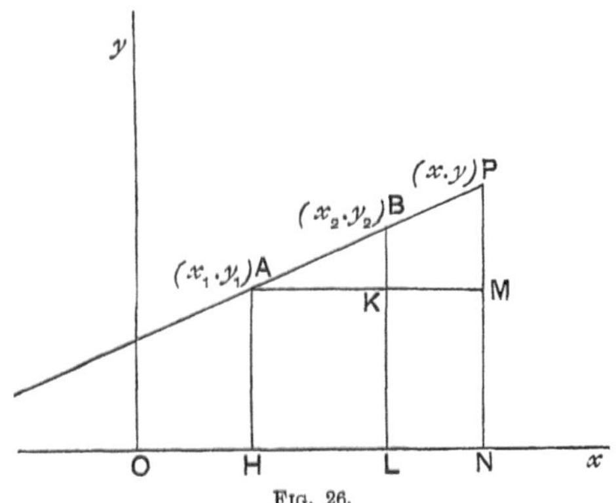

Fig. 26.

$$\therefore \text{ from (1) } \frac{y - y_1}{y_2 - y_1} = \frac{x - x_1}{x_2 - x_1},$$

and this is the reqd. equation.

Note. The slope of this line is $\frac{y_2 - y_1}{x_2 - x_1}$. This may be seen from the figure, or by writing the equation *in the form* $y = mx + c$.

29. The equation of the straight line in the preceding article might be found by the following method.

Suppose $y = mx + c$ is the equation of the straight line, *where m and c are unknown.*

The point (x_1, y_1) is on the straight line;

$$\therefore y_1 = mx_1 + c_1. \quad \dots\dots\dots\dots(1)$$

Also (x_2, y_2) is on the line; $\quad \therefore y_2 = mx_2 + c. \quad \dots\dots\dots\dots(2)$

And $\qquad\qquad\qquad\qquad y = mx + c. \quad \dots\dots\dots\dots(3)$

Subtracting (1) from (3), $y - y_1 = m(x - x_1)$.

   ,,   (1)   ,,   (2), $y_2 - y_1 = m(x_2 - x_1)$;

$\therefore$ by division, $\dfrac{y - y_1}{y_2 - y_1} = \dfrac{x - x_1}{x_2 - x_1}$ is the reqd. equation.

30. The following will be used a great deal in later chapters.

If (x_1, y_1) is a given point on a straight line, (x, y) any point on the line, θ the angle the line makes with the axis of x, and r the distance between the points (x, y), (x_1, y_1),

$$\frac{x - x_1}{\cos \theta} = \frac{y - y_1}{\sin \theta} = r.$$

[This will be found very useful in cases where we have to deal with the length (r) of any portion of a line.]

Let **P** be the point (x, y), **Q** the given point (x_1, y_1), so that **PQ** $= r$.

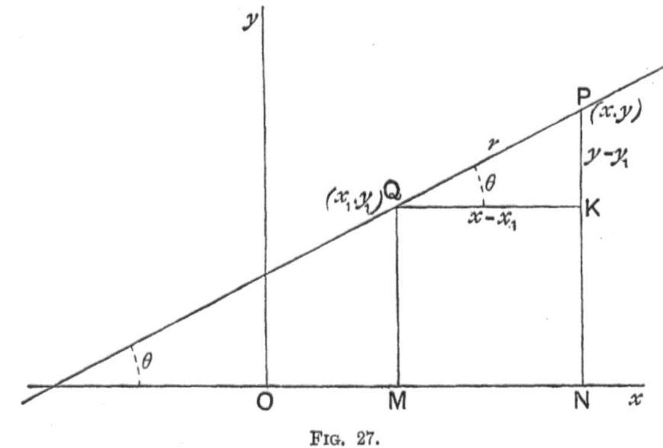

Fig. 27.

Draw the ordinates **PN**, **QM**, and draw **QK** perpendicular to **PN**.

From the $\triangle$ **PKQ**, **QK** = **PQ** $\cos \theta$ or $x - x_1 = r \cos \theta$ and $\dfrac{x - x_1}{\cos \theta} = r$.

Also **PK** = **PQ** $\sin \theta$, or $y - y_1 = r \sin \theta$ and $\dfrac{y - y_1}{\sin \theta} = r$.

$$\therefore \frac{x - x_1}{\cos \theta} = \frac{y - y_1}{\sin \theta} = r.$$

31. *The angle between the straight lines* $y = mx + c$, $y = m'x + c'$ *is equal to*

$$\tan^{-1}\left(\frac{m - m'}{1 + mm'}\right).$$

If AB, AC are the straight lines,

$$\angle \text{AB}x = \tan^{-1}m$$
$$= \theta \text{ suppose,}$$
$$\angle \text{AC}x = \tan^{-1}m'$$
$$= \theta' \text{ suppose.}$$

Then if $\phi = \angle\text{CAB}$, the angle required,

$$\phi = \theta - \theta';$$
$$\therefore \ \tan\phi = \tan(\theta - \theta')$$
$$= \frac{\tan\theta - \tan\theta'}{1 + \tan\theta\tan\theta'}$$
$$= \frac{m - m'}{1 + mm'};$$

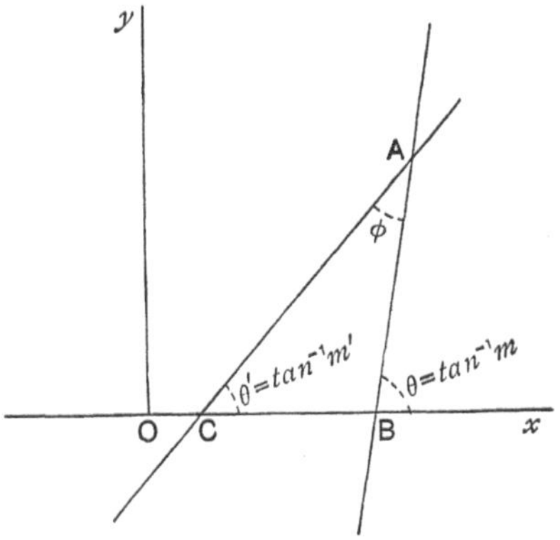

Fig. 28.

$$\therefore \ \phi = \tan^{-1}\left(\frac{m - m'}{1 + mm'}\right).$$

COROLLARY 1.   **If the two straight lines $y = mx + c$, $y = m'x + c'$ are parallel, $m = m'$.**

This is evident from a diagram, for the lines make equal angles with Ox, and therefore their slopes (or gradients) are equal.

COROLLARY 2.   **If the two straight lines $y = mx + c$, $y = m'x + c'$ are at right angles, $mm' = -1$.**

The tangent of the angle between them $= \dfrac{m - m'}{1 + mm'}$.

But $\tan 90° = \infty$; $\therefore \ \dfrac{m - m'}{1 + mm'} = \infty$; $\therefore \ mm' + 1 = 0$, or $mm' = -1$.

32. **If the two straight lines $a_1x + b_1y + c_1 = 0$, $a_2x + b_2y + c_2 = 0$ are at right angles, $a_1a_2 + b_1b_2 = 0$.**

The 'm' or slope of the first line $= -\dfrac{a_1}{b_1}$.

,,   ,,   ,,   second ,, $= -\dfrac{a_2}{b_2}$.

$\therefore$ by the preceding article $\left(-\dfrac{a_1}{b_1}\right) \times \left(-\dfrac{a_2}{b_2}\right) = -1,$ $\quad (mm' = -1)$

$$\text{or } a_1 a_2 + b_1 b_2 = 0.$$

Note carefully the following illustrations :

(1) $5x - 2y = 3,$

$5x - 2y = k$ are parallel lines for all values of k, for their slopes are equal.

(2) $5x - 2y = 3,$

$2x + 5y = k$ are perpendicular lines for all values of k, for $a_1 a_2 + b_1 b_2 = 5 \times 2 - 2 \times 5 = 0.$

(3) $\dfrac{x}{2} + \dfrac{y}{3} = 5,$

$2x - 3y = k$ are perpendicular lines for all values of k, for $a_1 a_2 + b_1 b_2 = \dfrac{1}{2} \times 2 - \dfrac{1}{3} \times 3 = 0.$

33. *To find the equations of the straight lines drawn through the point (h, k) and making an angle a with the given line $y = mx + c$.*

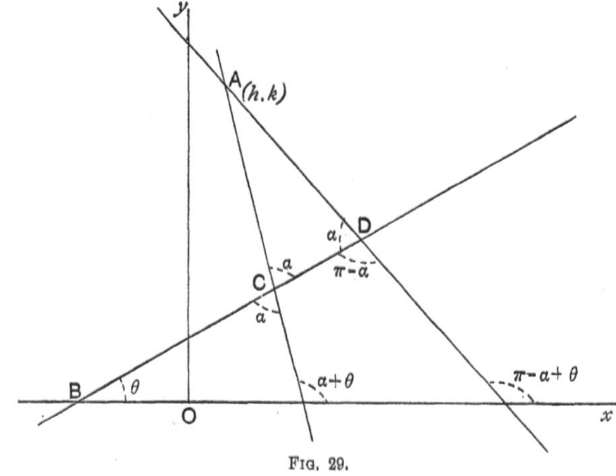

Fig. 29.

Let A be the point (h, k), BCD the given line $y = mx + c$. Then if AC, AD are the lines whose equations are required,

$$\angle \text{ACD} = \angle \text{ADC} = a.$$

Hence if BCD makes an angle θ with Ox, we see from the diagram that the slopes of the required lines are

$$\tan (a + \theta) \text{ and } \tan (\pi - a + \theta).$$

$$\tan (a + \theta) = \frac{\tan a + \tan \theta}{1 - \tan a \tan \theta} \qquad \tan (\pi - a + \theta) = -\tan (a - \theta)$$

$$= \frac{\tan a + m}{1 - m \tan a}; \qquad\qquad = -\frac{\tan a - \tan \theta}{1 + \tan a \tan \theta}$$

$$= \frac{m - \tan a}{1 + m \tan a};$$

$$\therefore\ y - k = \frac{m + \tan a}{1 - m \tan a}(x - h) \text{ and } y - k = \frac{m - \tan a}{1 + m \tan a}(x - h) \quad \text{(Art. 26.)}$$

are the required equations.

34. *If λ is any constant, $ax + by + c + \lambda(a'x + b'y + c') = 0$ ...(1) represents a straight line passing through the point of intersection of the straight lines $ax + by + c = 0$ ...(2), and $a'x + b'y + c' = 0$ ...(3).*

Equation (1) represents a straight line, for it is of the first degree.

Let (x_1, y_1) be the co-ordinates of the common point of (2) and (3).

The point (x_1, y_1) lies on (2) ; $\therefore\ ax_1 + by_1 + c = 0$ .....(4)

,,   ,,   ,,   (3) ; $\therefore\ a'x_1 + b'y_1 + c = 0$ .....(5)

Multiplying (5) by λ, and adding to (4),

$$ax_1 + by_1 + c + \lambda(a'x_1 + b'y_1 + c) = 0.$$

From this we see that the values x_1, y_1 satisfy equation (1) ;

$\therefore$ the straight line represented by (1) passes through (x_1, y_1), the common point of (2) and (3).   Q.E.D.

35. If, without transgressing the laws of Algebra, we combine the equations of two given straight lines and obtain a third equation of the first degree, this equation will represent a straight line passing through the point of intersection of the given straight lines.

Consider the straight lines

$$a_1 x + b_1 y + c_1 = 0, \quad \dots\dots\dots\dots\dots\dots\dots(1)$$

$$a_2 x + b_2 y + c_2 = 0. \quad \dots\dots\dots\dots\dots\dots\dots(2)$$

Multiplying (1) by (λ) and (2) by (μ), where λ and μ are any constants, and adding, $\lambda(a_1x + b_1y + c) + \mu(a_2x + b_2y + c_2) = 0$ ...(3).

This equation is of the first degree, and therefore represents a straight line. Also, any values of x and y, which satisfy *both* (1) and (2), satisfy (3) also.

$\therefore$ the line (3) passes through the point of intersection of (1) and (2).

Example i. *Find the equation of the straight line passing through the origin and through the point of intersection of $5x - 2y + 4 = 0$ and $4x - 3y + 5 = 0$.*

Multiplying the first equation by 5, $25x - 10y + 20 = 0$.

,, ,, ,, second ,, ,, 4, $16x - 12y + 20 = 0$.

Subtracting, $9x + 2y = 0$.

This equation is of the first degree and has no constant term ; therefore it represents a straight line through the origin.

Also it is formed by combining the given equations, and therefore passes through their point of intersection. It is therefore the equation required.

Example ii. *Find the equation of the straight line through the point $(0, -3)$, and the point of intersection of $3x - 7y + 9 = 0$ and $5x + 4y = 3$.*

The straight line $3x - 7y + 9 + \lambda(5x + 4y - 3) = 0$ passes through the intersection of the given lines for all values of λ.

If it also passes through the point $(0, -3)$, we have by substitution,

$$21 + 9 + \lambda(-12 - 3) = 0 \text{ and } \lambda = 2 ;$$

$$\therefore \ 3x - 7y + 9 + 2(5x + 4y - 3) = 0,$$

or $13x + y + 3 = 0$ is the required equation.

Example iii. *The straight line $2x - 5y + 3 + a(2x + 5y - 3) = 0$ passes through a fixed point for all values of a.*

The line passes through the intersection of

$$2x - 5y + 3 = 0, \text{ and } 2x + 5y - 3 = 0,$$

i.e. through the fixed point $(0, \frac{3}{5})$.

36. *To change the origin to the point (h, k), the directions of the axes being unchanged.*

Let O$'$ be the point (h, k), O$'x'$, O$'y'$ the new axes,

(x, y) the co-ordinates of P referred to the old axes,

(x', y') ,, ,, ,, ,, new ,,

Drawing the ordinate PMN,

$$x = \text{ON} = \text{OA} + \text{O'M} = h + x'.$$
$$y = \text{PN} = \text{O'A} + \text{PM} = k + y'.$$

Hence, if in any equation we write x + h for x and y + k for y, the resulting equation is the equation of the same locus referred to parallel axes through the point (h, k).

Example. The equation of the straight line

$$3x + 2y - 18 = 0,$$

when referred to parallel axes through the point (4, 3) is

$$3(x + 4) + 2(y + 3) - 18 = 0$$
$$\text{or } 3x + 2y = 0.$$

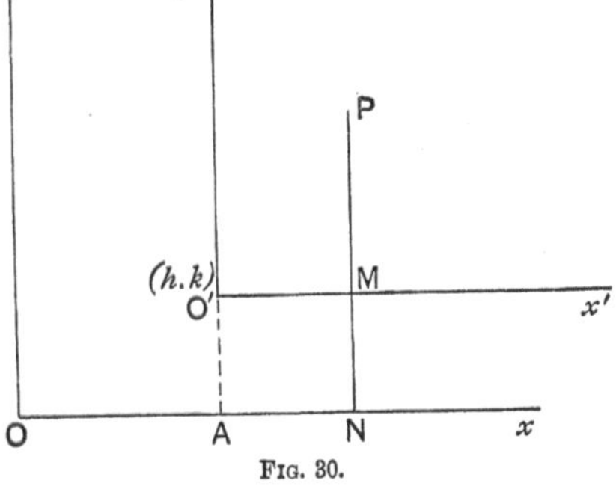

FIG. 30.

This proves that the straight line $3x + 2y - 18 = 0$ passes through the point (4, 3).

37. *As the point (x_1, y_1) moves from one side of the straight line* $\text{A}x + \text{B}y + \text{C} = 0$ *to the other, the expression* $\text{A}x_1 + \text{B}y_1 + \text{C}$ *changes its sign.*

Let P be the point (x_1, y_1), EF the straight line $\text{A}x + \text{B}y + \text{C} = 0$.

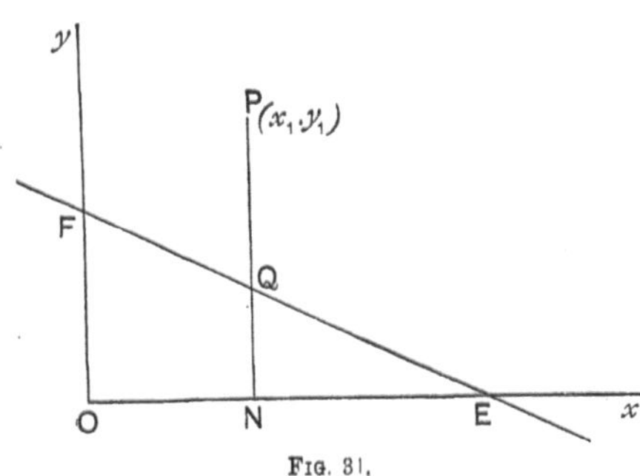

FIG. 31.

Draw the ordinate PN meeting EF at Q.

The point (x_1, QN) is on the line

$$\text{A}x + \text{B}y + \text{C} = 0 ;$$
$$\therefore \quad \text{A}x_1 + \text{B} \cdot \text{QN} + \text{C} = 0 ;$$
$$\therefore \quad \text{A}x_1 + \text{C} = - \text{B} \cdot \text{QN} ;$$
$$\therefore \quad \text{A}x_1 + \text{B}y_1 + \text{C}$$
$$= \text{B}y_1 - \text{B} \cdot \text{QN}$$
$$= \text{B}(\text{PN} - \text{QN}).$$

But as P moves to the other side of EF,

PN − QN changes its sign, which proves the proposition.

When P is at the origin, $x_1 = y_1 = 0$,

and then $Ax_1 + By_1 + C = C$.

$\therefore$ if C and the expression $Ax_1 + By_1 + C$ have the **same sign**, the point (x_1, y_1) and the origin are on the **same side of the line.**

If C and $Ax_1 + By_1 + C$ have **opposite signs,** the point (x_1, y_1) and the origin are on **opposite sides of the line.**

Consider the straight line $3x + 4y - 12 = 0$. Here $C = -12$.

Now when $x = 3$ and $y = 4$, $3x + 4y - 12 = 9 + 16 - 12 = 13$, which is positive, while $C(-12)$ is negative.

$\therefore$ the point $(3, 4)$ and the origin lie on opposite sides of the line $$3x + 4y - 12 = 0.$$

When $x = 2$ and $y = -1$, $3x + 4y - 12 = 6 - 4 - 12 = -10$.

$\therefore$ the point $(2, -1)$ and the origin lie on the same side of the line.

38. *The length of the perpendicular from the point* (x_1, y_1) *upon the straight line* $Ax + By + C = 0$ *is equal to*

$$\frac{Ax_1 + By_1 + C}{\sqrt{A^2 + B^2}}.$$

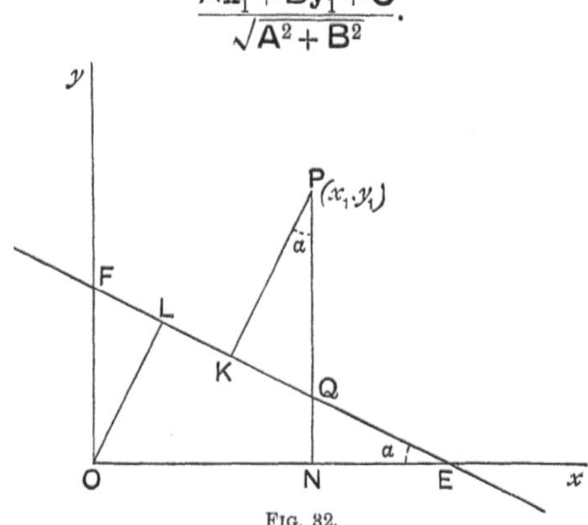

Fig. 32.

Let EF be the straight line, P the point (x_1, y_1).
Draw PK perpendicular to EF.
Also draw PN perpendicular to Ox, meeting EF at Q.
Let $\angle$ OEF $= a$.

At E, $y = 0$; $\therefore Ax + C = 0$ and $x = -\dfrac{C}{A}$, *i.e.* OE $= -\dfrac{C}{A}$.

In the same way at F, $x = 0$ and $OF = -\dfrac{C}{B}$;

$$\therefore \ \tan \alpha = \frac{OF}{OE} = \frac{A}{B} \ \text{ and } \ \sin \alpha = \frac{A}{\sqrt{A^2 + B^2}}, \ \cos \alpha = \frac{B}{\sqrt{A^2 + B^2}}.$$

Also $\angle KPQ = 90° - \angle KQP = 90° - \angle NQE = \alpha$;

$$\therefore \ PK = PQ \cos \alpha = (PN - QN) \cos \alpha = \frac{(y_1 - QN)B}{\sqrt{A^2 + B^2}} \ \dots\dots(1)$$

But Q is on the line $Ax + By + C = 0$, and $(x_1,\ QN)$ are its co-ordinates ;

$$\therefore \ Ax_1 + B \cdot QN + C = 0, \ \text{ whence } \ QN = -\frac{Ax_1 + C}{B}.$$

$$\therefore \ \text{from (1) } PK = \frac{\left(y_1 + \dfrac{Ax_1 + C}{B}\right)B}{\sqrt{A^2 + B^2}} = \frac{Ax_1 + By_1 + C}{\sqrt{A^2 + B^2}}.$$

COROLLARY. When P is at the origin, the expression for the perpendicular becomes

$$\frac{C}{\sqrt{A^2 + B^2}}.$$

But when P crosses the line, the expression $\dfrac{Ax_1 + By_1 + C}{\sqrt{A^2 + B^2}}$ changes its sign (Art. 37).

$\therefore$ if P and the origin are on the same side of the line, the expressions $\dfrac{C}{\sqrt{A^2 + B^2}}$ and $\dfrac{Ax_1 + By_1 + C}{\sqrt{A^2 + B^2}}$ have the same sign ; and when P and the origin are on opposite sides of the line, these expressions have opposite signs.

39. *Second Method.*

Draw OL perpendicular to the straight line.

First let us find the length of OL.

$$\tfrac{1}{2}OL \cdot EF = \triangle OEF = \tfrac{1}{2}OE \cdot OF ;$$

$$\therefore \ OL = \frac{OE \cdot OF}{EF}. \ \dots\dots \ \dots\dots\dots\dots\dots\dots(1)$$

At E, $y = 0$; $\therefore$ $Ax + C = 0$ and $x = -\dfrac{C}{A}$, *i.e.* $OE = -\dfrac{C}{A}$.

At F, $x = 0$; $\therefore$ $By + C = 0$ and $OF = -\dfrac{C}{B}$.

Also $EF = \sqrt{OE^2 + OF^2} = \sqrt{\dfrac{C^2}{A^2} + \dfrac{C^2}{B^2}} = \dfrac{C}{AB}\sqrt{A^2 + B^2}$;

$$\therefore \text{ from (1) } OL = \frac{C}{\sqrt{A^2 + B^2}}. \dotfill (2)$$

Next transfer the origin to the point (x_1, y_1).
The equation of the line is now $A(x + x_1) + B(y + y_1) + C = 0$,

$$\text{or } Ax + By + Ax_1 + By_1 + C = 0. \dotfill (3)$$

Also from (2) the length of the perpendicular from the new origin upon this line $= \dfrac{Ax_1 + By_1 + C}{\sqrt{A^2 + B^2}}$, for $Ax_1 + By_1 + C$ is the constant term in equation (3).

But the new origin is the point (x_1, y_1) referred to the old axes;

$\therefore$ $\dfrac{Ax_1 + By_1 + C}{\sqrt{A^2 + B^2}}$ is the length of the perpendicular required.

40. *The equations of the bisectors of the angles between the straight lines* $a_1x + b_1y + c_1 = 0$, $a_2x + b_2y + c_2 = 0$ *are*

$$\frac{a_1x + b_1y + c_1}{\sqrt{a_1{}^2 + b_1{}^2}} = \pm\frac{a_2x + b_2y + c_2}{\sqrt{a_2{}^2 + b_2{}^2}}.$$

If (x_1, y_1) is any point *on either bisector*, the lengths of the perpendiculars from (x_1, y_1) to the given lines are equal;

$$\therefore \frac{a_1x_1 + b_1y_1 + c_1}{\sqrt{a_1{}^2 + b_1{}^2}} = \pm\frac{a_2x_1 + b_2y_1 + c_2}{\sqrt{a_2{}^2 + b_2{}^2}}. \quad \left(p = \frac{Ax_1 + By_1 + C}{\sqrt{A^2 + B^2}} \text{ (Art. 38.)}\right)$$

But (x_1, y_1) is *any* point on either bisector.
Therefore, suppressing the suffixes in x_1 and y_1 the *two* equations

$$\frac{a_1x + b_1y + c_1}{\sqrt{a_1{}^2 + b_1{}^2}} = \pm\frac{a_2x + b_2y + c_2}{\sqrt{a_2{}^2 + b_2{}^2}} \dotfill (1)$$

are the equations of the two bisectors.

Corollary. If ABC, DBE are the given lines, and P a point on the bisector of the $\angle$ ABE,

the perpendicular PN is on the opp. side of AB to the origin,

and   ,,   ,,   PM   ,,   ,,   BE   ,,

$\therefore$ PN *and* PM *have the same sign.*

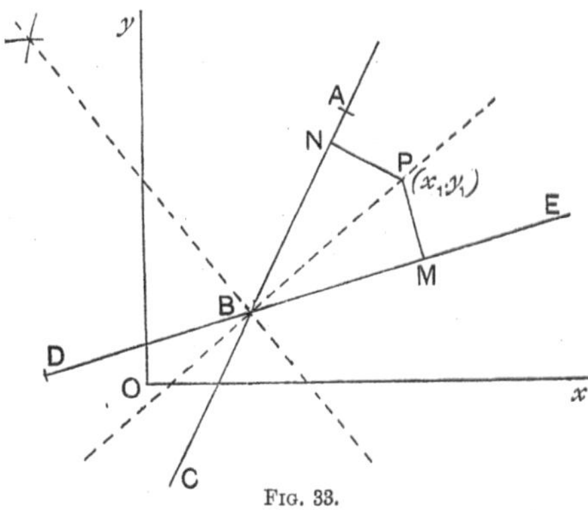

Fig. 33.

$\therefore$ the equation of BP (the bisector falling within the same angle DBC as the origin) is

$$\frac{a_1 x + b_1 y + c_1}{\sqrt{a_1{}^2 + b_1{}^2}} = + \frac{a_2 x + b_2 y + c_2}{\sqrt{a_2{}^2 + b_2{}^2}}.$$

In the same way it may be proved that the equation of the other bisector is

$$\frac{a_1 x + b_1 y + c_1}{\sqrt{a_1{}^2 + b_1{}^2}} = - \frac{a_2 x + b_2 y + c_2}{\sqrt{a_2{}^2 + b_2{}^2}}.$$

41. Note carefully the following :

A straight line passes through the origin if its equation contains no constant term.

The straight lines $x - y = 0$, $x + y = 0$ bisect the angles between the axes.

The straight line $y - y_1 = m(x - x_1)$ passes through the point (x_1, y_1) for all values of m.

The straight line $A(x - x_1) + B(y - y_1) = 0$ passes through the point (x_1, y_1) for all values of A and B.

The straight line $\lambda(ax+by+c)+\mu(a'x+b'y+c')=0$ passes through the intersection of $ax+by+c=0$, and $a'x+b'y+c'=0$ for all values of λ and μ.

The straight line $A(x-x_1)+B(y-y_1)=0$ passes through the point (x_1, y_1) and is parallel to $Ax+By+C=0$.

The straight line $B(x-x_1)-A(y-y_1)=0$ passes through the point (x_1, y_1) and is perpendicular to $Ax+By+C=0$.

Examples II. c.

[These questions may be taken orally with advantage.]

Write down, or mention, any special feature of the following straight lines:

  1. $Ax+By=0$.   **2.** $Ax+C=0$.   **3.** $By+C=0$.

  4. $x-y+2=0$.   **5.** $x+y-5=0$.

  6. What is the general equation of a straight line parallel to
$$3x-5y+7=0?$$

  7. What is the general equation of a straight line perpendicular to

    (1) $3x-5y=7$.   (2) $\dfrac{x}{2}+\dfrac{y}{3}=1$.   (3) $y=mx+c$.

    (4) $Ax+By+C=0$.   (5) $Ax-By=C$.   (6) $\dfrac{x}{a}-\dfrac{y}{b}=1$.

  8. What is the slope of the line $3x+6y-7=0$? What is the tangent of the acute angle it makes with the axis of y?

  9. What are the slopes of the lines joining the following pairs of points:

    (1) (0, 0), (3, 4).   (2) (0, 0), (h, k).   (3) (4, 5), (5, 6).

    (4) $(-2, 3)$, $(3, -5)$.   (5) (5, 3), $(-2, -4)$.

  10. What are the slopes of the straight lines perpendicular to the following:

    (1) $y=bx+c$.   (2) $y=-bx+c$.   (3) $by+x=c$.

    (4) $3x+4y+1=0$.   (5) $\dfrac{x}{2}-\dfrac{y}{3}=1$.   (6) $\dfrac{x}{4}+\dfrac{y}{5}=1$.

Write down, or read off, without simplifying, the length of the perpendicular from the point:

  11. (1, 1)   to the straight line $5x-y-1=0$.

  12. $(2, -1)$,,   ,, $3x-2y+4=0$.

  13. (0, 0)   ,,   ,, $4x-3y+5=0$.

  14. $(-2, 3)$,,   ,, $2y-3x-1=0$.

  15. (h, k) ,,   ,, $ay-bx=0$.

  16. $(-2, -3)$,,   ,, $y=3x$.

17.
$$5x - 3y - 7 = 0. \quad \dots\dots\dots\dots\dots\dots\dots\dots\dots (1)$$
$$3x + 8y + 7 = 0. \quad \dots\dots\dots\dots\dots\dots\dots\dots (2)$$
Adding,
$$8x + 5y = 0. \quad \dots\dots\dots\dots\dots\dots\dots\dots (3)$$
Interpret (3) in connection with (1) and (2).

18. What does the equation $3(y-2)+4(x-3)=0$ become when the origin is changed to the point $(3, 2)$?

19. What does the equation $2(y+3)-3(x+5)=0$ become when the origin is changed to the point $(5, 3)$?

Write down, or read off, without simplifying, the equations of the bisectors of the angles between the following pairs of straight lines, giving reasons for your answers:

20. $3x+4y-7=0$, $5x-12y+1=0$.   **21.** $x-3y=0$, $x+3y-5=0$.

22. $y=mx+c$, $y=m'x+c'$.

Examples. II. d.

[When you use a formula, quote it.]

Find the equation of the straight line:

1. Parallel to $3y-5x-7=0$, and passing through the origin.

2. Parallel to $y=3x+5$, and passing through the point $(2, -5)$.

3. Parallel to $3x+7y=5$, and passing through the point $(-1, 1)$.

4. Passing through the points $(3, 4)$, $(5, 6)$.   [Check your result.]

5.   ,,   ,, (a, b), $(-a, -b)$.   ,,   ,,

6.   ,,   ,, $(4, -5)$, $(0, 0)$.   ,,   ,,

7.   ,,   ,, $(4, 0)$, $(0, 0)$.   ,,   ,,

8.   ,,   ,, $(6, 5)$, $(6, -2)$.   ,,   ,,

9.   ,,   ,, $(3, -7)$, $(-2, 5)$.   ,,   ,,

10.   ,,   ,, $(0, 0)$, $(0, -3)$.   ,,   ,,

11. Find the acute angle between the straight lines $x=0$ and $y=\sqrt{3x}+4$.

12. Find the angle between the straight lines $x+y=7$ and $x-y=5$.

13. Find the tangent of the angle between the straight lines $y=5x-4$, $y=3x+6$.

14. Find the acute angle between the straight lines $4x-3y=12$, $5x+3y-2=0$.

15. Find the angle between the straight lines $3x-4y=12$, $4x+3y=2$.

16. What do you see about the inclinations of the lines $y-mx=c$, $y+mx=c$ to the axis of x?

17. Find the equation of the straight line through the point $(2, -3)$ and perpendicular to $y=2x+3$.

18. Find the equation of the straight line through the origin, and perpendicular to $3y-4x+5=0$.

19. Find the equations of the straight lines through the point $(3, -4)$ and making angles of $45°$ with the axis of x.

20. Prove that the following three points are collinear: $(-4, -5)$, $(1, -1)$, $(6, 3)$.

21. Find the equation of the straight line making an intercept 3 on the axis of x and at right angles to $y = 4x + 5$.

22. Find the equation of the straight line through the point $(5, 8)$ and making equal (positive) intercepts on the axes.

23. A straight line passes through the point $(2, 3)$ and the portion intercepted between the axes is bisected at that point: find its equation.

24. Find the equation of the straight line through the point $(2, 5)$ and making equal intercepts of opposite sign on the axes.

25. The equations $ax + by + c = 0$, $y = mx + k$ represent the same straight line: find the relations between the coefficients.

26. Find the equations of the straight lines through the point $(-2, -4)$ and making angles of $30°$ with the axis of x.

27. Find the equation of the straight line through the point $(-3, 4)$ and perpendicular to the line joining $(0, 0)$ and $(5, 6)$.

Find the perpendicular distance:

28. From the point $(3, 4)$ to the straight line $3x - 4y = 7$.

29. ,, ,, $(4, -2)$,, ,, $5x + 12y = 3$.

30. ,, ,, $(1, -1)$,, ,, $6x - 7y - 13 = 0$.

31. ,, ,, (a, b) ,, ,, $ax + by = 0$.

32. ,, ,, $(-2, -4)$,, ,, $\dfrac{x}{3} - \dfrac{y}{4} = 1$.

33. ,, ,, $(3, 6)$,, ,, $\dfrac{x}{12} - \dfrac{y}{5} = 1$.

34. From the line $y = 4x$ to the point of intersection of $y = 3x - 5$ and $y = 4x - 8$.

35. Between the parallels $3x - 4y = 12$, $3x - 4y = 3$.

36. ,, ,, $lx + my + n = 0$, $lx + my - n = 0$.

37. ,, ,, $\dfrac{x}{a} + \dfrac{y}{b} = 5$, $\dfrac{x}{a} + \dfrac{y}{b} + 9 = 0$.

[In numbers 38 to 42 the λ method of Art. 34 may be used with advantage.]

38. Find the equation of the straight line passing through the point $(1, 26)$ and the point of intersection of $5x - 7y + 6 = 0$ and $2y - 9x - 4 = 0$.

39. Find the equation of the straight line passing through the intersection of $3x - 5y + 6 = 0$ and $5x - 7y + 4 = 0$, and parallel to $13x - 19y + 7 = 0$.

40. Find the equation of the straight line through the point of intersection of $2x + y - 5 = 0$, and $7x - 9y + 1 = 0$, and perpendicular to the former.

41. A straight line passes through the origin and the point of intersection of $19x - 23y + 2 = 0$, and $3x + 5y - 1 = 0$: find its equation.

42. Find the equations of the straight lines passing through the intersection of $5x - 9y + 13 = 0$ and $9x - 5y + 11 = 0$, and making angles of $45°$ with the axis of x.

What is the area contained by the following straight lines?

43. $x=0,\ y=0,\ 4x-3y=24.$ **44.** $x=0,\ y=c,\ y=mx.$

45. $x-y=0,\ x+y=0,\ y=k.$ **46.** $x=0,\ y=0,\ x-5=0,\ y+6=0.$

47. $x-y=0,\ x+y=0,\ x-y+5=0,\ x+y-7=0.$

48. Show that the area of the triangle contained by the straight lines whose equations are $2y+x-5=0,\ y+2x-7=0,\ x-y+1=0,$ is $\frac{3}{2}$.

49. If $Ax+By=C$ and $y=mx+k$ represent the same straight line, prove that

$$\frac{C}{\sqrt{A^2+B^2}}=\frac{k}{\sqrt{1+m^2}}.$$

50. If $Ax+By=C$ and $x\cos a+y\sin a=p$ represent the same straight line, find p in terms of A, B, and C.

What are the equations of the straight lines bisecting the angles between?

51. $4x-3y=8$ and $3x+4y=10.$ **52.** $3x-y+4=0$ and $x+3y-5=0.$

53. $x-y=0$ and $x+y=0.$ **54.** $ax-by+c=0,\ bx-ay+c=0.$

55. $y-b=\dfrac{2m}{1-m^2}(x-a),\ \ y-b=\dfrac{2m'}{1-m'^2}(x-a).$

56. $y=x\tan\theta,\ y=x\tan 3\theta.$

57. In the equation $y-mx-c+\lambda(y-m'x-c')=0,$ λ is a variable quantity which may have any value: prove that the straight line represented by this equation passes through a fixed point.

58. Prove the same when the equation is $y-3x-11+\lambda(y-2x-9)=0,$ and find the co-ordinates of the point.

59. Prove that the straight line $y+\lambda x=5$ passes through a fixed point for all values of λ.

60. Prove that the straight line $x(1+\lambda)+y(2-\lambda)+5=0$ passes through a fixed point for all values of λ, and find the co-ordinates of the point.

LOCUS PROBLEMS ON THE STRAIGHT LINE.

42. Example 1. *A point moves so that the square of its distance from the point* $(6,\ -4)$ *is always greater than the square of its distance from the point* $(3,5)$ *by* 18 *: find its locus.*

Let (x,y) be any point on the locus.

The square of its distance from $(6,\ -4)=(x-6)^2+(y+4)^2.$

,,   ,,   ,, $(3,\ \ \ 5)=(x-3)^2+(y-5)^2.$

$\therefore$ by hypothesis, $(x-6)^2+(y+4)^2-(x-3)^2-(y-5)^2=18.$

Simplifying $-6x+18y=0$

$$\text{or}\quad y=\frac{x}{3}.$$

$\therefore$ the locus is a straight line through the origin, making an angle $\tan^{-1}\frac{1}{3}$ with the axis of x.

Example ii. A *and* B *being fixed points, the vertex* C *of the triangle* ABC *moves so that* $\cot A + \cot B = k$, *a constant: find the locus of* C.

[To find the equation of the locus we have to discover the algebraic equation connecting (x, y), the co-ordinates of any point on the locus, with the given constant and the length of AB.]

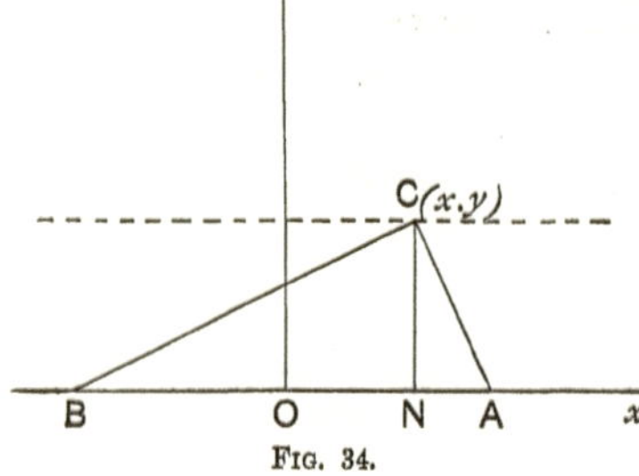

Fig. 34.

Let $AB = 2a$, and bisect it at O.

Take OA as axis of x, and Oy at rt. ∠s to OA as axis of y.

Let (x, y) be any point on the locus, and draw the ordinate CN.

$$\cot A = \frac{AN}{CN} = \frac{a - x}{y}$$

and

$$\cot B = \frac{BN}{CN} = \frac{a + x}{y};$$

$$\therefore k = \cot A + \cot B = \frac{2a}{y}.$$

$\therefore ky = 2a$ is the equation of the locus, a straight line parallel to AB the axis of x.

Note. **The choice of axes of co-ordinates is always an important step. A position of symmetry should be chosen when possible.**

Example iii. *A point moves so that its perpendicular distances from the straight lines* $3x + 4y - 5 = 0$, $3x - 4y - 7 = 0$ *are in the ratio of 1 to 2: find its locus.*

Let (x, y) be any point on the locus.

The perp. distance of (x, y) from

$$3x + 4y - 5 = 0 \text{ is } \frac{3x + 4y - 5}{5}. \qquad \left(p = \frac{Ax_1 + By_1 + C}{\sqrt{A^2 + B^2}} \right)$$

The perp. distance of (x, y) from

$$3x - 4y - 7 = 0 \text{ is } \frac{3x - 4y - 7}{5}. \qquad (\quad ,, \qquad ,, \quad)$$

$$\therefore \frac{3x + 4y - 5}{5} = \pm \frac{1}{2} \frac{3x - 4y - 7}{5}$$

or $\quad 6x + 8y - 10 = \pm (3x - 4y - 7).$

With the positive sign, $\quad 3x + 12y - 3 = 0$

$$\text{or} \quad x + 4y - 1 = 0. \dots\dots\dots\dots\dots\dots\dots\dots\dots\dots\dots(1)$$

With the negative sign, $\quad 9x + 4y - 17 = 0. \dots\dots\dots\dots\dots\dots\dots\dots\dots(2)$

The locus therefore is the two straight lines represented by (1) and (2).

Examples II. e.

1. Find the equation of, and draw the locus of a point equidistant from the points $(5, 0)$, $(-3, 0)$.

2. What is the equation of the locus of a point equidistant from the straight lines $x = -4$ and $x = 8$?   Draw the locus.

3. A point moves so that its distance from one straight line is always twice its distance from a perpendicular straight line.   Find the locus and draw it.

4. The difference of the squares of the distances of a point from the points $(3, 0)$, $(0, 3)$ is always equal to 6.   Find its locus and draw it.

5. The distance of a point from $(3, 2)$ is always equal to its distance from $(-4, 1)$.   Find the equation of, and draw its locus.

6. Find the equation of, and describe the locus of a point which moves at a perpendicular distance D from the straight line $Ax + By - C = 0$, and on the same side of the line as the origin.   Take A, B, and C as positive quantities.

7. Show that the locus of a point, which moves so that the sum of its perpendicular distances from two given intersecting straight lines is constant, is a straight line.

8. A point moves so that the sum of its distances from two perpendicular straight lines is equal to 6.   Find its locus and draw it.

9. P, any point on the line $Ax + By + C = 0$ is joined to the origin O. Find the equation of the locus of the middle point of OP and draw it.

10. OA, OB are two fixed perpendicular straight lines.   The straight line AB moves in such a manner that $OA + OB$ is always equal to 8.   Find the locus of the middle point of AB and draw it.

11. OA, OB are two fixed perpendicular straight lines, and perpendiculars PN, PM are drawn to them such that $PM - PN$ is always equal to 5. Find the locus of P and draw it.

12. P, any point on the line $Ax + By + C = 0$, is joined to the origin O, and OP is divided at Q in the ratio $1 : 2$.   Find the equation of the locus of Q and draw it.

13. Through a fixed point O any straight line is drawn, meeting two given parallel straight lines in P and Q; through P and Q straight lines are drawn in fixed directions, meeting in R.   Prove that the locus of R is a straight line.

[Take O as origin, and the axis of x at right angles to the parallel lines. If (x_1, y_1) are the co-ordinates of P, (x_2, y_2) the co-ordinates of Q, the equations of PR and QR may be taken as $y - y_1 = m_1(x - x_1)$, $y - y_2 = m_2(x - x_2)$, where m_1 and m_2 are constant.   Also $\dfrac{y_1}{x_1} = \dfrac{y_2}{x_2}$, and x_1, x_2 are constant.

From the above equations we now have to eliminate y_1 and y_2.]

14. The hypotenuse of a right-angled triangle slides between the axes of x and y, its ends always lying on the axes.   Prove that the locus of the vertex of the right angle is a straight line through the origin whose equation is $y = x \cot A$, where A is the angular point on the axis of x.

15. Show that the locus of a point which is equidistant from the lines $3x + 4y = 24$ and $4x + 3y = 12$ consists of two straight lines.   Find their equations, and draw a figure representing the four lines.

Examples II. f. Miscellaneous.

1. Find the point in the locus $y^2 = 4x$ where the ordinate is 8.

2. Find the co-ordinates of the points where the locus
$$x^2 + 3xy + 2y^2 - 5x - 8y + 6 = 0$$
cuts (1) the axis of x, (2) the axis of y.

3. What does the equation $(y-3)^2 = 4(x+5)$ become when the origin is changed to the point $(-5, 3)$?

4. Does the locus whose equation is $2x^2 - 4xy + y^2 - 10x - 5y - 14 = 0$ pass through the point $(1, -2)$?

5. With the equation $5x - 13y + 51 = 0$ change the origin to the point $(-5, 2)$. What do you deduce about the straight line represented by the equation?

6. Find the points on the locus $y^2 = 4x$ which are at a distance $2\sqrt{2}$ from the point $(3, 0)$.

7. If the origin is changed to the point $(-2, 3)$, what does the equation $x^2 + y^2 + 4x - 6y = 0$ become?

8. Find the length of the perpendicular from the origin upon the straight line joining the points $(a\cos a,\ a\sin a)$, $(a\cos\beta,\ a\sin\beta)$, and prove that the perpendicular bisects the distance between these points.

9. Prove, by using rectangular co-ordinates, that the straight line joining the middle points of two sides of a triangle is parallel to the third side and equal to half its length.

10. Find the equations of the medians of the triangle formed by joining the points $(3, 4)$, $(-5, 2)$, $(7, 6)$.

11. Prove, by using co-ordinates, that the straight lines joining the middle points of opposite sides of a quadrilateral bisect one another.

12. Find the perpendicular distance between the straight lines
$$y = 3(x-4), \quad y = 3(x-9).$$

13. Find the equation of the straight lines bisecting the angles between
$$y - b = (x-a)\tan 2\theta, \quad y - b = (x-a)\tan 4\theta.$$

14. Find the equations of the diagonals of the parallelogram formed by the straight lines $y = x$, $y = 3x$, $y = x - 4$, $y = 3x - 6$.

15. Prove that the area of the triangle formed by the three straight lines $y = mx$, $y = m'x$ and $y = c$ is equal to
$$\frac{c^2}{2}\left(\frac{1}{m} - \frac{1}{m'}\right).$$

16. The portion, **AB**, of a straight line intercepted between the axes passes through the fixed point (h, k). Find the equation of the locus of the middle point of **AB**.

17. Write down the general equation of a straight line perpendicular to $\dfrac{x}{a} + \dfrac{y}{b} = 1$, and hence determine the equation of a straight line perpendicular to $\dfrac{x}{a} + \dfrac{y}{b} = 1$, and through the point (a, b).

18. Find the equation of a straight line through the point (1, 4) and making with the axes in the first quadrant a triangle whose area is 8.

19. The portion of a straight line in the first quadrant cuts the axis of x at A and the axis of y at B. The point (10, 3) divides AB in the ratio of $3:5$. Find the equation of the line.

20. Prove that the straight line $3x(a+1)+4y(a-1)-8(a-1)=0$ passes through a fixed point for all values of a, and find its co-ordinates.

21. Find a point in the straight line $4x-2y+7=0$ equidistant from the points (2, 3), (-2, 4).

22. The co-ordinates of the points A, B, C, D are respectively (3, 6), (8, 10), (12, 6), (3, 4). Find the co-ordinates of the middle point of the line joining the middle points of AB and CD.

23. Find the co-ordinates of the centre of the circum-circle of the triangle formed by joining the points (-2, 2), (1, -2), (1, 3).

24. Find the co-ordinates of the points which divide internally and externally the line joining (-3, -4) and (-8, 7) in the ratio $7:5$.

25. Find the equation of the straight line perpendicular to the line $4x-3y=12$, and meeting it on the axis of x.

26. Find the equations of the straight lines parallel to the line $12x-5y=7$, and at a distance 2 from it.

27. Find the equation of the straight line parallel to the line $x+2y=5$, and at the same distance from the point (3, 2).

28. Find the equation of the straight line perpendicular to the line $\frac{x}{a}+\frac{y}{b}=1$, and through the point (b, a).

29. Find the co-ordinates of the points on the line $x+5y=13$, at a distance 2 from the line $12x-5y+26=0$.

30. Find the co-ordinates of the foot of the perpendicular from the point (a, 0) upon the line $y=mx+\dfrac{a}{m}$.

31. Find the perpendicular distance between the lines

$$y=(x-a)\tan a, \quad y=(x-a')\tan a.$$

32. Find the equations of the straight lines which are equally inclined to the lines $x+y=a$ and $x-y=a$, and form with them triangles whose areas are c^2.

33. Find the equations of the diagonals of the parallelogram whose sides have for their equations $x=a$, $x=b$, $x\cos a+y\sin a=p_1$, $x\cos a+y\sin a=p_2$.

34. Prove that the area of the triangle formed by joining the points $(am_1{}^2, am_1)$, $(am_2{}^2, am_2)$, $(am_3{}^2, am_3)$ is $\dfrac{a^2}{2}(m_1-m_2)(m_2-m_3)(m_3-m_1)$.

35. If O is the middle point of the line AB, and P any other point, prove that $PA^2+PB^2=2(AO^2+PO^2)$.

36. Find the equation of the straight line through the point (x_1, y_1) and perpendicular to the straight line $xx_1 + yy_1 = a^2$.

37. Find the co-ordinates of the orthocentre of the triangle formed by joining the points $(0, 0)$, $(8, 0)$, $(4, 6)$.

38. Find the equation of the straight line through the intersection of $3x - 4y = 7$ and $2x + 5y = 9$, and parallel to $x + y = 0$.

39. Find the equation of the straight line through the intersection of $ax + by + c = 0$ and $a'x + b'y + c' = 0$, and parallel to the axis of x.

40. Find the equation of the straight line through the intersection of $ax + by + c = 0$ and $a'x + b'y + c' = 0$, and parallel to the axis of y.

CHAPTER III.

PAIRS OF STRAIGHT LINES, Etc.

43. Let us consider the equation

$$y^2 - 5xy + 6x^2 = 0. \quad \dots\dots\dots\dots\dots\dots\dots(1)$$

It may be written $(y - 3x)(y - 2x) = 0.$

$\therefore$ if (x, y) is *any* point on the graph or locus whose equation is $y^2 - 5xy + 6x^2 = 0$, these values of x and y must satisfy one of the equations

$$y - 3x = 0, \quad y - 2x = 0.$$

In other words, the point (x, y) must lie on one of the two straight lines whose equations are $y - 3x = 0$, $y - 2x = 0$.

But each of these equations represents a straight line through the origin.

$\therefore$ equation (1) represents two straight lines passing through the origin, and their slopes are 3 and 2 respectively.

44. *Any homogeneous equation of the second degree in x and y represents two straight lines through the origin.*

$$ax^2 + 2hxy + by^2 = 0 \text{ is such an equation.}$$

It may be written $y^2 + \dfrac{2h}{b}xy + \dfrac{a}{b}x^2 = 0. \quad \dots\dots\dots\dots\dots\dots(1)$

Also if $y - m_1x$, $y - m_2x$ are the factors of the left-hand member, the equation may be written

$$(y - m_1x)(y - m_2x) = 0 ;$$

$\therefore$ if (x, y) is *any* point on the graph or locus whose equation is

$$ax^2 + 2hxy + by^2 = 0,$$

these values of x and y must satisfy one of the equations

$$y - m_1x = 0, \quad y - m_2x = 0,$$

and each of these equations represents a straight line through the origin.

$\therefore$ the equation $ax^2 + 2hxy + by^2 = 0$ represents two straight lines through the origin.

B.A.G.   D

COROLLARY. We know, by Algebra, that if the expression $ax^2 + 2hxy + by^2$ has *real* factors,

$$h^2 \not< ab.$$

If $h^2 < ab$, the factors are imaginary, and in such a case the straight lines are said to be imaginary.

If $h^2 = ab$, the expression $ax^2 + 2hxy + by^2$ is a perfect square, and the two straight lines are coincident.

Example i. The equation $y^2 - 10xy + 9x^2 = 0$
may be written $\qquad (y - 9x)(y - x) = 0.$
$\therefore$ the locus is the two straight lines $y = 9x$, $y = x$.

Example ii. The equation $y^2 - 4xy + 4x^2 = 0$
may be written $\qquad (y - 2x)^2 = 0.$
$\therefore$ in this case the locus is *two coincident* straight lines.

Example iii. Consider the equation $4y^2 + 3xy + x^2 = 0$.
Treating it as a quadratic for y,

$$y = \frac{-3 \pm \sqrt{9 - 16}}{8} x$$

$$= \frac{-3 \pm \sqrt{-7}}{8} x.$$

$\therefore$ the equation may be written

$$\left(y + \frac{3 - \sqrt{-7}}{8} x \right)\left(y + \frac{3 + \sqrt{-7}}{8} x \right) = 0,$$

and the straight lines are imaginary.

45. *Interpret the locus represented by the equation* $xy = 0$.

Here we see that if $x = 0$ or $y = 0$ (independently of one another), the given equation is satisfied.

$\therefore$ the locus is the two straight lines $x = 0$, $y = 0$, *i.e.* the axes of co-ordinates.

Or, we might look upon this as a particular case of Art. 44.

46. *Interpret the locus represented by the equation* $x^2 + y^2 = 0$.

A square is necessarily positive or zero, and here we have the sum of two squares equal to zero.

$\therefore$ the only real solution of this equation is

$$\left. \begin{array}{l} x = 0 \\ y = 0 \end{array} \right\} \text{ simultaneously.}$$

$\therefore$ the equation represents the point $(0, 0)$, the origin.

N.B. The expression $x^2 + y^2$ has imaginary factors $x + y\sqrt{-1}$ and $x - y\sqrt{-1}$, and therefore may be said to represent two imaginary straight lines.

Thus we have two *imaginary* straight lines passing through a *real* point $(0, 0)$.

47. *To find the angle between the two straight lines represented by the equation*
$$ax^2 + 2hxy + by^2 = 0.$$

Dividing through by x^2, the given equation may be written
$$b\left(\frac{y}{x}\right)^2 + 2h\left(\frac{y}{x}\right) + a = 0.$$

Hence, if the equations of the two lines are
$$y = m_1 x, \quad y = m_2 x,$$

m_1 and m_2 are the roots of this quadratic for $\dfrac{y}{x}$.

$$\therefore\ m_1 + m_2 = -\frac{2h}{b}, \quad\dotfill(1)$$

$$m_1 m_2 = \frac{a}{b}. \quad\dotfill(2)$$

Squaring (1), multiplying (2) by 4, and subtracting,
$$(m_1 - m_2)^2 = \frac{4h^2}{b^2} - \frac{4a}{b} = 4\left(\frac{h^2 - ab}{b^2}\right).$$

$$\therefore\ \text{the angle between the lines} = \tan^{-1}\frac{m_1 - m_2}{1 + m_1 m_2}$$

$$= \tan^{-1}\frac{\dfrac{2\sqrt{h^2 - ab}}{b}}{1 + \dfrac{a}{b}} = \tan^{-1}\frac{2\sqrt{h^2 - ab}}{a + b}. \quad\dotfill(3)$$

COROLLARY. If the straight lines represented by
$$ax^2 + 2hxy + by^2 = 0$$

are at **right angles**, we see from (3) that $a + b = 0$, for $\tan\dfrac{\pi}{2} = \infty$.

Hence, if the two straight lines represented by a homogeneous equation of the second degree are at right angles, the coefficients of x^2 and y^2 are equal, but of opposite sign.

When the actual values of a, h, and b are given, the angle between the lines $ax^2 + 2hxy + by^2 = 0$ is often best found by factorisation.

Example. *Find the angle between the straight lines* $2y^2 - xy - 3x^2 = 0$.
The equation may be written $(2y - 3x)(y + x) = 0$;

$$\therefore\ y = \frac{3x}{2} \text{ and } y = -x \text{ are the straight lines.}$$

$$\therefore\ \text{the angle reqd.} = \tan^{-1}\left(\frac{-1 - \frac{3}{2}}{1 - \frac{3}{2}}\right) \qquad \left(\tan^{-1}\frac{m - m'}{1 + mm'}\right)$$

$$= \tan^{-1}5 = 78°\ 41\tfrac{1}{2}' \text{ from 4-figure Tables.}$$

***48.** *To find the equation of the bisectors of the angles between the straight lines* $ax^2 + 2hxy + by^2 = 0$.

The given equation may be written $y^2 + \dfrac{2h}{b}xy + \dfrac{a}{b}x^2 = 0$.

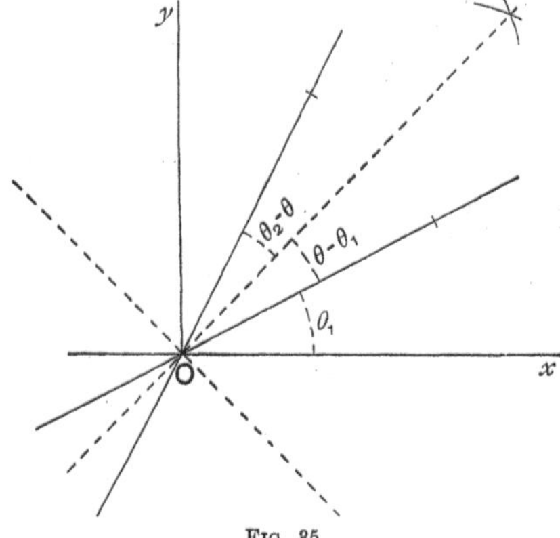

Fig. 35.

If these lines make angles θ_1, θ_2 with Ox, this equation is identical with $(y - \tan\theta_1 x)(y - \tan\theta_2 x) = 0$.

$$\therefore\ \tan\theta_1 + \tan\theta_2 = -\frac{2h}{b} \dots (1) \text{ and } \tan\theta_1 \tan\theta_2 = \frac{a}{b} \dots (2).$$

The two bisectors are at right angles; $\therefore$ if one makes an angle θ with Ox, as in the figure, the other makes an angle $\dfrac{\pi}{2} + \theta$ with Ox.

$\therefore$ the equation of the bisectors may be written

$$\left[y - x \tan \theta\right]\left[y - x \tan\left(\frac{\pi}{2} + \theta\right)\right] = 0$$

$$\text{or } (y - x \tan \theta)(y + x \cot \theta) = 0,$$

$$i.e. \ y^2 + (\cot \theta - \tan \theta)xy - x^2 = 0,$$

$$y^2 + 2 \cot 2\theta \, xy - x^2 = 0. \ \dots\dots\dots\dots\dots\dots(3)$$

Now from the figure we see that $\theta - \theta_1 = \theta_2 - \theta$;

$$\therefore \ 2\theta = \theta_1 + \theta_2,$$

$$\cot 2\theta = \cot(\theta_1 + \theta_2) = \frac{1 - \tan \theta_1 \tan \theta_2}{\tan \theta_1 + \tan \theta_2}$$

$$= \frac{1 - \dfrac{a}{b}}{-\dfrac{2h}{b}} \text{ from (1) and (2)}$$

$$= \frac{a - b}{2h} \, ;$$

$$\therefore \text{ from (3) } y^2 + \frac{a - b}{h}xy - x^2 = 0$$

$$\text{or } \frac{x^2 - y^2}{a - b} = \frac{xy}{h} \text{ is the reqd. equation.}$$

Examples III. a.

What do the following equations represent ?

1. $(x - a)(x - b) = 0.$
2. $(x - a)^2 + (y - b)^2 = 0.$
3. $(y - mx - c)(y - m'x - c') = 0.$
4. $(y - mx - c)^2 + (y - m'x - c')^2 = 0.$
5. $x^2 - y^2 = 0.$
6. $y^2 - xy = 0.$
7. $(x + y)^2 - a^2 = 0.$
8. $y^2 - xy - 6x^2 = 0.$
9. $(x + y - a)^2 + (x - y - a)^2 = 0.$
10. $xy - c(x + y) + c^2 = 0.$

Find the angles between the following pairs of straight lines :

11. $2x^2 - 7xy + 3y^2 = 0.$
12. $6x^2 + 5xy - 6y^2 = 0.$

13. Interpret the equation $ax^2 + by^2 = 0$, (1) when a and b have the same sign, (2) when they are of opposite sign.

14. Find the tangent of the angle between the straight lines represented by the equation $2y^2 + 4xy - 3x^2 = 0.$

15. Find the acute angle between the straight lines represented by $y^2 - xy - 6x^2 = 0$, and verify your result on squared paper.

16. Find the angle between the two straight lines represented by $ax^2 + bxy + cy^2 = 0$, and show that the one will make the same angle with the axis of x that the other does with the axis of y if $a = c$.

17. Show that the straight lines represented by $11y^2 + 16xy - x^2 = 0$ are inclined at an angle of $30°$ to the line $x + 2y = 1$.

18. Find the condition that the lines represented by

$$a^2x^2 + bcy^2 = a(b + c)xy$$

may be at right angles.

49. *To find the condition that an equation of the second degree may represent two straight lines.*

The most general equation of the second degree is

$$ax^2 + 2hxy + by^2 + 2gx + 2fy + c = 0. \quad \dots\dots\dots\dots(1)$$

(i) When a is not zero.

Treating the equation as a quadratic for x, and solving,

$$ax^2 + 2x(hy + g) + by^2 + 2fy + c = 0$$

and
$$x = \frac{-2(hy + g) \pm 2\sqrt{(hy + g)^2 - a(by^2 + 2fy + c)}}{2a}. \quad \dots\dots\dots(2)$$

$$\left[\frac{-b \pm \sqrt{b^2 - 4ac}}{2a}\right]$$

Now if equation (1) represents two straight lines, its left-hand side must break up into real factors of the first degree, or linear factors; therefore the quantity under the radical sign in (2) must be a perfect square.

This expression may be written

$$(h^2 - ab)y^2 + 2(hg - af)y + g^2 - ac\,;$$
$$\therefore\ \ 4(hg - af)^2 = 4(h^2 - ab)(g^2 - ac) \quad\quad (b^2 = 4ac)$$

if equation (1) represents two straight lines.

This therefore is the required condition.

It may be written $abc - af^2 - bg^2 - ch^2 + 2fgh = 0$.

The expression $abc - af^2 - bg^2 - ch^2 + 2fgh$ is called the **Discriminant**.

(ii) If a is zero, but b not zero, we shall obtain the required condition by treating the left-hand member of (1) as a quadratic for y.

The condition will be found to be

$$2fgh - bg^2 - ch^2 = 0,$$

which is the same as (3) when $a = 0$.

(iii) When both a and b are zero, the equation is

$$2hxy + 2gx + 2fy + c = 0$$

$$\text{or} \quad xy + \frac{g}{h}x + \frac{f}{h}y + \frac{c}{2h} = 0.$$

If the left-hand member has linear factors, they must be of the form
$$x + a, \; y + \beta \, ;$$

$\therefore$ the equation is identical with

$$(x + a)(y + \beta) = 0$$

$$\text{or} \quad xy + \beta x + ay + a\beta = 0 \, ;$$

$\therefore$ comparing coefficients,

$$\frac{g}{h} = \beta, \quad \frac{f}{h} = a, \quad \frac{c}{2h} = a\beta \, ;$$

$$\therefore \frac{fg}{h^2} = \frac{c}{2h} \quad \text{or} \quad 2fg - ch = 0,$$

the required condition.

Here again we observe that the condition is the same as (3) when $a = 0$ and $b = 0$.

Thus, whatever values a, h, and b may have, the condition that $ax^2 + 2hxy + by^2 + 2gx + 2fy + c = 0$ may represent two straight lines is
$$abc - af^2 - bg^2 - ch^2 + 2fgh = 0.$$

50. *If* $\qquad ax^2 + 2hxy + by^2 + 2gx + 2fy + c = 0$ $\quad$ ....(1)
represents two straight lines, they are parallel to the straight lines represented by $\qquad ax^2 + 2hxy + by^2 = 0.$ $\quad$ ....(2)

By hypothesis, the left-hand member of (1) can be resolved into two linear factors. Let this left-hand member be equal to

$$(l_1 x + m_1 y + n_1)(l_2 x + m_2 y + n_2).$$

Multiplying this out, we see that $(l_1 x + m_1 y)(l_2 x + m_2 y)$, the terms of the second degree, must be identical with

$$ax^2 + 2hxy + by^2.$$

$\therefore$ equation (2) is identical with $(l_1 x + m_1 y)(l_2 x + m_2 y) = 0$, and therefore represents the lines $l_1 x + m_1 y = 0$ and $l_2 x + m_2 y = 0$, which are respectively parallel to $l_1 x + m_1 y + n_1 = 0$ and $l_2 x + m_2 y + n_2 = 0$.

COROLLARY. *If the equation $ax^2 + 2hxy + by^2 + 2gx + 2fy + c = 0$ represents two* **perpendicular** *straight lines, $a + b = 0$, or the coefficients of x^2 and y^2 are equal but of opposite sign.*

The lines are parallel to the lines represented by

$$ax^2 + 2hxy + by^2 = 0 \; ;$$

$$\therefore \; a + b = 0. \quad \text{(Art. 47.)}$$

51. If two equations are combined in any way, without transgressing the laws of algebra, the resulting equation represents a locus which passes through the common points of the loci represented by the first two equations.

Take the equations

$$ax^2 + 2hxy + by^2 = c, \quad\dots\dots\dots\dots\dots\dots\dots(1)$$

$$lx + my = n. \quad\dots\dots\dots\dots\dots\dots\dots(2)$$

Consider the equation

$$n(ax^2 + 2hxy + by^2) = c(lx + my). \quad\dots\dots\dots\dots(3)$$

We have multiplied the members of (1) by the *equal* quantities n and $lx + my$.

Also whenever $lx + my = n$, (3) becomes $ax^2 + 2hxy + by^2 = c$, *i.e.* any values of x and y which satisfy (1) and (2) also satisfy (3).

In other words any point which lies on (1) and (2) lies also on (3), which proves the proposition in this case.

The proposition may be proved in the same way with any other equations.

52. *To find the equation of the straight lines joining the origin to the points of intersection of*

$$ax^2 + 2hxy + by^2 + 2gx + 2fy + c = 0 \quad\dots\dots\dots\dots(1)$$

$$and \;\; lx + my + n = 0. \quad\dots\dots\dots\dots(2)$$

The required equation will be homogeneous and of the second degree.

In equation (1),

multiply the terms of the second degree by $(-n)^2$,

      ,,   ,,   first   ,, $-(lx + my)n$,

      ,,   constant term by $(lx + my)^2$.

[*N.B.* These three expressions are equal for $lx + my = -n$.]

The result is

$$n^2(ax^2 + 2hxy + by)^2 - 2n(lx + my)(gx + fy) + c(lx + my)^2 = 0. \quad\dots(3)$$

Now any values of x and y which satisfy equations (1) and (2) satisfy (3) also, for it is formed by combining equations (1) and (2);

$\therefore$ the locus represented by (3) passes through the common points of (1) and (2).

Also equation (3) is homogeneous and of the second degree, and therefore represents two straight lines through the origin;

$\therefore$ it represents the lines required.

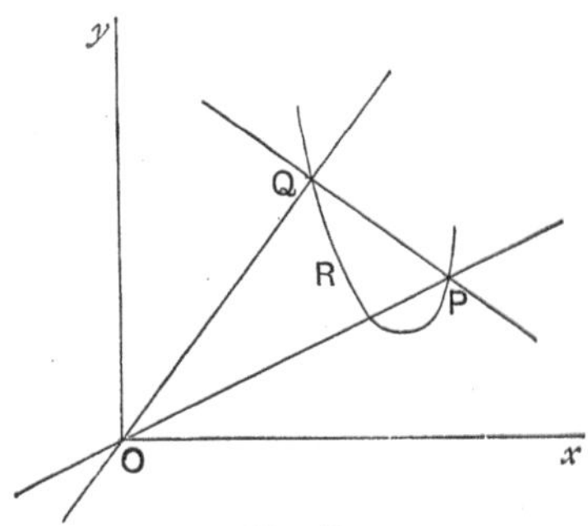

FIG. 36.

If PRQ is the curve represented by (1), PQ the straight line represented by (2), (3) is the equation of the straight lines OP, OQ.

53. *A homogeneous equation of the nth degree represents n straight lines, which all pass through the origin.*

When the coefficient of y^n, the highest power of y in the equation, is reduced to unity, the equation may be written

$$y^n + a_1 y^{n-1} x + a_2 y^{n-2} x^2 + \ldots + a_n x^n = 0.$$

Dividing through by x^n, this becomes

$$\left(\frac{y}{x}\right)^n + a_1 \left(\frac{y}{x}\right)^{n-1} + a_2 \left(\frac{y}{x}\right)^{n-2} + \ldots + a_n = 0. \ldots\ldots\ldots\ldots(1)$$

Treating this as an equation for $\frac{y}{x}$, it has n roots, for it is of the nth degree.

Let $m_1,\ m_2,\ m_3,\ \ldots m_n$ be the roots.

Then the equation (1) must be identical with

$$\left(\frac{y}{x} - m_1\right)\left(\frac{y}{x} - m_2\right)\left(\frac{y}{x} - m_3\right)\ldots\left(\frac{y}{x} - m_n\right) = 0. \ldots\ldots\ldots(2)$$

Hence if (x, y) is *any point on any one of the straight lines,* $y - m_1 x = 0,\ \ y - m_2 x = 0,\ \ldots y - m_n x = 0$, those values of x and y satisfy equation (1), for one of the factors of (2) will be zero;

$\therefore$ the given equation represents the n straight lines $y - m_1 x = 0$, $y - m_2 x = 0,\ \ldots y - m_n x = 0$, all of which pass through the origin.

COROLLARY. If some of the roots of equation (1) are imaginary, a corresponding number of the straight lines will be imaginary.

Example. *If* $12x^2 - 29xy + 15y^2 + 2x - 7y + \lambda = 0$ *represents two straight lines, find the value of* λ.

The terms of the second degree factorise into $(4x - 3y)(3x - 5y)$;

$\therefore$ if the equation represents two straight lines, the left-hand member must have factors of the form $(4x - 3y + m)(3x - 5y + n)$;

$\therefore$ the expression $12x^2 - 29xy + 15y^2 + 2x - 7y + \lambda$

is identical with

$$12x^2 - 29xy + 15y^2 + x(3m + 4n) - y(5m + 3n) + mn\ ;$$

$$\therefore\ \ 3m + 4n = 2, \ldots\ldots\ldots\ldots\ldots\ldots\ldots\ldots\ldots\ldots(1)$$
$$5m + 3n = 7, \ldots\ldots\ldots\ldots\ldots\ldots\ldots\ldots\ldots\ldots(2)$$
$$\text{and}\ \ mn = \lambda. \ldots\ldots\ldots\ldots\ldots\ldots\ldots\ldots\ldots\ldots(3)$$

Solving (1) and (2), $m = 2,\ n = -1$;

$$\therefore\ \lambda = -2,\ \text{from (3)}.$$

Examples III. b.

1. Solve the equation $2x^2 + 2y^2 + 5xy + 1 = 3x + 3y$ for y.   Draw the locus represented by the equation.

2. Solve the equation $x^2 - 3xy + 2y^2 + 4x - 5y + 3 = 0$ for x, and draw the locus represented by the equation.

3. Solve the equation $6y^2 - 17xy + 7x^2 = 0$ for y, and draw the locus represented by the equation.

In each of the following cases find a value of λ for which the equation will represent two straight lines :

4. $y^2 - 3xy + 2x^2 + x + 2y + \lambda = 0.$ **5.** $\lambda xy + 10x + 6y + 4 = 0$.

6. $\lambda y^2 - 7xy + 4x^2 - 11x + 8y - 3 = 0.$ **7.** $\lambda x^2 - y^2 - 4x + 8y - 15 = 0.$

8. $3x^2 + \lambda xy - 4y^2 - 3x + 11y - 6 = 0.$

9. Find the angle between the straight lines in Example 4 above.

10.   „   „   „   „   5   „

11. Prove that the equation $35x^2 - 24xy - 35y^2 + 2x + 12y - 1 = 0$ represents two straight lines at right angles, and find their equations.

12. Find the single equation which represents the two straight lines which pass through the point (3, 0) and make angles of 45° with the axis of x.

13. Show that $x^2 + 2xy + y^2 = 8ax + 8ay + 9a^2$ represents two parallel straight lines, and determine the perpendicular distance between them.

14. Find the condition that $ax^2 + by^2 + x + y = 0$ may represent two straight lines.

15. Find the condition that the straight lines $px^2 - qxy + ry^2 = 0$ should be (1) coincident, (2) perpendicular, (3) imaginary.

16. Given that $ax^2 + 2hxy + by^2 + 2gx + 2fy + c = 0$ represents two straight lines, find the angle between them.

17. Prove that if $ax^2 + 2hxy + by^2 + 2gx + 2fy + c = 0$ represents two straight lines, these lines are parallel if $h^2 = ab$.

18. Prove that $y^2 - 4xy + 4x^2 - 6x + 3y = 0$ represents two parallel straight lines, and draw them.

19. If λ is any positive quantity, prove that the equation
$$\lambda x^2 - y^2 + 8y - 16 = 0$$
represents two straight lines which intersect at the point (0, 4).

20. In the previous example, what will the locus be if λ is a negative quantity ?

21. Find the angle between the straight lines represented by
$$xy - 3x + 2y - 6 = 0.$$

22. If the equation $2hxy + 2gx + 2fy + c = 0$ represents two straight lines, prove that $2fg = ch$, and that these lines and the axes form a rectangle whose diagonals are $\dfrac{x}{f} - \dfrac{y}{g} = 0$ and $\dfrac{x}{f} + \dfrac{y}{g} + \dfrac{1}{h} = 0.$

23. Prove that the locus $(y - 2x)^2 + \lambda(y + 5x - 7) = 0$ passes through a fixed point for all values of λ, and find the co-ordinates of that point.

24. Find the equation of the two straight lines joining the origin to the points of intersection of $x^2 - 2xy + 3y^2 + 3x - 4y = 0$ and $3x + 4y = 1$.

25. Prove that the straight lines joining the origin to the points of intersection of $x^2 - xy + y^2 + 3(x + y) - 2 = 0$ and $x - y - \sqrt{2} = 0$ are at right angles.

26. Prove that, if the straight lines joining the origin to the points of intersection of $3x^2 - xy + 3y^2 + 2x - 3y + 4 = 0$ and $2x + 3y = \lambda$ are at right angles, $6\lambda^2 - 5\lambda + 52 = 0.$

27. Prove that the equation $y^2 - 2xy \sec \alpha + x^2 = 0$ represents two straight lines inclined at an angle α.

*CHAPTER IV.

OBLIQUE AXES. POLAR CO-ORDINATES.

54. When the axes xOx', yOy' are not at right angles, they are said to be **oblique.**

In this case the ordinate PN must be drawn parallel to Oy.

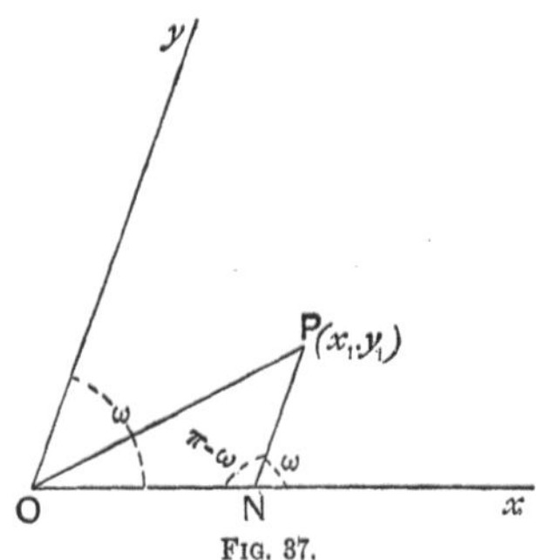

Fig. 87.

Thus in the figure (ON, NP) are the co-ordinates of the point P.

The same rule of signs is used as with rectangular axes.

The angle between the axes Ox and Oy is usually denoted by ω.

55. *To find the distance of the point (x_1, y_1) from the origin, when ω is the angle between the axes.*

Let P be the point (x_1, y_1), and draw the ordinate PN.

$$\angle ONP = \pi - \angle PNx = \pi - \omega \;;$$

$$\therefore \; OP^2 = ON^2 + PN^2 - 2 \cdot ON \cdot PN \cos(\pi - \omega) \quad (a^2 = b^2 + c^2 - 2bc \cos A)$$

$$= x_1^2 + y_1^2 + 2x_1 y_1 \cos \omega.$$

56. *To find the distance between the points (x_1, y_1), (x_2, y_2) when ω is the angle between the axes.*

Let P and Q be the points. Draw the ordinates QM, PN, and also draw QR parallel to Ox to meet PN at R.

From the $\triangle$ PQR,

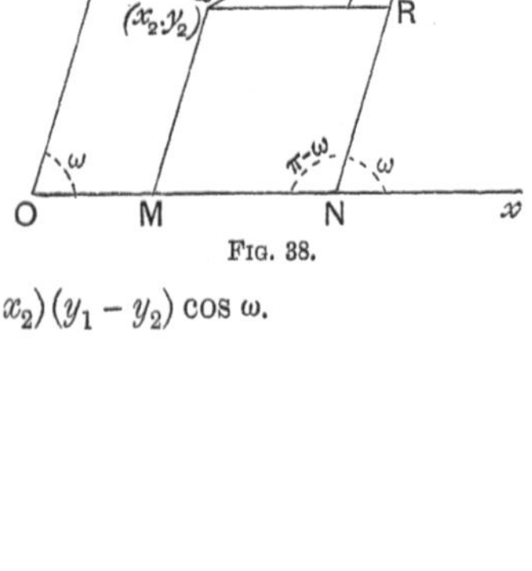

$$PQ^2 = QR^2 + PR^2 - 2QR \cdot PR \cos QRP$$

$$(a^2 = b^2 + c^2 - 2bc \cos A)$$

$$= (x_1 - x_2)^2 + (y_1 - y_2)^2$$

$$- 2(x_1 - x_2)(y_1 - y_2)\cos(\pi - \omega)$$

$$= (x_1 - x_2)^2 + (y_1 - y_2)^2 + 2(x_1 - x_2)(y_1 - y_2)\cos \omega.$$

Fig. 88.

57. *When ω is the angle between the axes, the co-ordinates of the middle point of the line joining the points (x_1, y_1), (x_2, y_2) are*

$$\frac{\mathbf{x}_1 + \mathbf{x}_2}{2}, \quad \frac{\mathbf{y}_1 + \mathbf{y}_2}{2}.$$

The proof of Art. 5 holds, for the $\triangle^s$ QRE, RPF are still similar.

58. *To find the co-ordinates of the point dividing in a given ratio the line joining two given points, when the axes are oblique.*

The proof and formulae of Art. 6 hold, for $\triangle^s$ QRE, RPF are still similar.

THE STRAIGHT LINE.　OBLIQUE AXES.

59. It will be assumed throughout that ω is the angle between the axes when the axes are oblique.

As in Art. 13, the equation of a straight line parallel to Oy, and making an intercept a on the axis of x, is $x = a$.

In the same way the equation of a straight line parallel to Ox, and making an intercept b on the axis of y, is $y = b$.

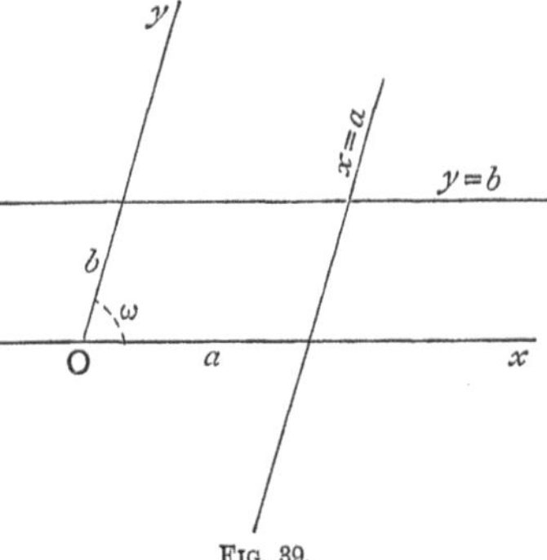

FIG. 39.

60. *To find the equation of a straight line through the origin and making an angle a with the axis of x.*

If P (x, y) is any point on the str. line, and PN its ordinate, *drawn parallel to* Oy, from the figure,

$$\frac{y}{x} = \frac{\text{PN}}{\text{ON}} = \frac{\sin a}{\sin (\omega - a)};$$

$$\therefore \ y = \frac{\sin a}{\sin (\omega - a)}x$$

is the required equation.

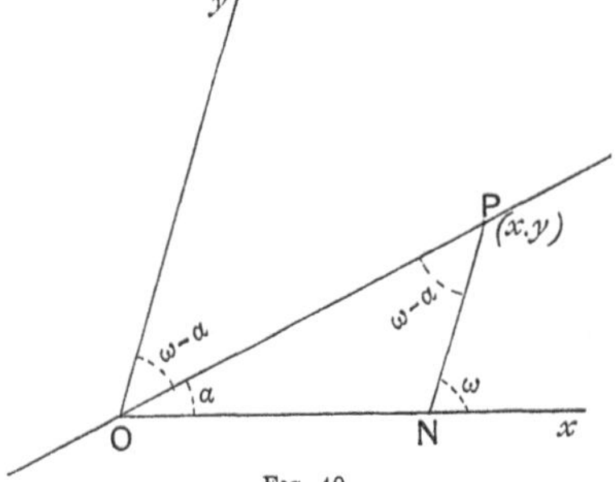

FIG. 40.

This may be written $y = mx$, where $m = \dfrac{\sin a}{\sin (\omega - a)}$.

N.B. **In all cases of oblique axes it will be found that 'm' is no longer the tangent of the angle which the line makes with Ox.**

61. *The equation of a straight line passing through the origin and through the point* (x_1, y_1) *is* $\dfrac{y}{y_1} = \dfrac{x}{x_1}$.

The method of proof in Art. 17 holds for this, for $\triangle^s$ PNO, QMO are still similar.

62. *To find the equation of a straight line making an intercept c on the axis of y and an angle α with the axis of x.*

Let P(x, y) be any point on the str. line PB which cuts Oy at B.

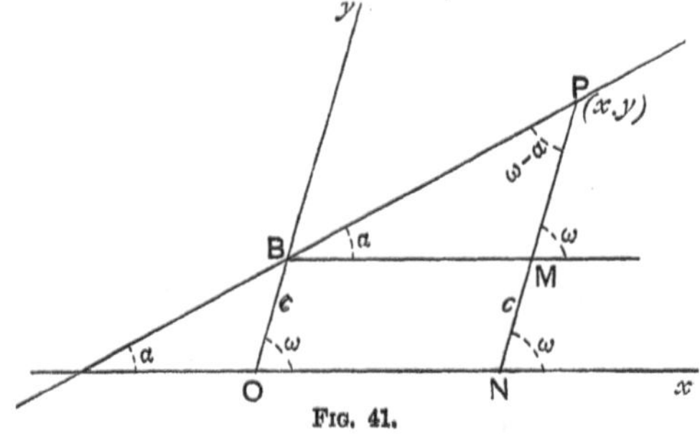

FIG. 41.

Draw PMN parallel to Oy, and BM parallel to Ox.

From the figure,
$$\frac{\text{PM}}{\text{BM}} = \frac{\sin \alpha}{\sin(\omega - \alpha)},$$

i.e.
$$\frac{y - c}{x} = \frac{\sin \alpha}{\sin(\omega - \alpha)};$$

$$\therefore \ y = \frac{\sin \alpha}{\sin(\omega - \alpha)} x + c \text{ is the reqd. equation.}$$

This may be written $y = mx + c$, where $m = \dfrac{\sin \alpha}{\sin(\omega - \alpha)}$.

63. *To find the angle which the straight line $y = mx + c$ makes with the axis of x.*

By the preceding Article, $m = \dfrac{\sin \alpha}{\sin(\omega - \alpha)}$;

$\therefore$ multiplying up and expanding,

$$m \sin \omega \cos \alpha - m \cos \omega \sin \alpha = \sin \alpha.$$

Dividing by cos α, and transposing,

$$\tan \alpha (1 + m \cos \omega) = m \sin \omega \,;$$

$$\therefore \; \alpha = \tan^{-1} \frac{m \sin \omega}{1 + m \cos \omega}.$$

64. *The equation of a straight line making intercepts a and b on the axes of x and y respectively is*

$$\frac{\mathbf{x}}{\mathbf{a}} + \frac{\mathbf{y}}{\mathbf{b}} = 1.$$

The method of proof in Art. 19 holds for this, for the $\triangle^s$ PNA, BOA are still similar.

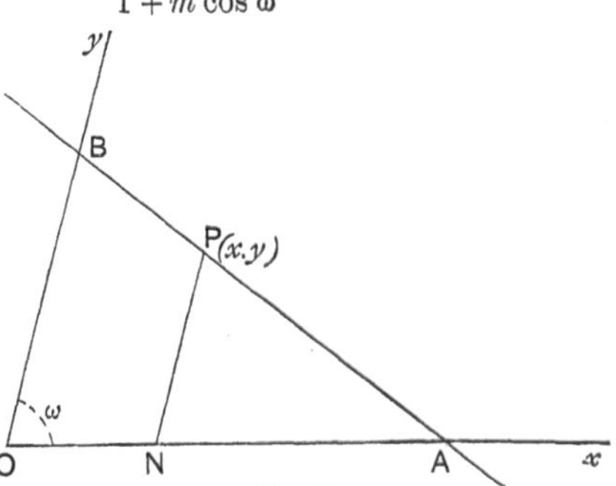

FIG. 42.

65. *If the perpendicular (p) from the origin upon a straight line makes angles α, β with the axes of x and y respectively, the equation of the line is*

$$\mathbf{x} \cos \alpha + \mathbf{y} \cos \beta = \mathbf{p}.$$

With the same construction as for rect. axes in Art. 20,

$$p = \mathsf{OQ} = \mathsf{OR} + \mathsf{RQ}$$
$$= \mathsf{OR} + \mathsf{PM}$$
$$= \mathsf{ON} \cos \alpha + \mathsf{PN} \cos \beta$$
$$= x \cos \alpha + y \cos \beta,$$

which proves the proposition.

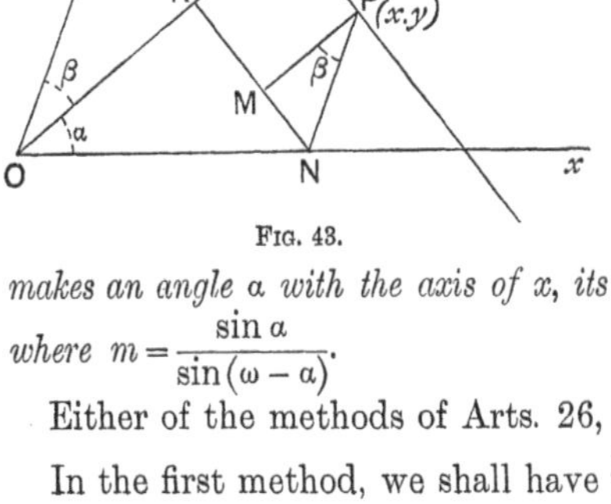

FIG. 43.

66. *If a straight line passes through the point (x_1, y_1) and makes an angle α with the axis of x, its equation is $y - y_1 = m(x - x_1)$, where $m = \dfrac{\sin \alpha}{\sin(\omega - \alpha)}$.*

Either of the methods of Arts. 26, 27 may be used.

In the first method, we shall have $\dfrac{y - y_1}{x - x_1} = \dfrac{\mathsf{PK}}{\mathsf{QK}} = \dfrac{\sin \alpha}{\sin(\omega - \alpha)} = m.$

67. *The equation of a straight line passing through the two points* (x_1, y_1), (x_2, y_2) *is* $\dfrac{y - y_1}{y_2 - y_1} = \dfrac{x - x_1}{x_2 - x_1}$.

Either of the methods of Arts. 28, 29 may be used to prove this.

68. *To find the co-ordinates of the point of intersection of two given straight lines.*

The method of Art. 23 holds good for oblique axes.

69. *To find the angle between the straight lines*
$$y = mx + c, \quad y = m'x + c'.$$

As in Art 31, if the lines make angles a, a' with Ox,

$$\tan \phi = \tan (a - a') = \frac{\tan a - \tan a'}{1 + \tan a \tan a'}.$$

Now
$$m = \frac{\sin a}{\sin (\omega - a)}. \quad \text{(Art. 62.)}$$

$\therefore$ as in Art. 63, multiplying up and expanding,
$$m \sin \omega \cos a - m \cos \omega \sin a = \sin a.$$

Dividing by $\cos a$ and transposing,
$$\tan a (1 + m \cos \omega) = m \sin \omega$$

and
$$\tan a = \frac{m \sin \omega}{1 + m \cos \omega}.$$

In the same way,
$$\tan a' = \frac{m' \sin \omega}{1 + m' \cos \omega};$$

$$\therefore \ \tan \phi = \frac{\dfrac{m \sin \omega}{1 + m \cos \omega} - \dfrac{m' \sin \omega}{1 + m' \cos \omega}}{1 + \left(\dfrac{m \sin \omega}{1 + m \cos \omega}\right)\left(\dfrac{m' \sin \omega}{1 + m' \cos \omega}\right)}$$

$$= \frac{(m - m') \sin \omega}{1 + (m + m') \cos \omega + mm'}, \text{ when simplified.}$$

COROLLARY 1.   The lines are parallel if $m - m' = 0$.

COROLLARY 2.   The lines are at right angles if
$$1 + (m + m') \cos \omega + mm' = 0.$$

70. *If (x_1, y_1) is a given point on a straight line, (x, y) any point on the line, α and β the angles the line makes with the axes of x and y respectively, and r the distance between the points (x, y), (x_1, y_1), the equation of the straight line may be written*

$$\frac{x - x_1}{l} = \frac{y - y_1}{m} = r,$$

where $l = \dfrac{\sin \beta}{\sin \omega} = \dfrac{\sin(\omega - \alpha)}{\sin \omega}$, and $m = \dfrac{\sin \alpha}{\sin \omega}$.

Let P be the point (x, y), Q the point (x_1, y_1), so that $PQ = r$.

Draw the ordinates QM, PN, and draw QK parallel to Ox to meet PN at K.

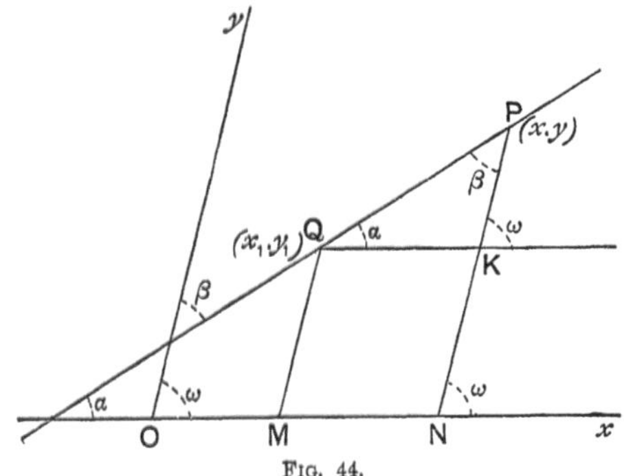

Fig. 44.

In the $\triangle$ PKQ, $\dfrac{QK}{\sin \beta} = \dfrac{PK}{\sin \alpha} = \dfrac{PQ}{\sin \omega}$,

i.e. $\dfrac{x - x_1}{\sin \beta} = \dfrac{y - y_1}{\sin \alpha} = \dfrac{r}{\sin \omega}$

or $\dfrac{x - x_1}{l} = \dfrac{y - y_1}{m} = r$, the reqd. equations.

71. *If λ is any constant $ax + by + c + \lambda(a'x + b'y + c') = 0$ represents a straight line passing through the intersection of the straight lines $ax + by + c = 0$ and $a'x + b'y + c' = 0$.*

The proof of Art. 34 holds good for oblique axes.

72. *If in any equation we write $x + h$ for x and $y + k$ for y, the resulting equation is the equation of the same locus referred to parallel axes through the point (h, k).*

The proof of Art. 36 holds good.

73. *As the point (x_1, y_1) moves from one side of the straight line $Ax + By + C = 0$ to the other, the expression $Ax_1 + By_1 + C$ changes its sign.*

The proof of Art. 37 holds good.

74. *To find the length of the perpendicular drawn from the point (x_1, y_1) to the straight line $Ax + By + C = 0$, when ω is the angle between the axes.*

Let the straight line cut the axes at E and F, and draw ON perpendicular to EF.　First let us find the length of ON.

$$\tfrac{1}{2}ON \cdot EF = \triangle OEF = \tfrac{1}{2}OE \cdot OF \sin \omega ;$$

$$\therefore \ ON = \frac{OE \cdot OF \sin \omega}{EF}. \quad\dots\dots\dots\dots\dots\dots(1)$$

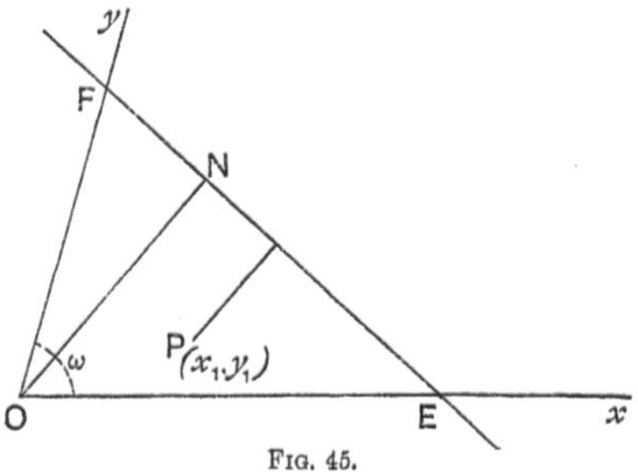

At E,　$y = 0$;

$\therefore \ Ax + C = 0$;

$\therefore \ OE = -\dfrac{C}{A}.$

At F,　$x = 0$;

$\therefore \ By + C = 0$;

$\therefore \ OF = -\dfrac{C}{B}.$

Fig. 45.

$$EF^2 = OE^2 + OF^2 - 2OE \cdot OF \cos \omega \quad (a^2 = b^2 + c^2 - 2bc \cos A)$$

$$= C^2\left(\frac{1}{A^2} + \frac{1}{B^2} - \frac{2 \cos \omega}{AB}\right) ;$$

$$\therefore \ EF = \frac{C}{AB}\sqrt{A^2 + B^2 - 2AB \cos \omega}.$$

$$\therefore \ \text{from (1)} \quad ON = \frac{C \sin \omega}{\sqrt{A^2 + B^2 - 2AB \cos \omega}}. \quad\dots\dots\dots\dots(2)$$

To find the length of the perpendicular from (x_1, y_1), first transfer the origin to the point (x_1, y_1).

The equation of the line is now

$$A(x + x_1) + B(y + y_1) + C = 0$$

$$\text{or}\quad Ax + By + Ax_1 + By_1 + C = 0. \ldots\ldots\ldots\ldots\ldots(3)$$

Also from (2), the length of the perpendicular from the new

origin upon this line $= \dfrac{(Ax_1 + By_1 + C)\sin \omega}{\sqrt{A^2 + B^2 - 2AB\cos \omega}}$, for $Ax_1 + By_1 + C$ is the

constant term in equation (3).

But the new origin is the point $(x_1,\ y_1)$ referred to the old axes.

$\therefore$ the length of the perpendicular from the point $(x_1,\ y_1)$ upon the straight line $Ax + By + C = 0$ is equal to

$$\frac{(Ax_1 + By_1 + C)\sin \omega}{\sqrt{A^2 + B^2 - 2AB\cos \omega}}.$$

Examples IV. a.

[ω is the angle between the axes of co-ordinates unless otherwise stated in the question.]

Find, without using the formula for the length of a line, the distances between the following pairs of points, when the axes are inclined at an angle of 60° :

1. $(0, 0)$, $(6, 9)$.   **2.** $(3, 5)$, $(7, 1)$.   **3.** $(-2, 3)$, $(2, -1)$.

From a diagram find the co-ordinates of the middle point of the line joining :

4. $(0, 0)$, $(4, 6)$.   **5.** $(-2, 4)$, $(2, 6)$.   **6.** $(1, 5)$, $(3, 3)$.

The angle between the axes being 60°, find the angles which the following straight lines make with the axis of x :

7. $y = 2x - 4$.   **8.** $y = (\sqrt{3} + 1)x$.

9. $y = x + 6$.   **10.** $y - 3x + 5 = 0$.

Without assuming a formula, find the equation of the straight line :

11. Through the origin and the point $(4, 6)$.

12.   ,,   ,,   ,,   ,, $(-4, 5)$.

13. Making intercepts 4, 4   on the axes.

14.   ,,   ,, $4, -3$,,

15. Through the points $(1, 1)$, $(4, 4)$.

16.   ,,   ,, $(3, 5)$, $(6, 8)$.

17. Through the point $(x_1,\ y_1)$ and perpendicular to the axis of x.

18.   ,,   ,,   ,,   ,,   ,, y.

19. Prove that the area of the triangle formed by joining the points $(x_1,\ y_1)$, $(x_2,\ y_2)$, $(x_3,\ y_3)$ is equal to

$$\frac{\sin \omega}{2}\left[x_1(y_2 - y_3) + x_2(y_3 - y_1) + x_3(y_1 - y_2)\right].$$

20. Prove that the straight lines $x-y=0$, $x+y=0$ bisect the angles between the axes.

21. Find the condition that the straight lines $y=mx+c$, $y=m'x+c'$ may make equal angles with the axis of x on opposite sides of it.

22. The condition that $lx+my+n=0$ and $l'x+m'y+n'=0$ may be equally inclined to the axis of x in opposite directions is $\dfrac{m}{l}+\dfrac{m'}{l'}=2\cos\omega$.

23. Find the value of a if the lines $y+3x=4$, $2y-ax=5$ are at right angles when the angle between the axes is $60°$.

24. Prove that the lines $2y-x=5$ and $y(2\cos\omega+1)+x(2+\cos\omega)=k$ are at right angles.

25. Prove that the straight lines $y=3x+4$, $7y+5x+4=0$ are at right angles when $60°$ is the angle between the axes.

26. Find the equation of the straight line through the origin and at right angles to $y+2x=5$, when $60°$ is the angle between the axes.

27. Find the equation of the straight line through the point $(2,-3)$ and perpendicular to $3x-y=4$, when $120°$ is the angle between the axes.

28. Find the angle between the axes if the straight lines $3x+4y=5$, $4x-3y=4$ are perpendicular.

POLAR CO-ORDINATES.

75. If Ox is a straight line in a given plane, drawn from a fixed given point O, the position of a point P in that plane is known if the angle POx (θ) and the distance OP (r) are known.

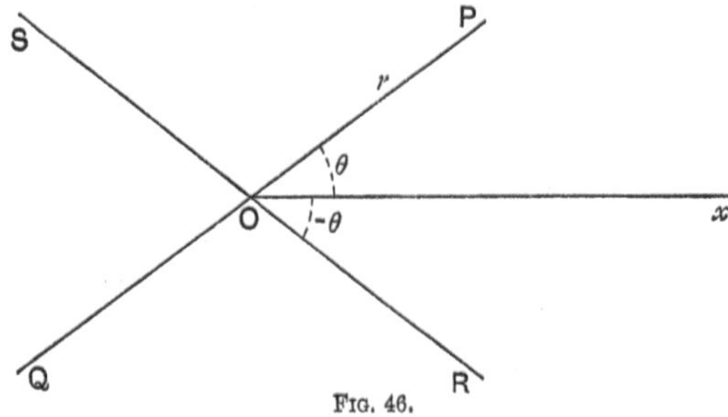

Fig. 46.

(r,θ) are called the **polar co-ordinates** of P.

r is called the **vectorial co-ordinate**, θ the **angular co-ordinate**.

O is called the **pole** or **origin**, and Ox the **initial line**.

The angle θ is measured thus :

Let OA revolve about O, in an anti-clockwise direction, from the position coinciding with Ox until it coincides with OP; θ is the angle it turns through.

The point $(-r, \theta)$ is the point Q, where OQ is measured in the opposite direction to OP, and equal to it in magnitude.

Thus we see that the points $(-r, \theta)$, $(r, \theta + \pi)$ are coincident.

In the same way, the points (r, θ), $(r, \theta + 2\pi)$ are coincident.

Also if R is the point $(r, -\theta)$, S is the point $(-r, -\theta)$, OS being drawn in an opposite direction to OR and equal to it in magnitude.

76. *If (x, y) are the co-ordinates of a point referred to rectangular axes Ox, Oy, and (r, θ) are the polar co-ordinates of the same point when O is the pole and Ox the initial line*

$$x = r \cos \theta, \quad y = r \sin \theta.$$

Let P be the point (x, y). Draw the ordinate PN.

$$OP = r, \quad \angle POx = \theta;$$

$$\therefore x = ON = OP \cos PON$$

$$= r \cos \theta$$

and $y = PN = OP \sin PON$

$$= r \sin \theta.$$

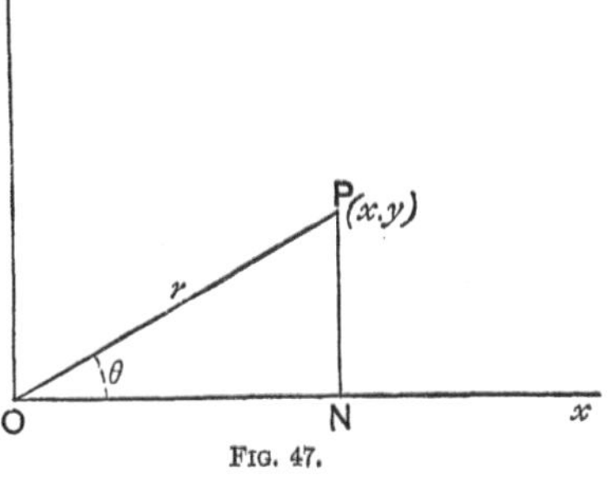

FIG. 47.

77. *To find the distance between the points (r_1, θ_1), (r_2, θ_2).*

Let P, Q be the points.　Join OP, OQ.

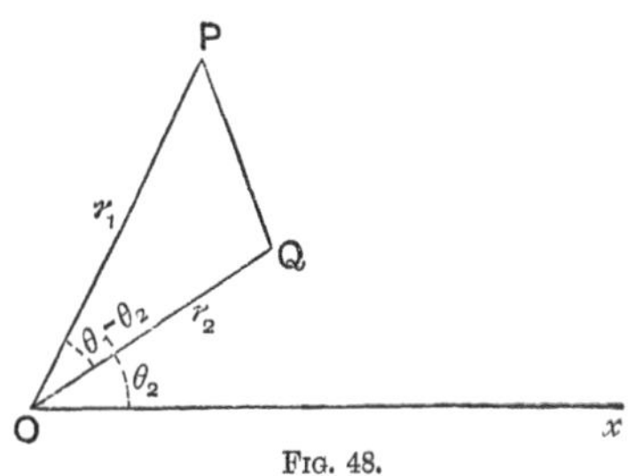

FIG. 48.

$$PQ^2 = OP^2 + OQ^2 - 2OP \cdot OQ \cos POQ \quad (a^2 = b^2 + c^2 - 2bc \cos A)$$

$$= r_1^2 + r_2^2 - 2r_1 r_2 \cos (\theta_1 - \theta_2);$$

$$\therefore PQ = \sqrt{r_1^2 + r_2^2 - 2r_1 r_2 \cos (\theta_1 - \theta_2)}.$$

78. *To find the area of the triangle formed by joining three points whose polar co-ordinates are given.*

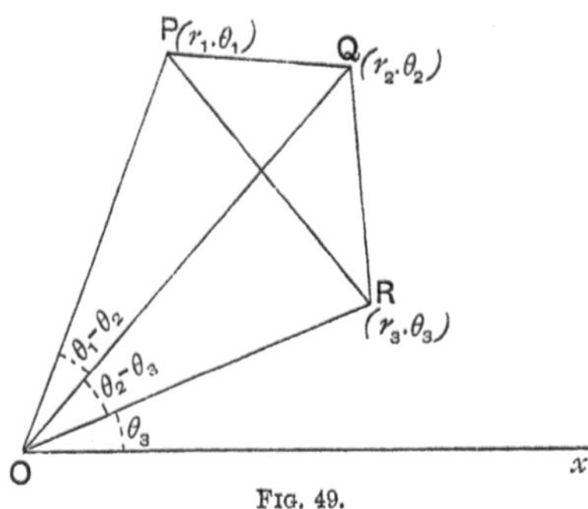

FIG. 49.

Let P, Q, R be the points whose polar co-ordinates are (r_1, θ_1), (r_2, θ_2), (r_3, θ_3) respectively.

Join OP, OQ, OR.

$\triangle$ PQR $= \triangle$ OPQ $+ \triangle$ OQR $- \triangle$ OPR

$\qquad = \tfrac{1}{2}$OP . OQ sin POQ $+ \tfrac{1}{2}$OQ . OR sin QOR $- \tfrac{1}{2}$OP . OR sin POR

$\qquad = \tfrac{1}{2}r_1 r_2 \sin(\theta_1 - \theta_2) + \tfrac{1}{2}r_2 r_3 \sin(\theta_2 - \theta_3) - \tfrac{1}{2}r_1 r_3 \sin(\theta_1 - \theta_3)$

$\qquad = \tfrac{1}{2}\left[r_1 r_2 \sin(\theta_1 - \theta_2) + r_2 r_3 \sin(\theta_2 - \theta_3) + r_3 r_1 \sin(\theta_3 - \theta_1) \right].$

79. *To find the equation of a straight line in polar co-ordinates.*

Let P(r, θ) be any point on the straight line PN.

Draw ON perpendicular to the line from the origin O, and let ON $= p$, and make an $\angle\, a$ with Ox, the initial line.

From the $\triangle$ OPN,

$\qquad$ ON $=$ OP cos PON,

$\qquad$ *i.e.* $\mathbf{p} = \mathbf{r}\cos(\theta - a)$,

the equation required.

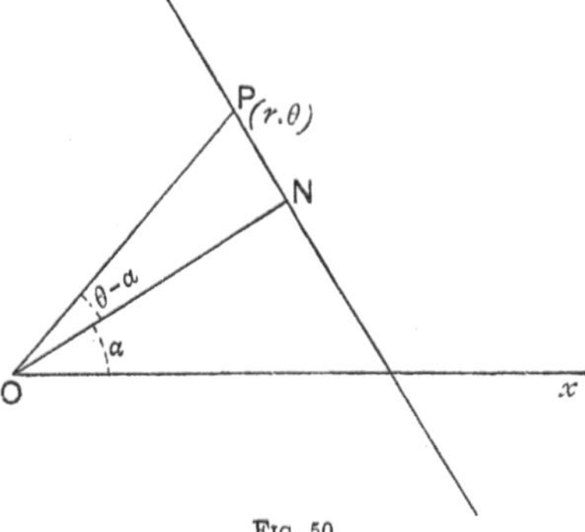

FIG. 50.

80. By transforming the polar to rectangular co-ordinates, we shall see that $p = r\cos(\theta - a)$ must represent a straight line.

The equation may be written

$$p = r\cos\theta\cos a + r\sin\theta\sin a,$$

which becomes in rectangular co-ordinates (Art. 76)

$$p = x \cos a + y \sin a,$$

an equation of the first degree in x and y.

81. *To find the polar equation of the straight line passing through the two given points*

$$(r_1,\ \theta_1),\ (r_2,\ \theta_2).$$

Let Q, R be the given points, P $(r,\ \theta)$ any point on the line. Join OP, OQ, OR.

$$\triangle\ POQ + \triangle\ QOR = \triangle\ POR ;$$

$$\therefore\ \tfrac{1}{2} rr_1 \sin(\theta - \theta_1)$$
$$+ \tfrac{1}{2} r_1 r_2 \sin(\theta_1 - \theta_2)$$
$$= \tfrac{1}{2} rr_2 \sin(\theta - \theta_2).$$

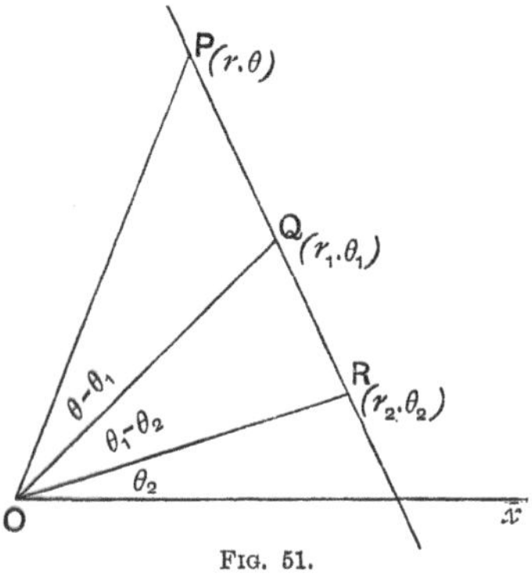

This may be written in the more symmetrical form

$$rr_1 \sin(\theta - \theta_1) + r_1 r_2 \sin(\theta_1 - \theta_2) + r_2 r \sin(\theta_2 - \theta) = 0.$$

FIG. 51.

82. *The equations* $p = r \cos(\theta - a)$, $p' = r \cos\left(\theta + \dfrac{\pi}{2} - a\right)$ *represent two perpendicular straight lines.*

(*First method of proof.*)

The equations may be written $\quad p = r \cos\theta \cos a + r \sin\theta \sin a$

and $p' = r \sin(a - \theta)$ or $p' = r \cos\theta \sin a - r \sin\theta \cos a.$

Putting these into rectangular co-ordinates, they become respectively,

$$p = x \cos a + y \sin a,$$
$$p' = x \sin a - y \cos a,$$

two perpendicular straight lines. $(aa' + bb' = 0)$

(*Second method of proof.*)

In the case of $p = r \cos(\theta - a)$, the perpendicular p makes an angle a with the initial line.

The second equation may be written $p' = r \cos\left[\theta - \left(a - \dfrac{\pi}{2}\right)\right];$

$\therefore$ the perpendicular p' makes an angle $a - \dfrac{\pi}{2}$ with the initial line.

The difference of the angles α and $\alpha - \dfrac{\pi}{2}$ is a right angle;

$\therefore$ the perpendiculars p, p' are at right angles;

$\therefore$ the lines are at right angles.

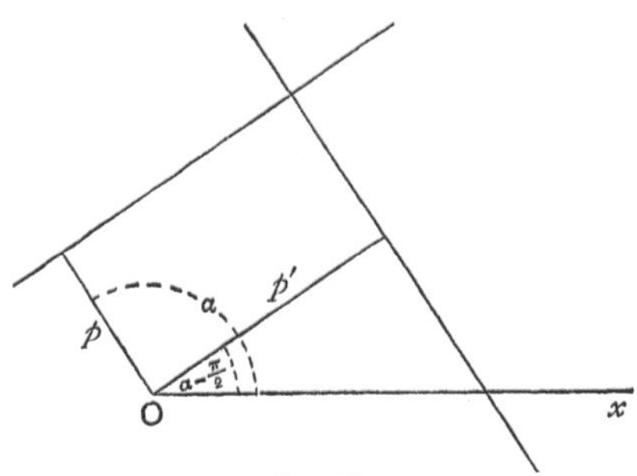

Fig. 52.

Note. If, in the polar equation of a straight line, we write $\theta + \dfrac{\pi}{2}$ for θ, we have the equation of a perpendicular straight line. By giving the proper value to the constant term, we can obtain the equation of any perpendicular straight line.

Examples IV. b.

Using a half inch unit of length, plot in one diagram the points whose polar co-ordinates are :

1. $\left(4, -\dfrac{\pi}{3}\right).$ 2. $\left(-3, \dfrac{\pi}{3}\right).$ 3. $\left(-2, -\dfrac{\pi}{6}\right).$

4. $(3, 0).$ 5. $(-5, \pi).$ 6. $\left(5, -\dfrac{\pi}{2}\right).$

Express the following equations in polar co-ordinates :

7. $x \cos \alpha + y \sin \alpha = p.$ 8. $x^2 + y^2 = a^2.$

9. $x^2 - y^2 = a^2.$ 10. $y^2 = 4ax.$ 11. $x^4 - y^4 = a^4.$

Express the following equations in rectangular co-ordinates :

(Interpret the equation in the first two cases.)

12. $r \cos \theta = a.$ 13. $r \cos (\theta + \alpha) = p.$

14. $A \cos \theta + B \sin \theta + \dfrac{C}{r} = 0.$ 15. $r = a \cos \theta.$

16. $r^2 = a \cos 2\theta.$ 17. $r = a \cos (\theta - \alpha).$

18. Find the polar equation of a straight line at right angles to the initial line and at a distance a from the origin.

19. Find the polar equation of a straight line parallel to the initial line and at a distance a from the origin.

20. What is the equation of a straight line through the origin and making an angle a with the initial line?

21. A regular hexagon has its centre at the origin and one side at right angles to the initial line. If each side $=a$, find the equations of its sides.

22. Find the co-ordinates of the point of intersection of the two straight lines $r \cos(\theta - 45°)=2$, $r \cos(\theta - 135°)=2$.

Draw the straight lines represented by the following equations:

23. $\theta = \dfrac{\pi}{3}$.　　　　**24.** $\theta = 0$.　　　　**25.** $1 = \sin(\theta - 30°)$.

26. $r \sin \theta = a$.　　　　**27.** $p = r \sin(\theta + a)$.　　**28.** $p = r \sin(\theta - a)$.

29. What is the equation of a straight line perpendicular to

$$p = r \sin(\theta + a)$$

and at the same perpendicular distance from the origin?

30. What is the equation of the straight line through the pole and at right angles to $p = r \cos(\theta - a)$?

31. Through the point whose polar co-ordinates are (r, θ), two straight lines are drawn to make equal angles, ϕ, with the radius vector; prove that their equations are $x \sin(\phi \pm \theta) \mp y \cos(\phi \pm \theta) = r \sin \phi$, the upper signs being taken together and the lower signs together.

CHAPTER V.

THE CIRCLE.   (RECTANGULAR AXES.)

83. *To find the equation of the circle whose radius is a and whose centre is at the origin.*

Let P be any point on the circle and (x, y) its co-ordinates.

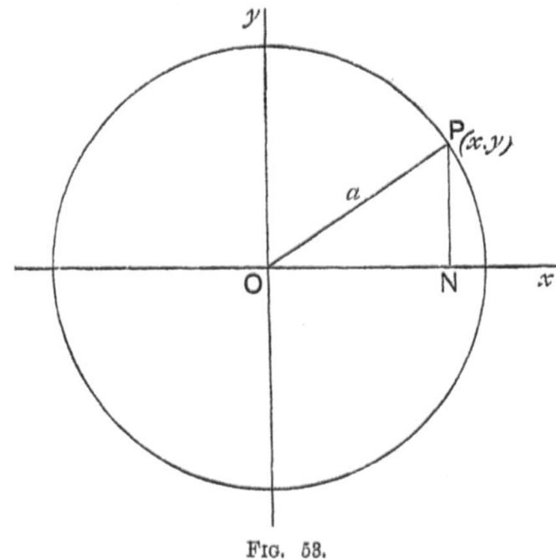

Fig. 58.

Draw the ordinate PN.

$$ON^2 + PN^2 = OP^2.$$

But $$ON = x, \ PN = y, \ OP = a \ ;$$

$$\therefore \ \mathbf{x}^2 + \mathbf{y}^2 = \mathbf{a}^2 \text{ is the required equation.}$$

84. *To find the equation of the circle whose centre is at the point (α, β) and whose radius is a.*

Let C be the centre, and take P any point on the curve, (x, y) being its ordinates.

Draw the ordinates CK, PM, and draw CN parallel to Ox to meet PM at N.

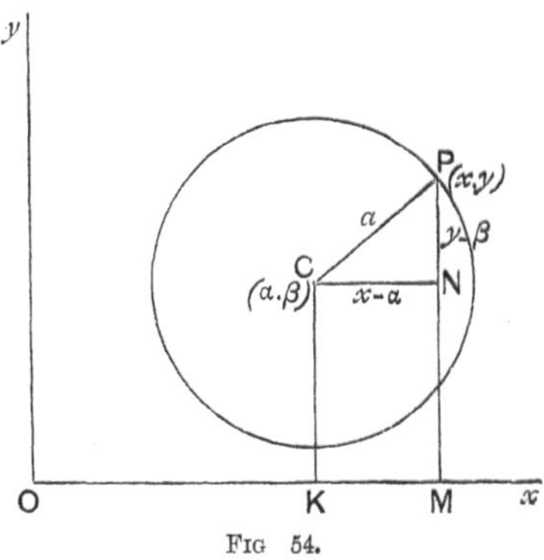

Fig 54.

In the right-angled $\triangle$ CNP,

$$CN^2 + PN^2 = CP^2.$$

But $$CN = KM = x - a,$$

$$PN = PM - CK = y - \beta \quad \text{and} \quad CP = a;$$

$$\therefore \ (\mathbf{x} - a)^2 + (\mathbf{y} - \beta)^2 = \mathbf{a}^2 \text{ is the required equation.}$$

85. The equation $(x - a)^2 + (y - \beta)^2 = a^2$

may be written in the form

$$x^2 + y^2 - 2ax - 2\beta y + a^2 + \beta^2 - a^2 = 0.$$

Hence we see that the equation of *any* circle may be written in the form

$$\mathbf{x^2 + y^2 + 2gx + 2fy + c = 0.}$$

This is called the **general equation of a circle.**

Note carefully its characteristics, remembering that **the axes are rectangular.** The coefficients of $\mathbf{x}^2$ and $\mathbf{y}^2$ are equal, and there is no term involving the product **xy.**

If $\mathbf{c} = 0$, the curve will pass through the origin, for its equation is satisfied by the values 0, 0 of **x** and **y.**

Example i. *Find the centre and radius of the circle $x^2 + y^2 + 2x - 2y - 2 = 0$, and draw the curve.*

The equation may be written $(x + 1)^2 + (y - 1)^2 = 4$.

$\therefore$ the centre is at the point $(-1, 1)$ and the radius $= 2$.

$$[(x - a)^2 + (y - \beta^2) = a^2]$$

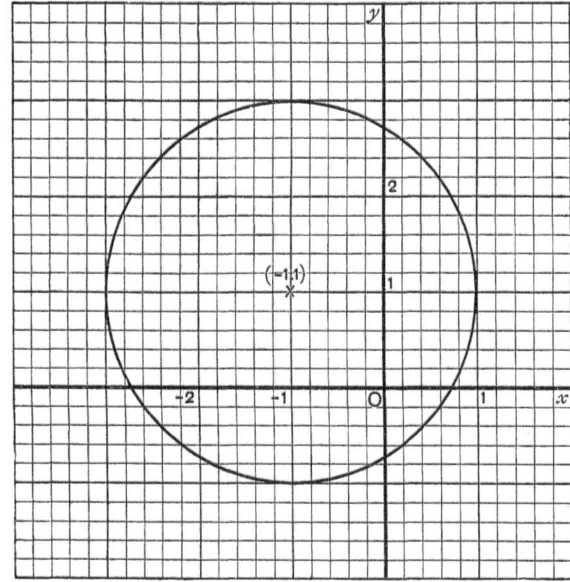

Fig. 55.

Example ii. *Find the co-ordinates of the centre and the length of the radius of the circle $x^2 + y^2 - 2gx - 2fy = 0$.*

The equation may be written $(x - g)^2 + (y - f)^2 = g^2 + f^2$.

$\therefore$ its centre is at the point (g, f),

and its radius $= \sqrt{g^2 + f^2}$.

Examples V. a.

Determine the radii of the circles :

1. $x^2 + y^2 = 4a^2$.

2. $(x + y)^2 + (x - y)^2 = 32a^2$.

Find the co-ordinates of the centre and the radius of each of the following circles. Draw the circle in each case :

3. $x^2 + y^2 - 2ax + 2ay = 2a^2$.

4. $x^2 + y^2 + 2x - 2y = 0$.

5. $x^2 + y^2 - 4x = 0$.

6. $x^2 + y^2 + 6y = 0$.

7. $x^2 + y^2 + 2gx + 2fy = c^2 - f^2 - g^2$.

8. $2(x^2 + y^2) - 3x + 4y = 0$.

9. $x^2 + y^2 - 5x + 7y = 0$.

10. $4x^2 + 4y^2 - 12x + 24y + 29 = 0$.

Find the equation of the circle :

11. Whose centre is at the point $(a, -a)$ and radius a.

12.   ,,   ,,   ,, (a, b) ,, $\sqrt{a^2 + b^2}$.

13. Which touches each axis at a distance 3 from the origin.

14. Which passes through the origin and makes positive intercepts each equal to a on the axes.

15. Which passes through the origin and the points $(a, 0)$, $(0, b)$.

16. Which touches both axes and passes through the point $(1, 2)$. [Two cases.]

17. Which touches the axis of x and has its centre at the point $(0, -4)$.

18. Which is described on the line joining $(0, 0)$ and (x_1, y_1) as diameter.

86. *Def.* If the two points in which a secant cuts the circumference of a circle move up to one another, the ultimate position of the secant, when these points coincide, is the tangent to the circle at the point of coincidence.

If (x, y) is any point on a curve and $(x + \Delta x, y + \Delta y)$ another point on it near (x, y), the slope (or gradient) of the tangent at the point (x, y) is equal to the limiting value of $\dfrac{\Delta y}{\Delta x}$ when the points move up to one another and coincide.

Let PQ be any curve, (x, y) the co-ordinates of P, $(x + \Delta x, y + \Delta y)$ the co-ordinates of Q, a point near to P.

Drawing

QR perpendicular to Ox

and

PR perpendicular to Oy,

Fɪɢ. 56.

we see from the diagram that $\dfrac{\Delta y}{\Delta x} = \dfrac{QR}{PR} = \tan \phi$, the slope of the chord PQ.

Now let the points move up to one another and ultimately coincide.

As they approach one another, the slope of the chord is always equal to $\dfrac{\Delta y}{\Delta x}$; therefore when they coincide the chord becomes a tangent, and its slope is the limiting value of $\dfrac{\Delta y}{\Delta x}$.

In using the above notation the student must remember that Δx is a small increment of x, and must therefore be regarded as a single quantity; the Δ cannot be separated from the x.

$\left[\right.$ **Differential Calculus.** If $y = f(x)$ is the equation of a curve, the slope of the tangent at the point (x, y) is $\left.\dfrac{dy}{dx}\right]$.

In finding the equations of tangents to curves the following will often be made use of.

If Δx and Δy are *small* quantities, $(\Delta x)^2$, $(\Delta y)^2$ and the product $\Delta x \cdot \Delta y$ are *very small*, and may generally be neglected in comparison with x and y.

Thus if $x = 1$ and $\Delta x = \dfrac{1}{1000} = \dfrac{1}{10^3}$, $\Delta x^2 = \dfrac{1}{10^6}$, and is *very* small compared with x.

Similarly, if $\Delta x = \dfrac{1}{10^3}$, $\Delta y = \dfrac{1}{10^5}$, $\Delta x \cdot \Delta y = \dfrac{1}{10^8}$, and may generally be neglected.

The student will doubtless have made use of the above in doing 'approximate' questions in Algebra and Arithmetic.

87. *To find the equation of the tangent to the circle* $x^2 + y^2 = a^2$ *at the point* (x_1, y_1).

Take a point $(x_1 + \Delta x_1, y_1 + \Delta y_1)$ on the curve and near to the point (x_1, y_1).

The point (x_1, y_1) is on the circle; $\quad \therefore \quad x_1^2 + y_1^2 = a^2$.

$\quad\quad$,, $\quad\quad (x_1 + \Delta x_1, y_1 + \Delta y_1)$ is on the circle;

$$\therefore \quad (x_1 + \Delta x_1)^2 + (y_1 + \Delta y_1)^2 = a^2.$$

$\therefore$ by subtraction,

$$2x_1\Delta x_1 + (\Delta x_1)^2 + 2y_1\Delta y_1 + (\Delta y_1)^2 = 0.$$

$\therefore$ in the limit, when the points approach one another,

$$2x_1\Delta x_1 + 2y_1\Delta y_1 = 0, \quad \text{neglecting} \ (\Delta x_1)^2 \ \text{and} \ (\Delta y_1)^2.$$

$$\therefore \frac{\Delta y_1}{\Delta x_1} = -\frac{x_1}{y_1} \text{ in the limit,}$$

i.e. $-\dfrac{x_1}{y_1}$ is the slope of the tangent at (x_1, y_1).

Also, the tangent passes through the point (x_1, y_1);

$$\therefore \text{ its equation is } \quad y - y_1 = -\frac{x_1}{y_1}(x - x_1) \quad [y - y_1 = m(x - x_1)]$$

$$\text{or } \quad yy_1 + xx_1 = x_1^2 + y_1^2,$$

$$\mathbf{xx_1 + yy_1 = a^2}.$$

Second Method by means of the Differential Calculus.

$$x^2 + y^2 = a^2.$$

Differentiating with respect to x,

$$2x + 2y\frac{dy}{dx} = 0 ; \qquad \therefore \frac{dy}{dx} = -\frac{x}{y}.$$

$\therefore \ -\dfrac{x_1}{y_1}$ is the slope of the tangent at the point (x_1, y_1);

$\therefore \ y - y_1 = -\dfrac{x_1}{y_1}(x - x_1)$ is the equation of the tangent,

$$\textit{i.e. } yy_1 + xx_1 = x_1^2 + y_1^2$$

$$\text{or } \mathbf{xx_1 + yy_1 = a^2}, \text{ as before.}$$

88. Normal. *Def.* A straight line drawn at right angles to a tangent to a curve and passing through its point of contact is said to be a *normal* to the curve.

To find the equation of the normal to the circle $x^2 + y^2 = a^2$ at the point (x_1, y_1).

The slope of the tangent $xx_1 + yy_1 = a^2$ is $-\dfrac{x_1}{y_1}$;

$$\therefore \quad \text{,,} \quad \text{,,} \quad \text{normal is } \frac{y_1}{x_1}. \quad (mm' = -1)$$

Also, the normal passes through the point (x_1, y_1);

$$\therefore \text{ its equation is } \quad y - y_1 = \frac{y_1}{x_1}(x - x_1) \text{ or } \frac{y}{y_1} = \frac{x}{x_1}.$$

Note. This is the equation of the radius drawn through the point (x_1, y_1), and we have therefore verified the well-known geometrical fact that a tangent to a circle is at right angles to the radius drawn through its point of contact.

89. Points of Intersection of Circles and Straight Lines.

Find the points of intersection of the circle

$$x^2 + y^2 = 25, \dots\dots\dots\dots\dots\dots(1)$$

and the straight line $\qquad x + y = 7. \dots\dots\dots\dots\dots\dots(2)$

We might plot the graphs of $x^2 + y^2 = 25$ and $x + y = 7$, as shown in the figure, and read off the co-ordinates of their

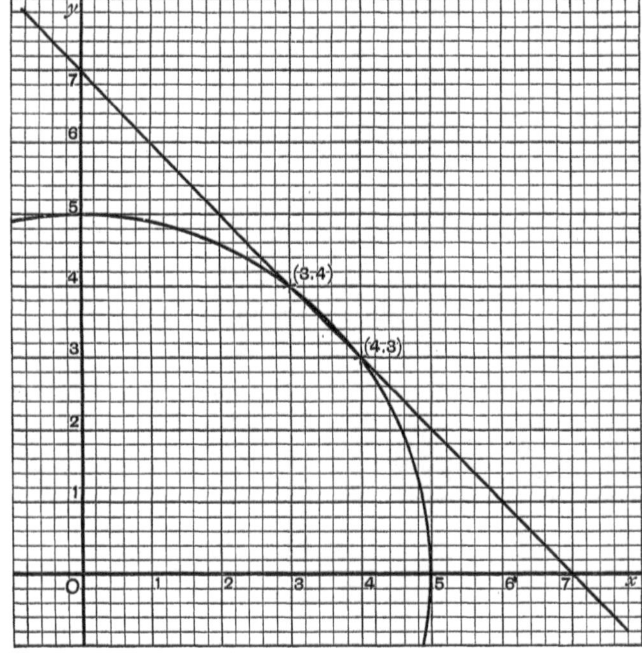

Fig. 57.

common points, and the beginner should do this, but let him also take note of the following carefully:

At the points where the straight line meets the circle, the corresponding values of x and y **satisfy both equations.**

The question might therefore be worded thus :
Find what values of x and y satisfy both the equations

$$x^2 + y^2 = 25 \text{ and } x + y = 7 :$$

in other words, **solve the equations.**

We therefore have to solve the equations in any way we please.
Squaring (2) and subtracting from (1), $-2xy = -24$;

$\therefore$ from (1), $\qquad x^2 - 2xy + y^2 = 25 - 24 = 1,$

$$x - y = \pm 1,$$
$$x + y = 7,$$

and $\qquad\qquad \left.\begin{array}{c} x = 4 \\ y = 3 \end{array}\right\}, \quad \left.\begin{array}{c} x = 3 \\ y = 4 \end{array}\right\}.$

$\therefore$ the straight line meets the circle at the points $(4, 3)$, $(3, 4)$. This can of course be verified from the figure.

90. The reasoning in the preceding article is quite general, and we thus deduce the following :

Given the equations of two curves, the points of intersection of the curves may be found by solving their equations, treating those equations as simultaneous.

91. *To find the condition that the straight line* $y = mx + c$ *may touch the circle* $x^2 + y^2 = a^2.$

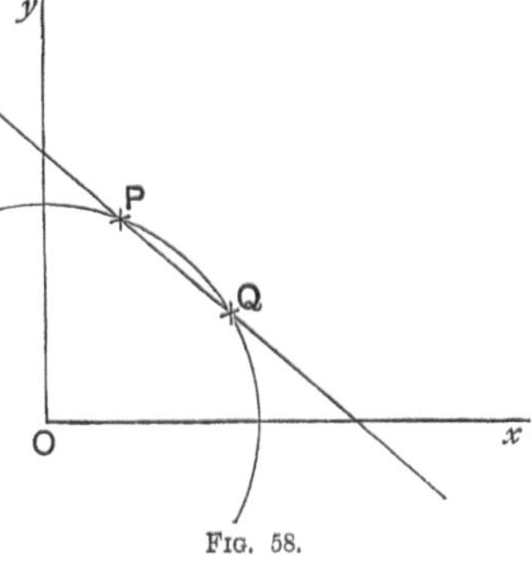

First Method. At the points of intersection, we have by substitution, $\qquad x^2 + (mx + c)^2 = a^2,$

$$x^2(1 + m^2) + 2mcx + c^2 - a^2 = 0.$$

If the straight line cut the curve at the points P and Q, the roots of this quadratic would give us the abscissae of the points P and Q. But when the line *touches* the circle, the points P and Q coincide, *i.e.* **the roots of the quadratic are equal.**

FIG. 58.

$\therefore \quad 4m^2c^2 = 4(1 + m^2)(c^2 - a^2), \quad [b^2 = 4ac]$

or $\quad c^2 = a^2(1 + m^2)$, the required condition.

Note. The above reasoning is quite general, and may therefore be applied in more general cases.

B.A.G.   F

Second Method. We know that if a straight line *touches* a circle, the length of the perpendicular from the centre of the circle upon the straight line is equal to the radius.

In this case the centre is at the origin.

$$\therefore \ \frac{c}{\sqrt{1+m^2}} = \pm a \qquad \left[p = \frac{Ax_1 + By_1 + C}{\sqrt{A^2 + B^2}} \right]$$

or $c^2 = a^2(1 + m^2)$, as before.

92. *To find the equation of a tangent to the circle* $x^2 + y^2 = a^2$, *in terms of its slope.*

Where the straight line $y = mx + c$ meets the circle, we have by substitution,
$$x^2 + (mx + c)^2 = a^2,$$

i.e. $x^2(1 + m^2) + 2mcx + c^2 - a^2 = 0.$

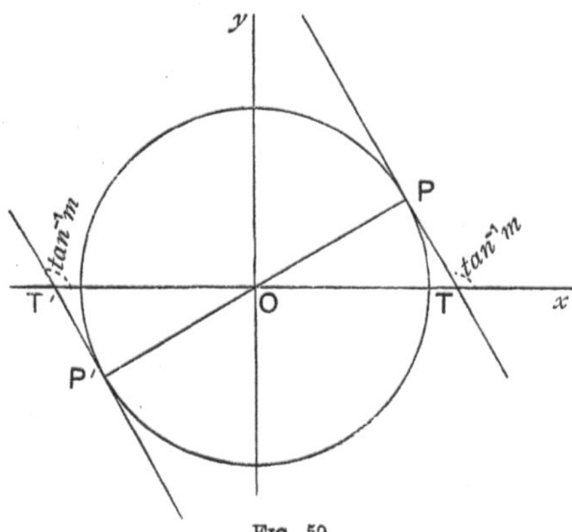

Fig. 59.

But if the straight line *touches* the curve, this quadratic for x will have equal roots

$$\therefore \ 4m^2c^2 = 4(c^2 - a^2)(1 + m^2), \quad [b^2 = 4ac]$$

i.e. $c^2 = a^2(1 + m^2),$

$$c = \pm a\sqrt{1 + m^2}.$$

We thus get two values for c. Therefore there are two tangents having the same slope, and their equations are

$$\mathbf{y = mx \pm a\sqrt{1 + m^2}.}$$

The tangents are parallel to one another and touch the circle at opposite ends of the diameter whose equation is $my + x = 0$.

In the figure, for the tangent PT we must take that sign which will make both intercepts on the axes positive, for P'T' that sign which will make both intercepts negative.

If we put $m = \tan \theta$, $\sqrt{1 + m^2} = \sec \theta$, and the equations of the tangents may be written $y = x \tan \theta \pm a \sec \theta$.

Examples V. b.

Without assuming the formula for a tangent, find the equation of the tangent to the circle:

1. $x^2 + y^2 = 25$ at the point $(3, 4)$.

2. $x^2 + y^2 = 5$,, ,, $(-1, 2)$.

3. $x^2 + y^2 = 4$, which makes an angle of $60°$ with the axis of x.

4. $x^2 + y^2 = 16$, ,, ,, $45°$,, ,,

5. $x^2 + y^2 = 64$, which has a slope equal to 3.

6. Find the radius of the circle which has its centre at the point $(-1, 2)$, and touches the straight line $12y + 5x + 7 = 0$.

7. Find the radius of the circle circumscribing the points $(4, 2)$, $(4, -2)$, $(1, -2)$.

Find the equation of the tangent to the circle $x^2 + y^2 = a^2$:

8. Which is parallel to $Ax + By + C = 0$.

9. Which is perpendicular to $Ax + By + C = 0$.

10. Which makes equal intercepts of positive sign on the axes.

11. Find the condition that the straight line $y = mx$ may touch the circle $x^2 + y^2 - ax - \beta y = 0$, and hence write down the equation of the tangent to the circle at the origin.

12. Find the co-ordinates of the points where the straight line $3y - 4x = 1$ meets the circle $x^2 + y^2 - 4x - 6y - 12 = 0$.

13. Find the length of the line $x = 2$ intercepted by the circle $x^2 + y^2 = 16$.

14. Prove that the straight line $y = x + a\sqrt{2}$ touches the circle $x^2 + y^2 = a^2$, and find its point of contact.

15. Find the co-ordinates of the centre of the circle circumscribing the triangle $(-2, 2)$, $(1, -2)$, $(1, 3)$.

16. Find the equation of the circle with radius $1·5$ and centre $(1·5, 2)$. Show that it touches the line $x = 0$. Draw the figure on squared paper, unit one inch.

17. Prove that the straight line $5x + 12y - 4 = 0$ touches the circle
$$x^2 + y^2 - 6x + 4y + 12 = 0.$$
Determine the co-ordinates of the point of contact.

18. Calculate the co-ordinates of the intersections of
$$x^2 + y^2 - 4x + 2y + 1 = 0$$
and $3x + 4y - 2 = 0$, and the length of the line between the points of inter-section. Check your results by a graphical solution.

19. Show that each of the circles
$$x^2 + y^2 - 2x\sqrt{3} + 2y = 0, \quad x^2 + y^2 + \frac{2x}{\sqrt{3}} - 2y = 0$$
circumscribes an equilateral triangle which has one side lying along an axis of co-ordinates.

Find the co-ordinates of the second point of intersection of these circles.

20. Prove that if the straight line $x\cos a + y\sin a = p$ touches the circle $(x - h)^2 + (y - k)^2 = r^2$, then $h\cos a + k\sin a - p = \pm r$, and explain the sign of the ambiguity.

93. *Tangents are drawn from the point (x_1, y_1) to the circle $x^2 + y^2 = a^2$, to find the equation of their chord of contact.*

Let P be the point (x_1, y_1), PA, PB the tangents. It is required to find the equation of AB. Let (h_1, k_1) be the co-ordinates of A, (h_2, k_2) the co-ordinates of B.

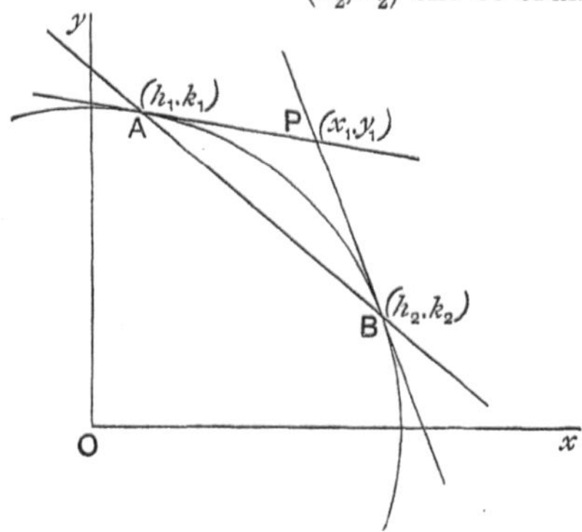

Fig. 60.

The equation of PA, the tangent at A, is
$$xh_1 + yk_1 = a^2.$$
The equation of PB is $xh_2 + yk_2 = a^2$.
But the point P(x_1, y_1) lies on both these lines;
$$\therefore \quad x_1h_1 + y_1k_1 = a^2 \quad \dots\dots\dots\dots\dots\dots(1)$$
and
$$x_1h_2 + y_1k_2 = a^2. \quad \dots\dots\dots\dots\dots\dots(2)$$

$\therefore\ x_1x + y_1y = a^2$ or $xx_1 + yy_1 = a^2$ is the equation required.

For firstly it represents a straight line.

Also from (1), we see that $(h_1,\ k_1)$ A lies on this line,

and   ,,   (2)   ,, $(h_2,\ k_2)$ B   ,,   ,,

$\therefore$ it is the equation of AB.

94. *Def.* **Pole and Polar.** If tangents are drawn at the extremities of any chord of a circle which passes through a fixed point, the locus of their intersection is called the **polar** of the point.   The fixed point is called the **pole** of the polar.

This locus will be proved to be a straight line.

To find the polar of the point (x_1, y_1) with respect to the circle $x^2 + y^2 = a^2$.

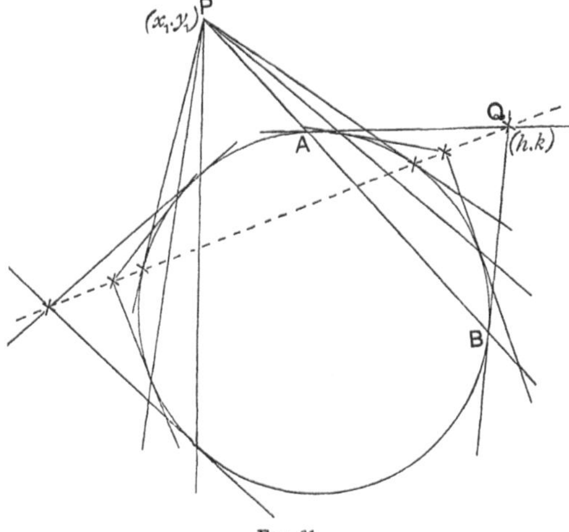

Fig. 61.

Let P be the point (x_1, y_1), and PAB any chord through P ; AQ, BQ the tangents at A and B.

It is required to find the locus of Q.

Let $(h,\ k)$ be the co-ordinates of Q.

Then the equation of its chord of contact AB is   (Art. 93)

$$xh + yk = a^2.$$

But (x_1, y_1) is on this line;
$$\therefore \quad x_1 h + y_1 k = a^2.$$
But (h, k) is *any point* on the locus;
$\therefore$ the equation of the locus is
$$x_1 x + y_1 y = a^2$$
or $\quad xx_1 + yy_1 = a^2$, a straight line.

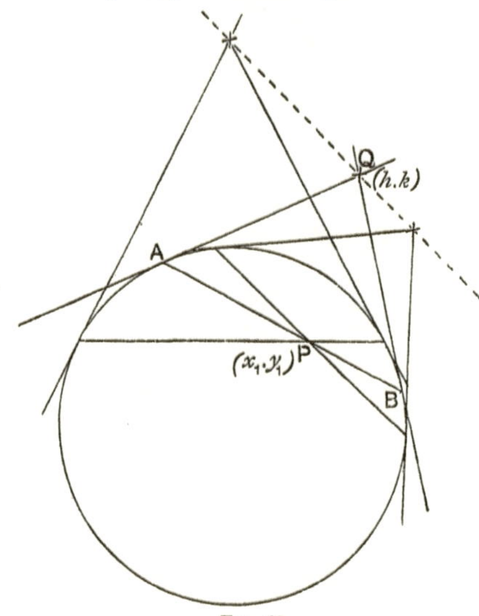

Fig. 62.

N.B. *When* (x_1, y_1) *is outside the circle, the polar is the same as the chord of contact of tangents drawn from* (x_1, y_1).

For the geometrical construction of the polar of a given point, see Baker and Bourne's *Geometry*, p. 371.

95. *If the polar of the point* P *passes through the point* Q, *the polar of the point* Q *passes through* P.

Let (x_1, y_1) be the co-ordinates of P, (x_2, y_2) those of Q.
The equation of the polar of P with respect to the circle
$$x^2 + y^2 = a^2$$
is $\quad xx_1 + yy_1 = a^2$.

This passes through the point Q; $\therefore x_1 x_2 + y_1 y_2 = a^2.$ ....(1)

But $xx_2 + yy_2 = a^2$ is the polar of Q, and by (1), this straight line passes through $(x_1, y_1)\mathsf{P}$, which proves the proposition.

96. *Given that* $\mathsf{A}x + \mathsf{B}y + \mathsf{C} = 0$ *is a tangent to the circle* $x^2 + y^2 = a^2$, *find the co-ordinates of its point of contact.*

Let (x_1, y_1) be the co-ordinates required.

Then $xx_1 + yy_1 = a^2$ is the equation of the tangent;

$\therefore$ this equation must be *identical* with

$$\mathsf{A}x + \mathsf{B}y = -\mathsf{C},$$

for the two equations represent the same straight line;

$\therefore$ comparing coefficients,

$$\frac{x_1}{\mathsf{A}} = \frac{y_1}{\mathsf{B}} = -\frac{a^2}{\mathsf{C}},$$

whence
$$x_1 = -\frac{a^2\mathsf{A}}{\mathsf{C}}, \quad y_1 = -\frac{a^2\mathsf{B}}{\mathsf{C}}.$$

COROLLARY. Since (x_1, y_1) is on the circle,

$$\frac{a^4\mathsf{A}^2}{\mathsf{C}^2} + \frac{a^4\mathsf{B}^2}{\mathsf{C}^2} = a^2;$$

$$\therefore a^2(\mathsf{A}^2 + \mathsf{B}^2) = \mathsf{C}^2 \quad \text{and} \quad \mathsf{C} = \pm a\sqrt{\mathsf{A}^2 + \mathsf{B}^2}.$$

97. *Find the co-ordinates of the pole of the straight line*

$$\mathsf{A}x + \mathsf{B}y + \mathsf{C} = 0,$$

with respect to the circle $x^2 + y^2 = a^2.$

Let (x_1, y_1) be the pole.

Then $xx_1 + yy_1 = a^2$ is the equation of the polar.

Now proceed as in the preceding article.

98. *If a straight line* PAB *cuts a circle at* A *and* B, *and* PC *is a tangent to the circle, to prove (analytically) that* $\mathsf{PA} \cdot \mathsf{PB} = \mathsf{PC}^2.$

Let $\qquad x^2 + y^2 + 2gx + 2fy + c = 0$ ....(1)

be the equation of the circle, (x_1, y_1) the co-ordinates of P, and let the equation of PAB be

$$\frac{x - x_1}{\cos\theta} = \frac{y - y_1}{\sin\theta} = r, \quad ....(2) \quad \text{(Art. 30)}$$

$$x = x_1 + r \cos \theta, \quad y = y_1 + r \sin \theta \,;$$

$\therefore$ where (1) and (2) meet we have by substitution,

$$(x_1 + r \cos \theta)^2 + (y_1 + r \sin \theta)^2 + 2g(x_1 + r \cos \theta) + 2f(y_1 + r \sin \theta) + c = 0,$$

$i.e.$ $r^2 + 2r(x_1 \cos \theta + y_1 \sin \theta + g \cos \theta + f \sin \theta) + x_1^2 + y_1^2 + 2gx_1 + 2fy_1 + c = 0.$

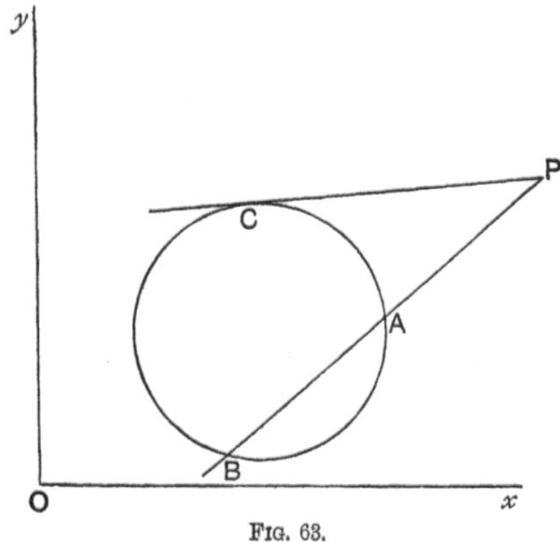

Fig. 63.

Now PA, PB are the roots of this equation;

$$\therefore \quad \mathrm{PA} \cdot \mathrm{PB} = x_1^2 + y_1^2 + 2gx_1 + 2fy_1 + c, \quad \dots\dots\dots\dots(2)$$

and is constant in value, for it is independent of θ.

Let the straight line PAB turn about P until the points A and B move up to one another and ultimately coincide.

The line becomes a tangent, and PA.PB becomes PC^2;

$\therefore$ since PA.PB is constant in value for all directions,

$$\mathrm{PA} \cdot \mathrm{PB} = \mathrm{PC}^2.$$

COROLLARY. $\mathrm{PC}^2 = \mathrm{PA} \cdot \mathrm{PB} = x_1^2 + y_1^2 + 2gx_1 + 2fy_1 + c$ from (2);

$$\therefore \quad \mathrm{PC} = \sqrt{x_1^2 + y_1^2 + 2gx_1 + 2fy_1 + c}.$$

99. *To prove that the* **length of the tangent** *drawn from the point* (x_1, y_1) *to the circle* $(x - a)^2 + (y - \beta)^2 = a^2$ *is equal to*

$$\sqrt{(x_1 - a)^2 + (y_1 - \beta)^2 - a^2}.$$

If P is the point (x_1, y_1), PQ a tangent, and C the centre of the circle,
$$PC^2 = (x_1 - a)^2 + (y_1 - \beta)^2, \quad \text{(Art. 4)}$$
$$CQ = a \; ;$$

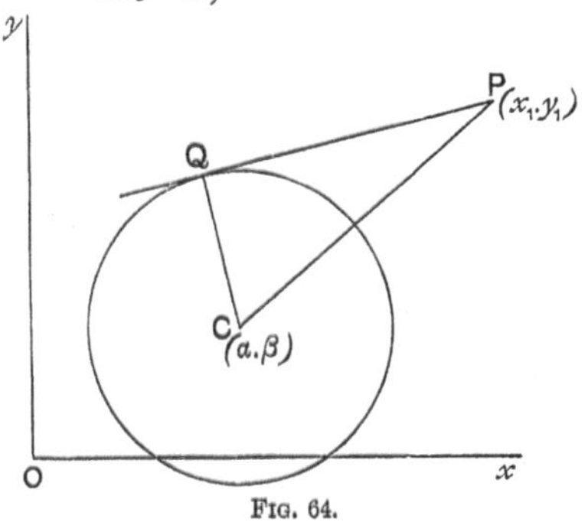

Fig. 64.

$$\therefore \ PQ^2 = PC^2 - CQ^2$$
$$= (x_1 - a)^2 + (y_1 - \beta)^2 - a^2$$
$$\text{and} \quad PQ = \sqrt{(x_1 - a)^2 + (y_1 - \beta)^2 - a^2}.$$

100. *Find the equation of the tangent to the circle*
$$x^2 + y^2 + 2gx + 2fy + c = 0$$
at the point (x_1, y_1).

Take the two points (x_1, y_1), $(x_1 + \Delta x_1, y_1 + \Delta y_1)$ on the circle, and near to one another.

Because (x_1, y_1) is on the curve,
$$x_1^2 + y_1^2 + 2gx_1 + 2fy_1 + c = 0.$$
Because $(x_1 + \Delta x_1, y_1 + \Delta y_1)$ is on the curve,
$$(x_1 + \Delta x_1)^2 + (y_1 + \Delta y_1)^2 + 2g(x_1 + \Delta x_1) + 2f(y_1 + \Delta y_1) + c = 0.$$
$\therefore$ subtracting,
$$2x_1\Delta x_1 + (\Delta x_1)^2 + 2y_1\Delta y_1 + (\Delta y_1)^2 + 2g\Delta x_1 + 2f\Delta y_1 = 0.$$
$\therefore$ in the limit when the points approach one another,
$$\Delta x_1(x_1 + g) + \Delta y_1(y_1 + f) = 0,$$
and the limiting value of $\dfrac{\Delta y_1}{\Delta x_1} = -\dfrac{x_1 + g}{y_1 + f}.$

$\therefore \ -\dfrac{x_1 + g}{y_1 + f}$ is the slope of the tangent at (x_1, y_1).

$\therefore$ the equation of the tangent is

$$y - y_1 = -\frac{x_1 + g}{y_1 + f}(x - x_1)$$

or $\quad xx_1 + yy_1 + gx + fy = x_1{}^2 + y_1{}^2 + gx_1 + fy_1$

$$= -gx_1 - fy_1 - c, \text{ for } (x_1, y_1) \text{ is on the circle.}$$

$$\therefore \quad xx_1 + yy_1 + g(x + x_1) + f(y + y_1) + c = 0$$

is the reqd. equation of the tangent.

Note. Here, and in Art. 87, the following rules hold for **writing down** the equation of a tangent to a circle :

We obtain the equation of the tangent at the point (x_1, y_1) to a circle by taking the equation of the circle, and writing

xx_1 instead of x^2,

$\qquad$ **yy_1** $\qquad$,, $\qquad$ **y^2,**

$\qquad$ **$x + x_1$** $\qquad$,, $\qquad$ **2x,**

$\qquad$ **$y + y_1$** $\qquad$,, $\qquad$ **2y, leaving the constant term unchanged.**

101. Second method by means of the Differential Calculus.

Differentiating with respect to x,

$$2x + 2y\frac{dy}{dx} + 2g + 2f\frac{dy}{dx} = 0,$$

whence $$\frac{dy}{dx} = -\frac{x + g}{y + f}.$$

$\therefore$ the slope of the tangent at the point (x_1, y_1) is $-\dfrac{x_1 + g}{y_1 + f}$.

We now proceed as in the first method.

102. *If tangents are drawn to the circle* $x^2 + y^2 + 2gx + 2fy + c = 0$, *the equation of their chord of contact is*

$$xx_1 + yy_1 + g(x + x_1) + f(y + y_1) + c = 0.$$

This is proved by the method of Art. 93.

103. *The polar of the point* (x_1, y_1) *with respect to the circle* $x^2 + y^2 + 2gx + 2fy + c = 0$ *is the straight line*

$$xx_1 + yy_1 + g(x + x_1) + f(y + y_1) + c = 0.$$

This is proved by the method of Art. 94.

104. *To find the equation of the pair of tangents drawn from the point $(x_1,\ y_1)$ to the circle $x^2 + y^2 = a^2$.*

Let $\mathsf{Q}(x,\ y)$ be any point on either of the tangents PA, PB drawn from the point P to the circle.

Let OP cut AB at N, and draw QM perpendicular to AB.

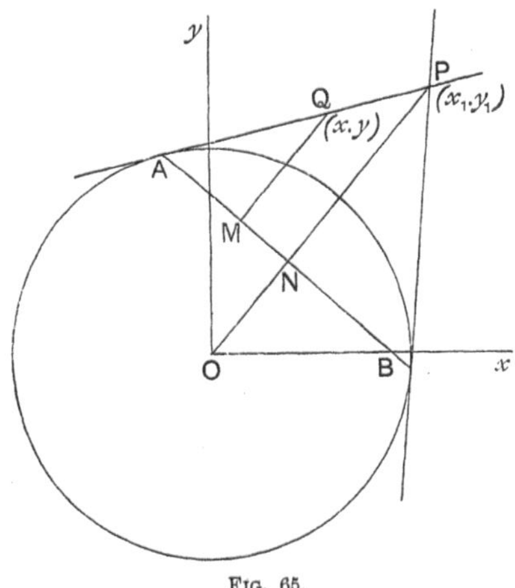

Fig. 65.

From the similar $\triangle^s$ AMQ, ANP,

$$\frac{\mathsf{QM}}{\mathsf{QA}} = \frac{\mathsf{PN}}{\mathsf{PA}}. \quad \dots\dots\dots\dots\dots\dots\dots\dots(1)$$

Now the equation of AB is $xx_1 + yy_1 - a^2 = 0$;

$$\therefore\ \mathsf{QM} = \frac{xx_1 + yy_1 - a^2}{\sqrt{x_1^2 + y_1^2}}, \quad \left(p = \frac{\mathsf{A}x_1 + \mathsf{B}y_1 + \mathsf{C}}{\sqrt{\mathsf{A}^2 + \mathsf{B}^2}}\right)$$

$$\mathsf{QA} = \sqrt{\mathsf{QO}^2 - \mathsf{AO}^2} = \sqrt{x^2 + y^2 - a^2},$$

$$\mathsf{PA} = \sqrt{\mathsf{PO}^2 - \mathsf{AO}^2} = \sqrt{x_1^2 + y_1^2 - a^2},$$

$$\mathsf{PN} = \frac{x_1^2 + y_1^2 - a^2}{\sqrt{x_1^2 + y_1^2}},$$

for PN is perpendicular to AB. $\left(p = \dfrac{\mathsf{A}x_1 + \mathsf{B}y_1 + \mathsf{C}}{\sqrt{\mathsf{A}^2 + \mathsf{B}^2}}\right)$

$$\therefore \text{ from (1) } \frac{xx_1 + yy_1 - a^2}{\sqrt{x_1^2 + y_1^2}\sqrt{x^2 + y^2 - a^2}} = \frac{x_1^2 + y_1^2 - a^2}{\sqrt{x_1^2 + y_1^2}\sqrt{x_1^2 + y_1^2 - a^2}}$$

$$\text{or } (xx_1 + yy_1 - a^2)^2 = (x^2 + y^2 - a^2)(x_1^2 + y_1^2 - a^2).$$

But (x, y) is any point on either tangent ;

$\therefore$ this is the equation of the two tangents PA, PB.

Example i. *To find the equations of the tangents drawn from the point* $(4, 2)$ *to the circle* $x^2 + y^2 = 4$.

Let $y - 2 = m(x - 4)$ be a tangent to the circle.

Since the line is a tangent, the length of the perpendicular from the centre $(0, 0)$ of the circle upon the line is equal to the radius, 2.

$$\therefore \frac{-2 + 4m}{\sqrt{1 + m^2}} = 2. \quad \left(p = \frac{Ax_1 + By_1 + C}{\sqrt{A^2 + B^2}} \right)$$

$$2m - 1 = \sqrt{1 + m^2}.$$

Squaring and solving for m, we have

$$m = 0 \text{ or } \tfrac{4}{3}.$$

$$\therefore \ y - 2 = 0, \text{ and } y - 2 = \tfrac{4}{3}(x - 4), \text{ or } 4x - 3y = 10,$$

are the tangents required.

Example ii. *To find the equation of a tangent to the circle*

$$(x - a)^2 + (y - \beta)^2 = r^2$$

in terms of its slope, m.

$y = mx \pm r\sqrt{1 + m^2}$ touches the circle $x^2 + y^2 = a^2$ for all values of m.

$\therefore$ transferring the origin to the point $(-a, -\beta)$,

$$y - \beta = m(x - a) \pm r\sqrt{1 + m^2}$$

is a tangent to the circle $(x - a)^2 + (y - \beta)^2 = r^2$ for all values of m.

Examples V. c.

Find with respect to the circle $x^2 + y^2 = a^2$ the poles of the following straight lines :

1. $y = mx + c.$ **2.** $xx_1 + yy_1 = b^2.$ **3.** $y = 2a.$ **4.** $2x = a.$

5. AB, AC are tangents to the circle $x^2 + y^2 = a^2$ drawn from the point $A(x_1, y_1)$. $Q(x, y)$ is any point on the chord of contact BC. OA meets BC at N and is perpendicular to it. Therefore

$$AB^2 - AQ^2 = BN^2 - QN^2 = OB^2 - OQ^2.$$

$$AB^2 = x_1^2 + y_1^2 - a^2 \text{ (Art. 99).} \quad AQ^2 = (x - x_1)^2 + (y - y_1)^2.$$

Using the equation $AB^2 - AQ^2 = OB^2 - OQ^2$ deduce the equation of BC.

6. Find the condition that the straight line $6x + 8y = k$ may touch the circle $x^2 + y^2 - 10x = 0$.

7. Find the equation of the straight line which touches the circle
$$x^2 + y^2 - 3x - 5y = 0$$
at the origin.

8. The equation of a straight line is $ax + by + a^2 + b^2 = 0$, and that of a circle, $x^2 + y^2 + ax + by = 0$.   Adding these equations, we have
$$(x + a)^2 + (y + b)^2 = 0.$$
What does this prove?

9. Prove that the lines $x = 1$, $y = 2$ each touch the circle
$$x^2 + y^2 - 4x - 2y + 4 = 0,$$
and find the other co-ordinates of the points of contact.

10. Find the equation of the circle which has its centre at the point $(9, 4)$ and passes through the point $(1, -2)$.
If tangents are drawn to this circle from the origin, find the equation of their chord of contact.

11. Find the equations of the tangents to the circle $x^2 + y^2 = a^2$ drawn from the point $(0, k)$.

12. Find, by the second method of Art. 91, the condition that the straight line $Ax + By + C = 0$ should touch the circle $x^2 + y^2 = a^2$.

13. Find the quadratic equation which will give the abscissae of the points where $y = mx + c$ meets the circle $x^2 + y^2 = a^2$.
Deduce the condition that (1) the straight line may *cut* the circle, (2) the straight line should not meet the circle.

14. Find the quadratic equation which will give the abscissae of the points where the straight line $xx_1 + yy_1 = a^2$ meets the circle $x^2 + y^2 = a^2$.
Deduce the condition that (1) the straight line may *cut* the circle, (2) the straight line may *touch* the circle, (3) the straight line should not meet the circle.
Give the geometrical interpretation to each result.

15. Prove that the straight line $gx + fy = 0$ touches the circle
$$x^2 + y^2 + gx + fy = 0$$
at the origin.

16. Find the condition that $Ax + By + C = 0$ may touch the circle
$$(x - h)^2 + (y - k)^2 = r^2.$$

Note. The point $(a \cos \alpha . a \sin \alpha)$ lies on the circle $x^2 + y^2 = a^2$ for all values of α, for $a^2 \cos^2\alpha + a^2 \sin^2\alpha \equiv a^2$.

17. The equation of a tangent to the circle $x^2 + y^2 = a^2$ may be written in the form $x \cos \alpha + y \sin \alpha = a$.

18. If (x_1, y_1) is the middle point of a chord of the circle $x^2 + y^2 = a^2$, find the equation of the chord.

19. Find the length of the common chord of the circles
$$(x - a)^2 + (y - b)^2 = c^2, \quad (x - b)^2 + (y - a)^2 = c^2 \ ;$$
and hence prove that the condition that these two circles should touch one another is $2c^2 = (a - b)^2$.

20. Find the condition that the chord of contact of tangents from the point (x_1, y_1) to the circle $x^2 + y^2 = a^2$ should subtend a right angle at the centre.

21. Tangents are drawn from the point (h, k) to the circle $x^2 + y^2 = a^2$. Prove that the area of the triangle formed by them and their chord of contact is

$$\frac{a(h^2 + k^2 - a^2)^{\frac{3}{2}}}{h^2 + k^2}.$$

22. Find the angle between the straight lines which join the points of intersection of $\frac{x}{r} + \frac{y}{s} = 1$ with $x^2 + y^2 = r^2$ to the origin, (1) direct from a diagram, (2) by using the method of Art. 52.

23. Find the condition that the circle $x^2 + y^2 + 2gx + 2fy + c = 0$ may

(1) touch the axis of x.

(2)   ,,   ,, y.

(3) have its centre in the axis of x.

(4)   ,,   ,,   ,, y.

24. Where the circles $x^2 + y^2 + 2gx + 2fy + c = 0$, $x^2 + y^2 + 2g'x + 2f'y + c' = 0$ meet, we have by subtraction $2(g - g')x + 2(f - f')y + c - c' = 0$. Interpret the last equation.

25. Add the equations $x^2 + y^2 - 2ax - 2by = c^2$, $x^2 + y^2 + 2ax + 2by = d^2$, and interpret the result.

26. Giving reasons, interpret the connection between the circles

$$x^2 + y^2 - c^2 = 0, \quad x^2 + y^2 - 2ax - 2by - c^2 = 0,$$

and the straight line $ax + by = 0$.

27. Find the equation of the two straight lines joining the points of intersection of the circle $x^2 + y^2 + 2gx + 2fy = 0$ and the straight line $ax + by = 1$ to the origin.

28. Find the equation of the two straight lines joining the points of intersection of the circle $x^2 + y^2 - 8x - 6y + 21 = 0$ and the straight line $x + y = 8$ to the origin.

29. Find the equations of the tangents drawn to the circle $x^2 + y^2 = 36$ from the point $(8, 6)$.

30. Find the pole of the straight line $3x + 4y - 45 = 0$ with respect to the circle $x^2 + y^2 - 6x - 8y + 5 = 0$.

[If (x_1, y_1) is the pole, the equations $xx_1 + yy_1 - 3(x + x_1) - 4(y + y_1) + 5 = 0$ and $3x + 4y - 45 = 0$ represent the same line and are therefore identical.]

31. Find the equation of the circle which has its centre at the point $(4, 3)$ and touches the straight line $5x - 12y - 10 = 0$.

32. Draw the circle $x^2 + y^2 - 8x - 4y + 16 = 0$, and find the equations of the tangents to it which pass through the origin.

33. Prove that the equation of the chord joining the points (x_1, y_1), (x_2, y_2) on the circle $x^2 + y^2 = a^2$ may be written

$$(y - y_1)(y_2 + y_1) + (x - x_1)(x_2 + x_1) = 0.$$

Deduce the equation of the tangent at the point (x_1, y_1).

34. Prove that the common tangents of the circles $x^2 + y^2 + 2x = 0$, $x^2 + y^2 - 6x = 0$ form an equilateral triangle.

35. Form the equation of the circle which passes through the origin, has the line $y = mx$ for a diameter, and touches the line $my + x = a$.

36. Prove that the product of the abscissae of the points when the straight line $y = mx$ meets the circle $x^2 + y^2 + 2gx + 2fy + c = 0$ is equal to

$$\frac{c}{1 + m^2}.$$

37. A straight line is drawn through the point (x_1, y_1), A, making an angle θ with the axis of x, and meets the circle $x^2 + y^2 - 2gx - 2fy + c = 0$ at the points P and Q. If B is the middle point of PQ, prove that

$$\mathsf{AB} = (g - x_1)\cos\theta + (f - y_1)\sin\theta.$$

38. Find the equations of the tangents to the circle $x^2 + y^2 - 6x - 6y + 9 = 0$ which are parallel to $y = x$.

39. Draw the circle $x^2 + y^2 - 2cy = 0$. Take a point $\mathsf{P}(2a, 0)$ and bisect OP, O being the origin, at Q, and draw a straight line QAB through the centre of the circle and cutting it at A and B. Find the equation of the circle passing through the points P, A, B, and show that it touches the axis of x at P.

$[x^2 + y^2 - 2cy + \lambda(cx + ay - ac) = 0$ is a circle passing through the intersection of $x^2 + y^2 - 2cy = 0$ and $cx + ay - ac = 0.]$

40. Two circles are drawn to touch both the straight lines $y = 0$, $y = x\tan 2a$, and to pass through the point (h, k). Prove that the distances of the points of contact from the origin are the roots of the equation

$$x^2 - 2(h + k\tan a)x + h^2 + k^2 = 0.$$

Find the equation of the common chord of the two circles.

41. Find the equation of the circle which passes through the two points $(a, 0)$, $(-a, 0)$, and whose radius is $\sqrt{a^2 + b^2}$.

Write down the equation of the polar of the point (a, β) with respect to this circle. If (a, β) is a fixed point, a a constant, and b a variable, prove that this polar passes through another fixed point.

LOCUS PROBLEMS ON THE CIRCLE.

105. Example 1. *Pairs of straight lines are drawn through the points $(a, 0)$, $(-a, 0)$, so that the angle between each pair is constant and equal to a. Find the locus of their intersection.*

The equations of the lines may be written

$$y = m(x - a), \quad \dots\dots\dots\dots\dots\dots\dots\dots\dots\dots(1)$$
$$y = m'(x + a). \quad \dots\dots\dots\dots\dots\dots\dots\dots\dots(2)$$

Since a is the angle between these lines,

$$\tan a = \frac{m - m'}{1 + mm'}$$

or $(1 + mm')\tan a = m - m'. \quad \dots\dots\dots\dots\dots\dots\dots\dots(3)$

[We have to find the algebraic relation between (x, y), the co-ordinates of the intersection of (1) and (2), and the angle a. We must therefore eliminate m and m' between equations (1), (2), and (3).]

Substituting for m and m' from (1) and (2) in equation (3), we have

$$\tan a \left[1 + \frac{y^2}{x^2 - a^2} \right] = \frac{y}{x - a} - \frac{y}{x + a},$$

i.e. $\tan a [x^2 + y^2 - a^2] = 2ay,$

i.e. $x^2 + y^2 - 2ay \cot a = a^2$, the equation of the locus.

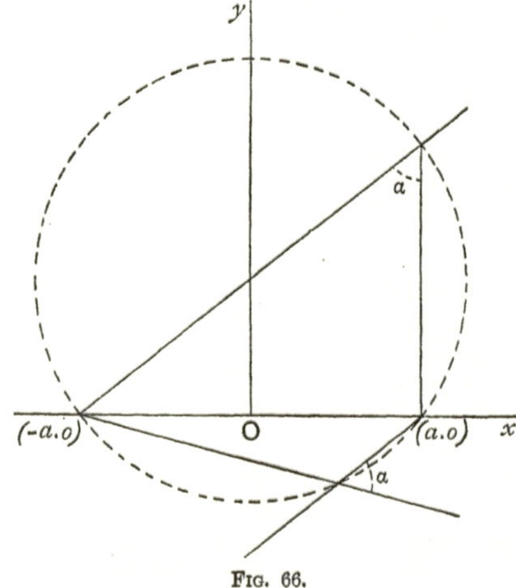

Fig. 66.

Thus we see that the locus is a circle whose centre is in the axis of y. The circle passes through the points $(a, 0)$, $(-a, 0)$.

We have here an analytical proof of the geometrical fact that the angle in a segment of a circle is constant.

Example ii. *Find the locus given by the equations* $x = a + r \cos \theta$, $y = b + r \sin \theta$, *where θ is a variable.*

To find the equation of the locus, we must eliminate θ.

$$x - a = r \cos \theta, \quad y - b = r \sin \theta.$$

Squaring and adding these equations, we have

$$(x - a)^2 + (y - b)^2 = r^2 (\cos^2 \theta + \sin^2 \theta),$$

i.e. $(x - a)^2 + (y - b)^2 = r^2$;

$\therefore$ the locus is a circle whose radius is r, and whose centre is at the point (a, b).

Examples V. d.

LOCUS PROBLEMS ON THE CIRCLE.

1. Chords of a circle are drawn through a fixed point : find the locus of their middle points.

2. Find the locus of the middle point of a line of constant length which moves so that its ends always lie on two fixed perpendicular lines.

3. The length of the tangent drawn from a point to the circle $x^2 + y^2 - 4x - 6y = 0$ is always equal to $2\sqrt{3}$. Find the equation of, and draw the locus of the point.

4. A point moves so that the sum of the squares of its distances from the angular points of a square is constant. Prove that its locus is a circle.

5. A point P moves so that the length of the tangent from P to the circle $x^2 + y^2 + 2gx + 2fy + c = 0$ is twice the distance of P from the origin. Find the equation of, and interpret the locus.

6. Find the locus of the intersection of the straight lines

$$(x - a) = m(y - \beta + c) \text{ and } m(x - a) = -(y - \beta - c),$$

m being a variable quantity, a, β, and c constants.

[The equation of the locus is found by eliminating m.]

7. If $x = a(1 + \cos a)$ and $y = a(1 + \sin a)$, find the locus of the point (x, y) when a is a variable angle.

8. A point moves so that the sum of the squares of its distances from a number of given points is constant. Prove that its locus is a circle.

9. Prove that the locus of a point from which the tangents to two given circles are in a constant ratio is a circle.

10. A and B being fixed points, a point P moves so that $PA = m \cdot PB$, where m is constant. Prove that the locus of P is a circle.

11. Two straight lines are drawn through the points $(a, 0)$, $(-a, 0)$ at right angles to one another. Find the locus of their point of intersection. What geometrical property of the circle is thus proved ?

[The equations of the lines may be written $y = m(x - a), \quad y = -\dfrac{1}{m}(x + a).$

If we now eliminate m between these equations we have a locus on which the two straight lines meet for all values of m.]

12. Two straight lines are drawn through the points (x_1, y_1), (x_2, y_2) at right angles to one another. Find the equation of the locus of their intersection.

13. Show that the locus of the intersection of the straight lines $x \cos a + y \sin a = a$, $x \sin a - y \cos a = b$ for all values of a is the circle $x^2 + y^2 = a^2 + b^2$.

14. A point moves so that the square of its distance from the base of an isosceles triangle is equal to the rectangle contained by its distances from the equal sides. Prove that its locus is a circle.

[Take the middle point of the base as origin and the altitude as axis of y.

The equations of the equal sides may then be written $\dfrac{x}{a} + \dfrac{y}{b} = 1, \quad -\dfrac{x}{a} + \dfrac{y}{b} = 1.$]

15. The straight lines $y - mx = \sqrt{a^2m^2 + b^2}$, $my + x = \sqrt{a^2 + b^2m^2}$ are at right angles to one another. Find the locus of their intersection when m is a variable quantity, a and b being constants.

[To find the locus we have to eliminate m. This can be done by squaring the equations, and adding.]

16. Find the locus of the intersection of the perpendicular straight lines $y - mx = \sqrt{a^2m^2 + b^2}$, $my + x = \sqrt{a^2 - b^2}$, m being a variable, a and b constants.

17. Tangents are drawn to the circles $x^2 + y^2 = a^2$, $x^2 + y^2 = b^2$ at right angles to one another. Find the locus of their point of intersection.

[Use the 'm' equation of a tangent.]

18. A point moves so that the sum of the squares of its distances from the sides of an equilateral triangle is constant. Prove that its locus is a circle.

19. O is a fixed point and P a point which moves on a fixed straight line; in OP a point Q is taken such that the rectangle OP . OQ is constant. Prove that the locus of Q is a circle.

Revision Questions on the Circle.

[These may be taken orally, or the answers may be written down without any working.]

What is the equation of the circle whose centre is at

1. The origin, and whose radius is 2 ?

2. The point (1, 2)　　,,　　　,,　　　3 ?

3.　　　,,　　　(0, -1) ,,　　　,,　　　1 ?

4.　　　,,　　　(3, 0)　　,,　　　,,　　　4 ?

What is the general equation of a circle

5. Which passes through the origin ?

6. Whose centre lies in the axis of x ?

7.　　　,,　　　　　,,　　　　　　,,　　　y ?

8. Whose centre lies at the point $(-a, -b)$?

9. Whose radius is a ?

What are the co-ordinates of the centre of the circle whose equation is

10. $x^2 + y^2 - 4x - 6y = 0$?　　　　**11.** $x^2 + y^2 + 3x - 8y = 1$?

12. $x^2 + y^2 + x + y = \frac{7}{8}$?

What is the equation of *any* diameter of the circle

13. $x^2 + y^2 = a^2$?　　　　**14.** $(x - 1)^2 + (y - 2)^2 = 4$?

15. $(x + 2)^2 + (y - 1)^2 = 9$?

16. What is the equation of the diameter of the circle

$$(x - 1)^2 + (y - 4)^2 = 16$$

which passes through the origin ?

17. What is the equation of the diameter of the circle $x^2 + y^2 = a^2$ which passes through the point (3, 1) ?

18. What is the slope, or gradient, of the diameter of the circle $(x-1)^2+(y-2)^2=9$ which passes through the point $(5, 3)$?

What is the equation of the tangent to the circle

19. $x^2+y^2=5$ at the point $(1, 2)$?

20. $4x^2+4y^2=1$ at the point (x_1, y_1)?

21. $x^2+y^2+2fx+2gy=0$ at the point (x_1, y_1)?

22. $x^2+y^2+4x-6y+3=0$ at the point $(1, 2)$?

23. $2x^2+2y^2-6x+8y-4=0$ at the point (x_1, y_1)?

What is the equation of the chord of contact of tangents to the circle $x^2+y^2=9$:

24. Drawn from the point $(5, 4)$? **25.** Drawn from the point $(-4, 1)$?

26. ,, ,, $(6, 0)$? **27.** ,, ,, $(0, -4)$?

Give the conditions that the following pairs of equations may be identical :

28. $Ax+By+C=0$, $lx+my+n=0$.

29. $3x+4y-1=0$, $Ax+By+C=0$.

30. $Ax+By+C=0$, $lx-m=0$.

31. $Ax+By+1=0$, $3x-5y+1=0$.

32. $Ax+By+C=0$, $Ax-3y-4=0$.

Give, without simplifying, the length of the tangent to the circle

33. $x^2+y^2+fx+gy+c=0$ from the point (x_1, y_1).

34. ,, ,, ,, $(0, 0)$.

35. $x^2+y^2-4x-5y-9=0$,, $(3, 5)$.

36. $(x-1)^2+(y-2)^2=4$,, $(5, 6)$.

37. $(x+1)^2+(y-3)^2=1$,, $(6, 0)$.

What is the equation which gives the abscissae of the points where the circle $x^2+y^2+gx+fy+c=0$ meets

38. The axis of x? **39.** The axis of y?

40. The straight line $x=y$? **41.** The straight line $y=2x$?

What is the condition that the circle $x^2+y^2+2gx+2fy+c=0$ should touch

42. The axis of x? **43.** The axis of y?

CHAPTER VI.

REVISION PAPERS.

Revision Paper VI. a.

1. The straight line joining the points (2, 4), (10, 12) is divided into four equal parts. Find the co-ordinates of the points of division.

2. Find the area of the triangle joining the points (0, 0), (6, 4), (8, 2) when unity represents one foot.

3. On squared paper plot the points (6, 8), (2, 4), unit half an inch. Join the points and read off the values of the intercepts this line makes on the axes of co-ordinates.

Deduce the equation of the straight line, and check your result.

4. Show on squared paper (unit 0·1 inch) the lines $\dfrac{x}{45}+\dfrac{y}{40}=1$, and $y=0\cdot 3x-30$, and measure the angle between them.

Also find the angle to the nearest degree by calculation.

5. A circle is described in the first quadrant touching the straight line $y=x\tan 2a$, and also touching the axis of x at the point (4, 0). Prove that its equation is $x^2+y^2-8x-8y\tan a+16=0$.

6. Determine which is the greater of the circles whose equations are $x^2+y^2-4ax-4ay+4a^2=0$ and $x^2+y^2-3ax-4ay=0$. Prove that these circles touch one another.

Revision Paper VI. b.

1. Determine graphically, or by calculation, the co-ordinates of the points of trisection of the line joining the points (8, 3), (1, 7).

2. Draw on squared paper the line bisecting at right angles the line which joins the points (5, 1) and (2, 4), unit 1 inch.

By inspection of your figure, write down the equation of the line so found, and verify your result by working out the equation.

3. In the equation $Ax+By=C$, interpret the constants when (i) $A=1$, (ii) $B=1$, (iii) $C=1$, (iv) $A^2+B^2=1$.

4. In the circle $x^2+y^2=a(5x-12y)$ find the equation of the diameter through the origin, and the equation of the tangent at the origin.

5. Find the equation of a circle (radius a) referred to a diameter and the tangent at its extremity as axes.

Show that the straight line $3x-4y+2a=0$ touches the circle, and find its point of contact.

6. Find the co-ordinates of the middle point of the chord which the circle $x^2+y^2-2x+2y=2$ cuts off on the line $y=x-1$.

Find also the equation of the locus of the middle points of all chords of the circle which are parallel to the line $y=x-1$.

Revision Paper VI. c.

1. Place the points (1, 2), (-3, 2), (2, -1), (-2, -3) on squared paper, taking rectangular axes, and an inch as unit.

State the co-ordinates of the middle points P and Q of the lines joining the first two points and the last two, and find the abscissa of the point where PQ cuts the axis of x.

2. Graphically, or by calculation, find the co-ordinates of the centre of gravity of three equal masses at the points (0, 1), (1, 2), and (3, 0).

3. Find the equation to a line passing through the point (-3, -2) and perpendicular to the line $2x + 3y = 3$.

Find also the equation to a line passing through the origin and the intersection of the two mutually perpendicular lines.

Verify the equations obtained by showing the lines on squared paper, explaining the nature of your verification.

4. Show that if $a^2 m^2 + 2al = 1$, the straight line $lx + my = 1$ will touch the circle $x^2 + y^2 - 2ax = 0$.

5. Find the equations of the two tangents which can be drawn from the origin to the circle $x^2 + y^2 + 10(x + y) + 40 = 0$, and determine the angle between them. Also find the equation of the concentric circle which passes through the origin.

6. Find the length of the tangent from the origin to the circle

$$x^2 + y^2 - 6x + 8y + 4 = 0.$$

What is the radius of this circle? Determine whether the point (1, 2) is inside or outside it.

Revision Paper VI. d.

1. Draw on squared paper the line $2x + 0 \cdot 7y = 2 \cdot 5$ inches.

Find to one decimal place the length of the perpendicular to this line from the origin, and check your result.

2. ABCD is a rectangle having AB 5 inches long and BC 3 inches long. The diagonal AC is drawn, and BO let fall perpendicular to it and produced to X. With OX and OA as axes of x and y, and unit one inch, find by drawing and measurement, or by calculation, the equations of the sides AB, BC, CD, each in the form $\frac{x}{a} + \frac{y}{b} = 1$, giving a and b in each case to two significant figures.

3. A parallel being drawn to the base of a triangle so as to cut the sides, find the locus of the intersection of perpendiculars to the sides through the points of cutting.

4. Find to one decimal place the length of the tangent from the point (5, 6) to the circle $x^2 + y^2 - 4x + 4y = 0$.

5. A circle is described to pass through the origin and to touch the lines $x = 1$, $x + y = 2$. Prove that the radius r is a root of the equation

$$r^2(3 - 2\sqrt{2}) - 2r\sqrt{2} + 2 = 0.$$

6. Find the equation of the common tangents to the circles

$$x^2 + y^2 = ax, \quad x^2 + y^2 = ay.$$

Revision Paper VI. e.

1. On squared paper draw a quadrilateral with its corners at $(3, 5)$, $(1, 4)$, $(0, 3)$, $(3, 1)$, the unit being one inch.

Find the area of the quadrilateral in any way you please, stating your method.

2. Draw on squared paper the line $3x + 4y = 10$, unit one inch. Find (a) by measurement, (b) by calculation, the length of the perpendicular from the origin on the straight line and the co-ordinates of the foot of the perpendicular.

3. The ends A, B of a straight line AB, of constant length a, slide upon the fixed rectangular axes OX, OY respectively. If the rectangle OAPB be completed, show that the co-ordinates of the foot of the perpendicular drawn from P to AB are $(a \cos^3\theta, a \sin^3\theta)$, where θ is the acute angle AB makes with Ox.

4. If the equation of a circle is $x^2 + y^2 + 2gx + 2fy + c = 0$, find the condition that must be satisfied by the parameters g, f, c, in order that it may be possible to find a point, or points, on the circle at equal perpendicular distances from the axes.

5. If the point (h, k) do not lie on the perimeter of the circle

$$(x - a)^2 + (y - \beta)^2 = r^2,$$

interpret the expression $(h - a)^2 + (k - \beta)^2 - r^2$ geometrically.

6. Describe the circle $x^2 + y^2 = c^2$, and draw the lines

$$(x\sqrt{3} + y - c)(-x\sqrt{3} + y - c) = 0.$$

Find the equation of the line joining the feet of the perpendiculars upon these lines from any point (a, b) on the circle.

Prove that this line also passes through the foot of the perpendicular from the point (a, b) upon the line $2y + c = 0$.

Revision Paper VI. f.

1. Determine the equation of the straight line which passes through the origin and the point of intersection of the lines $5x - 3y = 11$, $x + 2y = 10$.

2. Find the cosine of the angle between the straight lines whose equations are $ax + by + c = 0$, $a'x + b'y + c' = 0$.

3. If the axis of y is measured vertically upwards, prove that $y' - mx' - b$ represents the vertical height of the point (x', y') above the line whose equation is $y = mx + b$. Deduce the length of the perpendicular from the point (x', y') on the line $y = mx + b$.

4. Show that the circles $x^2 + y^2 + 2x - 8y + 8 = 0$, $x^2 + y^2 + 10x - 2y + 22 = 0$ touch each other, and find the point of contact.

5. Show that the distances of two points from the centre of a circle are to one another in the same ratio as the distances of each point from the polar of the other with regard to the circle.

6. Prove that the equation of the circle circumscribing the quadrilateral formed by the four straight lines $y = 2x$, $x = 2y$, $x + y = a$, $x + y = 2a$ is

$$9(x^2 + y^2) - 15a(x + y) + 10a^2 = 0.$$

Find the equations of the tangents drawn to it from the origin.

Revision Paper VI. g.

1. What is the general nature of the family of straight lines obtained from the equation $y = mx + 2 \cdot 2$, by giving various numerical values to m?

Show on squared paper the three lines of the family for which m has the values $1 \cdot 8$, $-1 \cdot 8$, $6 \cdot 4$.

2. Find the area of the triangle whose sides are formed by the straight lines $y = x$, $y = 2x$, $x + y - 6 = 0$.

3. Find the equations of the straight lines bisecting the angles between
$$3x + 4y = 5, \quad 12x - 5y = 41.$$

4. Solve the equation $2x^2 + 2y^2 + 5xy + 1 = 3x + 3y$ for y.

Give the values of y when $x = 0 \cdot 1$, $0 \cdot 2$, $0 \cdot 5$.

Draw the locus represented by the equation.

5. Having given $y = \sqrt{a^2 - x^2}$, where $a = 2$ inches, find (by putting $x = a \sin \theta$, or otherwise) a number of sets of corresponding values of x and y; and hence trace the curve represented by the equation.

6. AP, any chord of a circle (radius a) through a fixed point A, is produced to Q so that $AQ = m \cdot AP$. Find the locus of Q.

Revision Paper VI. h.

1. What is the general equation to a straight line through the point of intersection of the given straight lines
$$ax + by = 1, \quad lx + my = 1?$$

Find the equation of the straight line passing through their point of intersection and also through the origin.

2. Find the equations of the straight lines parallel to $12x - 5y = 26$ and at a distance of one unit from it.

3. Straight lines parallel to $y = kx$ are drawn to cut the given lines $y = mx$, $y = nx$. Find the equation of the locus of the middle points of the parallel intercepts.

4. Prove that the equation $x^2 + 3xy + y^2 - 7x - 3y + 1 = 0$ represents two straight lines.

Find the angle between these lines, and prove that the co-ordinates of their point of intersection are -1 and 3.

5. Find the equations of the tangents to the circle $x^2 + y^2 - 2ax = 0$ which are parallel to the line $5x + 12y = 0$.

6. Find the equations of all the common tangents of the circles
$$x^2 + y^2 + 2x = 0 \quad \text{and} \quad x^2 + y^2 - 6x = 0,$$
and draw a figure.

CHAPTER VII.

ORTHOGONAL CIRCLES.

106. The angle between two curves at their point of intersection is the angle between their tangents at that point.

When two circles intersect at a point so that the tangents at that point are at right angles, the circles are said to cut one another **orthogonally**.

Such circles are often called **orthogonal circles**.

If two circles cut one another orthogonally, the sum of the squares on their radii is equal to the square on the line joining their centres.

This is at once seen from a diagram, for a tangent to one circle will gram, for a tangent to one circle will pass through the centre of the other.

FIG. 67.

Examples VII. a.

1. Two circles of radii a and b intersect at an acute angle θ. Prove that the length of their common chord is

$$\frac{2ab \sin \theta}{\sqrt{a^2 + b^2 + 2ab \cos \theta}}.$$

2. Find the angle at which the circles

$$(x - a)^2 + (y - \beta)^2 = r^2, \quad (x - a')^2 + (y - \beta')^2 = r'^2 \text{ intersect.}$$

3. Prove that the circles $x^2 + y^2 - 2ax + c^2 = 0$, $x^2 + y^2 - 2by - c^2 = 0$ cut one another orthogonally.

4. Prove that the circles

$$x^2 + y^2 + 2gx + 2fy + c = 0, \quad x^2 + y^2 + 2g'x + 2f'y + c' = 0$$

cut one another orthogonally if $2gg' + 2ff' = c + c'$.

5. Find the equation of the circle which cuts the circles

$$x^2 + y^2 - 8y + 12 = 0 \text{ and } x^2 + y^2 - 4x - 6y - 3 = 0$$

orthogonally and passes through the origin.

6. Find the equation of the circle which cuts $x^2 + y^2 = a^2$ orthogonally and has its centre at the point (h, k).

7. Three circles are described of equal radii (r), having their centres at the points $(0, 0)$, $(0, b)$, $(a, 0)$. Find the equation of the circle which cuts them all orthogonally.

8. Find the general equation of the circles which cut the circle
$$x^2 + y^2 - 2x + 2y - 2 = 0$$
at right angles at the point $(1, 1)$.

9. Find the condition that the circles
$$x^2 + y^2 - 2ax - c^2 = 0, \quad x^2 + y^2 + 2bx - c^2 = 0$$
may cut orthogonally.

10. Find the equation of the circle which passes through the origin and cuts orthogonally the circles whose equations are
$$x^2 + y^2 - 6x + 8 = 0, \quad x^2 + y^2 - 2x - 2y = 7.$$

11. Find the locus of the centres of circles which cut the circles
$$x^2 + y^2 + 4x - 6y + 9 = 0, \quad x^2 + y^2 - 4x + 6y + 4 = 0$$
orthogonally.

12. A circle, whose centre is in the axis of x, cuts the circles
$$x^2 + y^2 - 6y + 5 = 0, \quad x^2 + y^2 + 6x - 31 = 0$$
orthogonally.   Find its equation.

RADICAL AXIS OF TWO CIRCLES.

107. *Def.* The locus of a point, which moves so that the tangents drawn from it to two given fixed circles are equal, is a

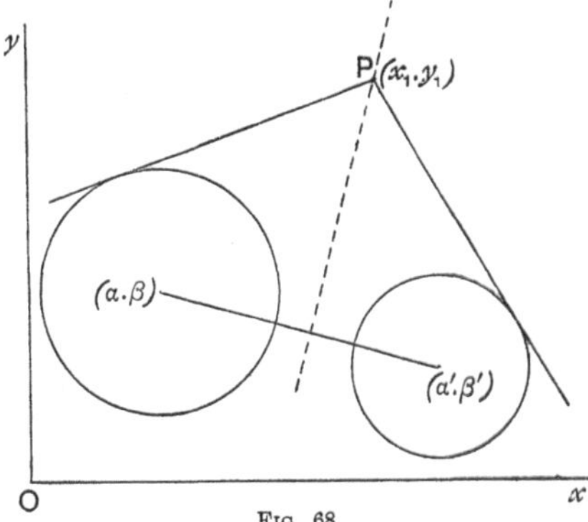

Fig. 68.

straight line perpendicular to the line of centres of the circles, and is called their **Radical Axis.**

Let $(x - \alpha)^2 + (y - \beta)^2 = a^2$, $(x - \alpha')^2 + (y - \beta')^2 = a'^2$ be the equations of the circles.

Then if $\mathsf{P}(x_1, y_1)$ is any point on the locus,

$$(x_1 - \alpha)^2 + (y_1 - \beta)^2 - a^2 = (x_1 - \alpha')^2 + (y_1 - \beta')^2 - a'^2 \quad \text{(Art. 99)}$$

or $\quad 2x_1(\alpha' - \alpha) + 2y_1(\beta' - \beta) + \alpha^2 + \beta^2 - \alpha'^2 - \beta'^2 - a^2 + a'^2 = 0.$

$\therefore$ suppressing suffixes,

$$2x(\alpha' - \alpha) + 2y(\beta' - \beta) + \alpha^2 + \beta^2 - \alpha'^2 - \beta'^2 - a^2 - a'^2 = 0$$

is the equation of the locus.

Being of the first degree, this is a straight line.

Its slope $= -\dfrac{\alpha' - \alpha}{\beta' - \beta}.$

The slope of the line of centres $= \dfrac{\beta' - \beta}{\alpha' - \alpha}.$

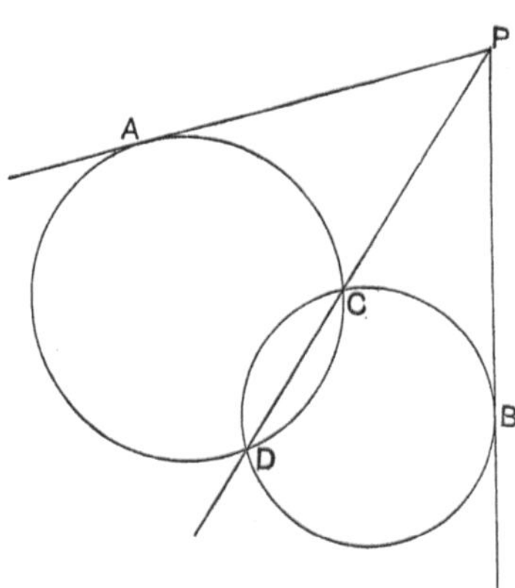

Fig. 69.

The product of these slopes

$$= -1; \quad (mm' = -1)$$

$\therefore$ the radical axis is perpendicular to the line of centres.

Note 1. When the coefficients of x^2 and y^2 are unity in the equations of two circles, the equation of their radical axis is at once obtained by subtraction.

Note 2. When the circles intersect, **their common chord is their radical axis.**

This is easily proved geometrically, for from the figure, if P is any point on the common chord,

$$\mathsf{PA}^2 = \mathsf{PC} . \mathsf{PD} = \mathsf{PB}^2;$$

$\therefore$ PCD is the radical axis.

108. For the sake of brevity, let $\mathsf{S} \equiv x^2 + y^2 + 2gx + 2fy + c$

and $\quad \mathsf{S}' \equiv x^2 + y^2 + 2g'x + 2f'y + c',$

so that $\mathsf{S} = 0$, $\mathsf{S}' = 0$ are the equations of circles.

Let us consider the equation $\mathsf{S} - \lambda\mathsf{S}' = 0,$ ....(1)

λ being any constant.

In this equation the coefficients of x^2 and y^2 are equal, being $(1 - \lambda)$, and there is no term involving the product xy;

$\therefore$ $\mathsf{S} - \lambda\mathsf{S}' = 0$ represents a circle.

Again, any values of x and y which make $S = 0$ and $S' = 0$, also make $S - \lambda S' = 0$.

$\therefore$ the circle $S - \lambda S' = 0$ passes through the two common points of $S = 0$ and $S' = 0$.

Also, by giving λ different values, this circle $S - \lambda S' = 0$ may be made to pass through any third point.

$\therefore$ $S - \lambda S' = 0$ **is the general equation of a circle passing through the common points of the circles** $S = 0$, $S' = 0$.

109. $S - \lambda(Ax + By + C) = 0$ *is a circle through the common points of the circle* $S = 0$ *and the straight line* $Ax + By + C = 0$, λ *being any constant.*

The method of the previous article proves this at once. It is left as an exercise for the student.

Notice that $Ax + By + C = 0$ is the radical axis of the circles $S = 0$, $S - \lambda(Ax + By + C) = 0$. (Note to Art. 107.)

110. The previous articles will be seen to be applications of the following important general theorem :

If $S = 0$, $S' = 0$ represent any two curves,

$S - \lambda S' = 0$ represents a curve passing through their common points, λ having any constant value.

For any values of x and y which make $S = 0$ and $S' = 0$, also make $S - \lambda S' = 0$.

111. *The three radical axes of the circles* $S_1 = 0$, $S_2 = 0$, $S_3 = 0$ *are concurrent.*

$S_1 - S_2 = 0$ is the radical axis of $S_1 = 0$ and $S_2 = 0$.

$S_2 - S_3 = 0$,,   ,, $S_2 = 0$ and $S_3 = 0$.

$\therefore$ by the above theorem,

$$(S_1 - S_2) + (S_2 - S_3) = 0,$$

i.e. $S_1 - S_3 = 0$, passes through the common point of $S_1 - S_2 = 0$ and $S_2 - S_3 = 0$.

But $S_1 - S_3 = 0$ is the radical axis of $S_1 = 0$ and $S_3 = 0$.

$\therefore$ the three radical axes are concurrent.

Def. The common point of the three radical axes of three circles taken in pairs is called their **Radical Centre.**

112. λ *having any value and a being constant, $x^2 + y^2 + \lambda x - a^2 = 0$ represents a system of circles having the axis of y for a common radical axis.*

The equation represents a circle, for the coefficients of x^2 and y^2 are equal, and there is no term involving the product xy.

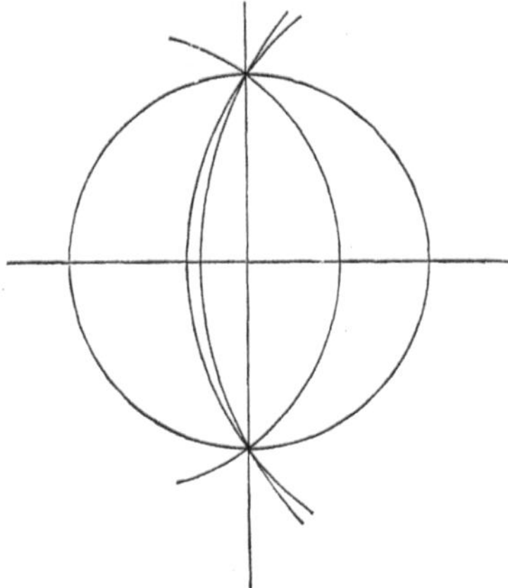

Also, whatever be the value of λ, it passes through the common points of

$$x^2 + y^2 - a^2 = 0 \quad \ldots\ldots\ldots(1)$$

and $\qquad x = 0. \ldots\ldots\ldots(2)$

The common points of (1) and (2) are the fixed points $(0, \pm a)$.

$\therefore$ the line joining these points, *i.e.* the axis of y, is the radical axis of every circle in the system.

CoROLLARY. In the same way it may be shown that

$$x^2 + y^2 + \lambda y - a^2 = 0$$

is a system of circles having the axis of x for a common radical axis.

Fig. 70.

Def. Such a system of circles is said to be **co-axal**.

113. Example i. *Find the equation of a system of circles which all have the straight line $3x - 5y = 7$ for their radical axis, one circle of the system having its centre at the origin and radius 4.*

$x^2 + y^2 - 16 + \lambda(3x - 5y - 7) = 0$ is the required equation, where λ may have any value; for it represents a circle passing through the common points of the circle $x^2 + y^2 - 16 = 0$, and the straight line $3x - 5y - 7 = 0$. Also $x^2 + y^2 - 16 = 0$ is one of the circles of the system, for this is what the equation of the system becomes when we take λ equal to zero.

Example ii. *Find the equation of the circle which passes through the points of intersection of $x^2 + y^2 + 4x - 8y + 4 = 0$ and $x^2 + y^2 + 2x + 2y - 2 = 0$, and also through the origin.*

The general equation of a circle through the intersections of the given circles is
$$x^2 + y^2 + 4x - 8y + 4 + \lambda(x^2 + y^2 + 2x + 2y - 2) = 0. \quad \ldots\ldots\ldots\ldots(1)$$

But the required circle passes through the origin $(0, 0)$.

$\therefore$ the values $(0, 0)$ must satisfy equation (1).

$$\therefore\ 4-2\lambda=0\ \text{ and }\ \lambda=2.$$

Substituting this value of λ in equation (1), we have for the required equation
$$3x^2+3y^2+8x-4y=0.$$

Examples VII. b.

1. Find the equation of the radical axis of the circles
$$x^2+y^2=1,\ \ x^2+y^2-4x-6y+11=0.$$
Draw a figure on squared paper, unit one inch.

2. Prove that the radical axis of two circles bisects their common tangents.

3. Find the radical axis of the circles
$$x^2+y^2-2x-2y+1=0\ \text{ and }\ 2x^2+2y^2-4x-5y-4=0.$$

4. Find the radical axis of the circles
$$x^2+y^2+2ax-c^2=0\ \text{ and }\ x^2+y^2+2bx-c^2=0,$$
and prove that the distance of any point on the first circle from the radical axis bears a constant ratio to the square of the length of the tangent drawn from that point to the second circle.

5. Find the equation of a system of circles which have the line $x-y=0$ for their radical axis.

6. All circles represented by the equation
$$(x-a_1)^2+(y-b_1)^2-c_1{}^2=\lambda[(x-a_2)^2+(y-b_2)^2-c_2{}^2]$$
have a common radical axis for all values of λ.

7. Find the radical axes and the radical centre of the circles
$$(x-1)^2+(y-2)^2=6,\ \ \ (x-2)^2+(y-3)^2=8,\ \ \ (x-3)^2+(y-1)^2=10.$$

8. Find the equation of the circle passing through the intersection of $x^2+y^2=1$ and $x^2+y^2=6x$ and through the point $(-1,1)$.

9. Find the equation of the circle passing through the intersection of $x^2+y^2=20$ and $2x-y=4$, and also through the point $(-2,0)$.

10. Show that the square of the tangent which can be drawn from any point on one circle to another is proportional to the perpendicular from the point on their radical axis.

11. OAB, OCD are two perpendicular straight lines, and AC, BD intersect at E. Prove that the circles drawn on AD, BC, OE as diameters have a common radical axis.

12. Find the radical axis of the system of circles represented by
$$x^2+y^2-x(\lambda-2)-y(\lambda-4)+1=0,$$
where λ may have any value. Draw two circles of the system on squared paper.

13. $2x-y=0$ is the common radical axis of a system of circles. One of the circles has its centre at the origin and a radius c: find an equation to represent the system.

14. Interpret the equation $x^2+y^2+x(\lambda-4)-2y(\lambda+3)+9-\lambda=0$, where λ may have any value.

15. Find the equation of the system of circles having the same radical axis as $x^2+y^2=25$, and $(x-1)^2+(y-1)^2=27$.

16. Find the condition that two circles of the system $x^2+y^2-2\lambda x-c^2=0$ may cut orthogonally, c being constant and λ variable.

17. Find the equation of the circle through the intersection of the circle $x^2+y^2=8$ and the straight line $2x+3y=5$, and also passing through the point $(6, 0)$.   Also find the co-ordinates of its centre.

18. Find the equation of the circle which passes through the origin and through the points of intersection of $x^2+y^2+6x-8y+4=0$ and $3x-2y+2=0$.

19. λ being a variable quantity, prove that $x^2+y^2+\lambda(x+2y)-20=0$ represents a system of circles passing through two fixed points.   Find the co-ordinates of the points.

20. Find the equations of the circles of radius 7 which pass through the common points of the circle $x^2+y^2=24$ and the straight line $3x-4y=0$.

21. λ being a variable, prove that each circle of the system represented by $x^2+y^2+\lambda x-2\lambda y-16=0$ passes through two fixed points.   Illustrate with a figure.

THE CIRCLE. OBLIQUE AXES.

***114.** *To find the equation of the circle whose radius is a and whose centre is at the origin, ω being the angle between the axes of co-ordinates.*

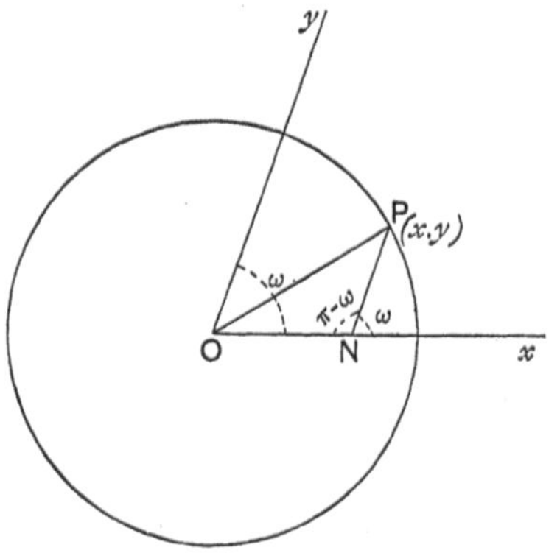

Let **P** be any point on the circumference and (x, y) its co-ordinates.

Draw the ordinate **PN** parallel to Oy.

In the $\triangle$ **OPN**,

$$\angle \mathsf{ONP} = \pi - \angle xOy = \pi - \omega\,;$$

$$\therefore \mathsf{OP}^2 = \mathsf{PN}^2 + \mathsf{ON}^2$$
$$- 2\,.\,\mathsf{PN}\,.\,\mathsf{ON}\cos(\pi-\omega),$$
$$\left[a^2 = b^2 + c^2 - 2bc\cos\mathsf{A}\right]$$

i.e. $\quad a^2 = y^2 + x^2$
$$- 2xy\cos(\pi-\omega)\,;$$

$$\therefore \mathbf{x^2 + y^2 + 2xy\cos}\,\omega = \mathbf{a}^2 \text{ is the reqd. equation.}$$

FIG. 71.

* **115.** *To find the equation of the circle whose radius is a and whose centre is at the point (α, β), ω being the angle between the axes of co-ordinates.*

Let P be any point on the circumference and (x, y) its co-ordinates; C the centre of the circle.

Draw the ordinates at C and P, CM and PN, parallel to Oy, and also draw CK parallel to Ox to meet PN at K.

$$\angle CKP = \angle ONP = \pi - \omega;$$

$\therefore$ in $\triangle$ CKP,

$$CP^2 = CK^2 + KP^2$$
$$- 2CK \cdot KP \cos(\pi - \omega),$$
$$[a^2 = b^2 + c^2 - 2bc \cos A]$$

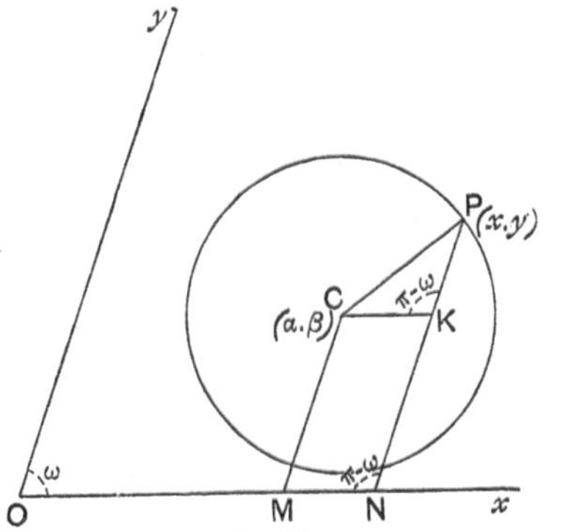

FIG. 72.

i.e. $a^2 = (x - \alpha)^2 + (y - \beta)^2 + 2(x - \alpha)(y - \beta) \cos \omega$;

$$\therefore \ (\mathbf{x} - \alpha)^2 + (\mathbf{y} - \beta)^2 + 2(\mathbf{x} - \alpha)(\mathbf{y} - \beta) \cos \omega = \mathbf{a}^2$$

is the reqd. equation.

The above equation may be written

$$x^2 + y^2 + 2xy \cos \omega - 2x(\alpha + \beta \cos \omega)$$
$$- 2y(\beta + \alpha \cos \omega) + \alpha^2 + \beta^2 + 2\alpha\beta \cos \omega - a^2 = 0.$$

We therefore see that, with oblique axes, the general equation of a circle is

$$x^2 + y^2 + 2xy \cos \omega + 2gx + 2fy + c = 0.$$

* **116.** *Equation of a tangent when the axes are oblique.*

If (x_1, y_1), $(x_1 + \Delta x_1, y_1 + \Delta y_1)$ are two points on a curve near to one another, the equation of the chord joining them is

$$\frac{y - y_1}{y_1 + \Delta y_1 - y_1} = \frac{x - x_1}{x_1 + \Delta x_1 - x_1} \quad \text{or} \quad \frac{y - y_1}{\Delta y_1} = \frac{x - x_1}{\Delta x_1},$$

i.e. $y - y_1 = \frac{\Delta y_1}{\Delta x_1}(x - x_1).$

Hence, as with rectangular axes, we have to determine the limiting value of $\frac{\Delta y_1}{\Delta x_1}$ when the points approach one another, and ultimately coincide. If this value is m, the equation of the tangent at the point (x_1, y_1) is

$$y - y_1 = m(x - x_1).$$

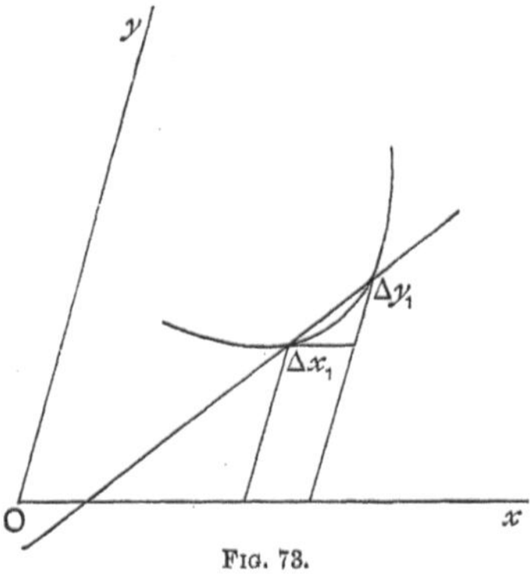

We see from the figure that with oblique axes the value of $\dfrac{\Delta y_1}{\Delta x_1}$ is not the slope of the chord as defined with rectangular axes.

*117. *To find the equation of the tangent to the circle*

$$x^2 + y^2 + 2xy \cos \omega = a^2$$

at the point (x_1, y_1).

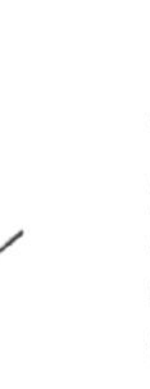

Fig. 73.

If (x_1, y_1), $(x_1 + \Delta x_1, y_1 + \Delta y_1)$ are two points on the circle near to one another, the equation of the chord joining them is

$$\frac{y - y_1}{\Delta y_1} = \frac{x - x_1}{\Delta x_1}. \quad \dots\dots\dots\dots\dots\dots\dots(1)$$

Since these points are on the circle,

$$x_1{}^2 + y_1{}^2 + 2x_1 y_1 \cos \omega = a^2$$

and $(x_1 + \Delta x_1)^2 + (y_1 + \Delta y_1)^2 + 2(x_1 + \Delta x_1)(y_1 + \Delta y_1) \cos \omega = a^2 \, ;$

$\therefore$ by subtraction,

$$2x_1 \Delta x_1 + (\Delta x_1)^2 + 2y_1 \Delta y_1 + (\Delta y_1)^2$$
$$+ 2(y_1 \Delta x_1 + x_1 \Delta y_1 + \Delta x_1 \Delta y_1) \cos \omega = 0.$$

But when the two points are very near to one another, $(\Delta x_1)^2$, $(\Delta y_1)^2$, and the product $\Delta x_1 . \Delta y_1$ are all very small;

$\therefore$ neglecting these quantities,

$$x_1 \Delta x_1 + y_1 \Delta y_1 + (y_1 \Delta x_1 + x_1 \Delta y_1) \cos \omega = 0,$$

whence $\Delta y_1 (y_1 + x_1 \cos \omega) = - \Delta x_1 (x_1 + y_1 \cos \omega);$

$\therefore$ by multiplication with equation (1),

$$(y - y_1)(y_1 + x_1 \cos \omega) = - (x - x_1)(x_1 + y_1 \cos \omega)$$

is the equation of the tangent.

It may be written

$$xx_1 + yy_1 + (x_1 y + y_1 x) \cos \omega = x_1{}^2 + y_1{}^2 + 2x_1 y_1 \cos \omega$$

or $xx_1 + yy_1 + (x_1 y + y_1 x) \cos \omega = a^2.$

Examples VII. c.

1. Draw the circle whose equation is
$$x^2 + y^2 + xy = a^2.$$

2. Find the equation of the circle which passes through the origin and cuts off intercepts h and k from the axes of co-ordinates (oblique).

3. Given that the straight line $lx + my = n$ touches the circle
$$x^2 + y^2 + 2xy \cos \omega = a^2,$$
find the co-ordinates of its point of contact.

4. If the equation $25(x + y - a)^2 = 18xy$ represents a circle, find the angle between the axes of co-ordinates.

5. Prove that the circle $x^2 + y^2 + 2xy \cos \omega - 2ax - 2ay + a^2 = 0$ touches both axes, and find the co-ordinates of its centre.

6. Find the condition that the straight line $y = mx$ may touch the circle $x^2 + y^2 + 2xy \cos \omega + 2gx + 2fy + c = 0$.

7. Find the angle between the axes when the equation
$$x^2 + y^2 = a^2 + \sqrt{3}xy + ax$$
represents a circle. Also find the radius and the co-ordinates of the centre.

8. Find the equation of the circle circumscribing the triangle formed by the axes of co-ordinates and the straight line $x + 2y = 6$, when $60°$ is the angle between the axes.

9. Prove that the straight line $2x + y = 2a$, touches the circle
$$x^2 + y^2 + xy = a^2,$$
and find the co-ordinates of its point of contact.

10. Find the equations of the tangents to the circle $x^2 + y^2 + xy = a^2$ at the points where it cuts the axis of y.

11. Find the equation of the straight line which touches the circle $x^2 + y^2 + xy - 4x - 5y = 0$ at the origin.

12. Find the equation of the tangent to the circle
$$x^2 + y^2 + 2xy \cos \omega + 2gx + 2fy = 0$$
at the origin, and the equation of the radius which passes through the origin.

POLAR EQUATION OF A CIRCLE.

118. *The polar equation of a circle of radius a, whose centre is at the origin, is $r = a$.*

This is evident from the fact that every point on the circumference of the circle is at a distance a from the origin.

B.A.G.　　　　　　　　　H

119. *The equation of the tangent at the point* (a, a) *on the circle* $r = a$ *is* $a = r \cos(\theta - a)$.

Let Q be the point (a, a), and P(r, θ) any point on the tangent at Q.

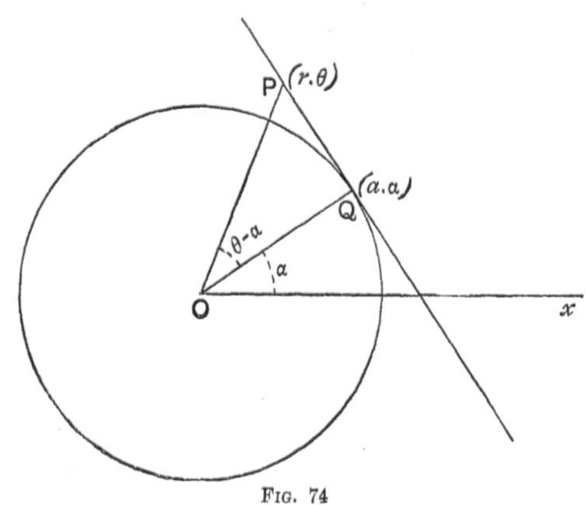

Fig. 74

$$\angle POQ = \angle POx - \angle QOx = \theta - a\ ;$$
$$\therefore\ \ OQ = OP \cos(\theta - a),$$

i.e. $a = r \cos(\theta - a)$ is the equation of the tangent at Q.

120. *To find the polar equation of a circle of radius a whose circumference passes through the origin and whose centre lies in the initial line.*

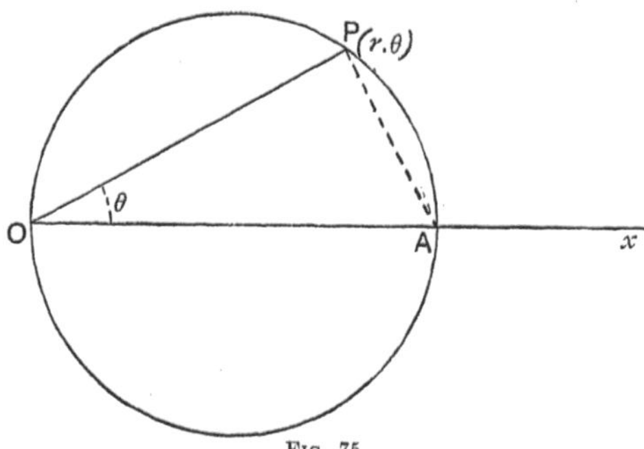

Fig. 75.

Let P$(r,\ \theta)$ be any point on the circumference and OA the diameter through the origin.

$\angle$ OPA is a rt. $\angle$;

$$\therefore\ \ OP = OA \cos \theta,$$

i.e. $r = 2a \cos \theta$ is the required equation.

121. *To find the equation of the tangent to the circle $r = 2a \cos \theta$ at the point (r_1, θ_1).*

Let P be the point (r_1, θ_1), C the centre, PK the tangent, OK the perpendicular on the tangent from the origin.

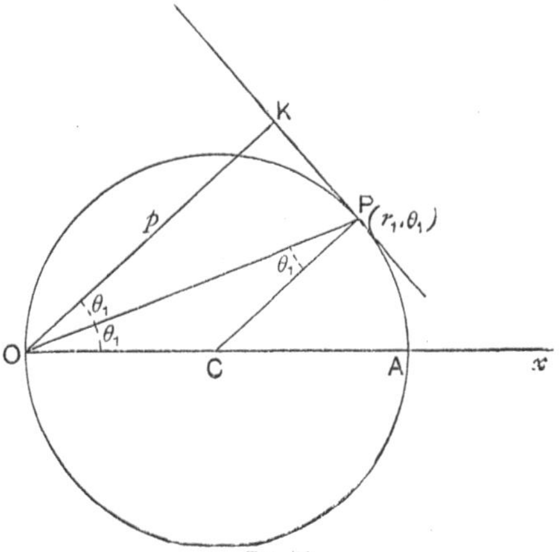

$$\angle KOP = \angle OPC = \angle POC$$
$$= \theta_1 ;$$
$$\therefore \ OK = OP \cos \theta_1$$
$$= r_1 \cos \theta_1$$
$$= 2a \cos^2 \theta_1 .$$

Also $\angle KOx = 2\theta_1$;

$\therefore$ the equation of PK is

$$2a \cos^2 \theta_1 = r \cos(\theta - 2\theta_1). \quad [p = r \cos(\theta - a)]$$

Fig. 76.

122. *To find the polar equation of a circle which passes through the origin and has its centre at the point (a, a).*

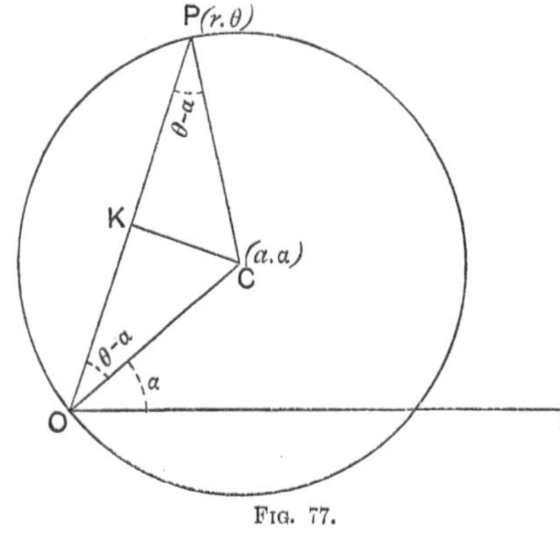

Fig. 77.

Let $P(r, \theta)$ be any pt. on the circle, C its centre, and draw CK perpendicular to OP.

$$\angle OPC = \angle POC = \theta - a ;$$
$$\therefore \ OP = 2OK$$
$$= 2OC \cos(\theta - a);$$
$$\therefore \ r = 2a \cos(\theta - a)$$

is the required equation.

Alternative Method. If OB is the diameter through O,

$$OP = OB \cos POB ;$$
$$\therefore \ r = 2a \cos(\theta - a), \text{ as before.}$$

123. *To find the polar equation of the tangent at the point* $(r_1,\ \theta_1)$ *to the circle whose equation is*

$$r = 2a\cos(\theta - \alpha).$$

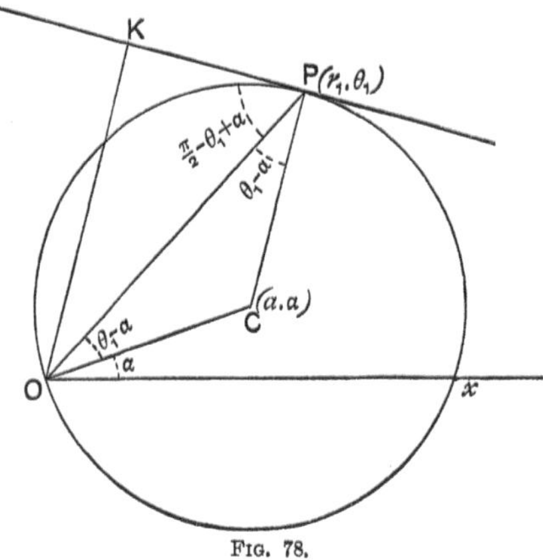

FIG. 78.

Let PK be the tangent at the point $P(r_1,\ \theta_1)$, and draw OK perpendicular to it.

If $C(a,\ \alpha)$ is the centre of the circle,

$$\angle \text{KPO} = \frac{\pi}{2} - \angle \text{OPC}$$

$$= \frac{\pi}{2} - \angle \text{COP}$$

$$= \frac{\pi}{2} - (\theta_1 - \alpha)\ ;$$

$$\therefore\ \ \text{OK} = \text{OP}\sin\left[\frac{\pi}{2} - (\theta_1 - \alpha)\right] = r_1\cos(\theta_1 - \alpha).$$

Also $\angle \text{KO}x = \angle \text{KOP} + \angle \text{PO}x = (\theta_1 - \alpha) + \theta_1 = 2\theta_1 - \alpha\ ;$

$\therefore$ the equation of the tangent PK is

$$r_1\cos(\theta_1 - \alpha) = r\cos(\theta - 2\theta_1 + \alpha). \quad \left[p = r\cos(\theta - \alpha)\right]$$

124. *To find the polar equation of the circle whose centre is at the point* $(c, 0)$ *in the initial line and whose radius is* a.

Let $P(r,\ \theta)$ be any point on the circle and C the centre of the circle.

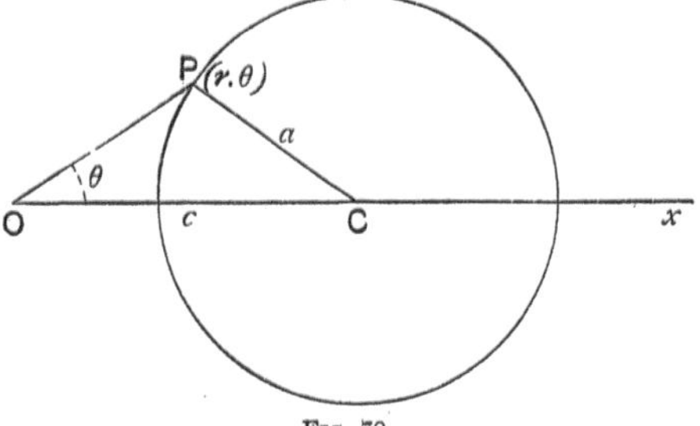

FIG. 79.

From the $\triangle$ OPC,

$$\text{OP}^2 + \text{OC}^2 - 2\text{OP}\,.\,\text{OC}\cos\text{POC} = \text{CP}^2\ ; \quad \left[b^2 + c^2 - 2bc\cos A = a^2\right]$$

$$\therefore\ \ r^2 + c^2 - 2cr\cos\theta = a^2 \text{ is the equation required.}$$

125. *To find the polar equation of the circle whose centre is at the point $(c,\ a)$ and whose radius is a.*

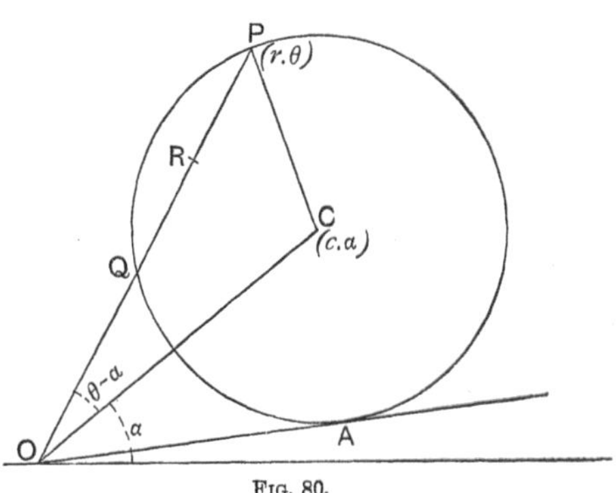

Fig. 80.

Let $C(c,\ a)$ be the centre of the circle, and $P(r,\ \theta)$ any point on its circumference.

From the $\triangle\,OPC$,

$$OP^2 + OC^2 - 2OP\,.\,OC\cos POC = CP^2\,;\quad [b^2 + c^2 - 2bc\cos A = a^2]$$

$$\therefore\ r^2 + c^2 - 2cr\cos(\theta - a) = a^2 \text{ is the required equation.}$$

126. In the preceding article, if OP meets the circle again at Q, we see that $OP,\ OQ$ are the roots of the equation

$$r^2 - 2cr\cos(\theta - a) + c^2 - a^2 = 0,$$

considered as a quadratic for r;

$\therefore\ OP\,.\,OQ =$ the product of the roots $= c^2 - a^2$, which is constant.

Also if OA is a tangent from O to the circle,

$$c^2 - a^2 = OC^2 - CA^2 = OA^2,$$

$\therefore\ OP\,.\,OQ = OA^2$, a well-known geometrical fact.

Again, if R is the middle point of PQ,

$$2OR = OP + OQ = \text{the sum of the roots of the above equation}$$
$$= 2c\cos(\theta - a)\,;$$

$\therefore$ the polar equation of the locus of R is

$$2r = 2c\cos(\theta - a)\ \text{ or }\ r = c\cos(\theta - a),$$

which represents a circle on OC as diameter (Art. 122).

Examples VII. d.

Draw the following loci :

1. $r = 2a \sin \theta$.

2. $r = 2a \sin (\theta + \alpha)$.

3. $r - 4 (\cos \theta + \sin \theta) = 0$.

4. $r = 2a \sin (\theta - \alpha)$.

5. Find the condition that the line $pr \cos \theta + qr \sin \theta = 1$ may touch the circle $r = 2a \cos \theta$.

6. Find the co-ordinates of the centre of the circle $r = 2 \cos \theta + 4 \sin \theta$.

7. What does the following equation represent ?

$$r = A \cos (\theta - \alpha) + B \cos (\theta - \beta) + C \cos (\theta - \gamma)\dots .$$

8. The polar equation of a circle being $r = 2c \cos \theta$, show that the equation $2c \cos \beta \cos \alpha = r \cos (\beta + \alpha - \theta)$ represents a chord such that the radii drawn to its extremities from the pole make angles α, β with the initial line.

9. Show that the polar equation $r = a \cos \theta + b \sin \theta$ represents a circle through the pole, and find the polar co-ordinates of its centre.

10. Find the polar equations of the circles passing through the points whose polar co-ordinates are $\left(a, \dfrac{\pi}{2} \right)$, $\left(b, \dfrac{\pi}{2} \right)$, and touching the straight line $\theta = 0$.

11. Draw the circle $r = 2 \cos \theta$, unit one inch. From first principles find the equation of the tangent at the point on the circle whose angular co-ordinate is $\dfrac{\pi}{6}$.

12. Through the point on the circle $r = 2a \cos \theta$ whose angular co-ordinate is $\dfrac{\pi}{6}$ a radius is drawn: find its equation.

13. In the circle $r = 2a \cos (\theta - \alpha)$, find the equation of the tangent at the origin.

14. In the circle $r^2 - 2cr \cos \theta + c^2 - a^2 = 0$, find the equations of the tangents which are perpendicular to the initial line.

15. OQ is drawn from the fixed point O to meet a fixed straight line AQ at Q. In OQ a point P is taken so that $OP \cdot OQ = a^2$. Find the locus of P, using polar co-ordinates.

16. O is a fixed point on a circle OPB of radius a, and OP is produced to Q so that $OQ = k \cdot OP$: find the locus of Q.

CHAPTER VIII.

THE PARABOLA.

127. Definition. A **Conic Section**, or more shortly a **Conic**, is the locus of a point which moves so that its distance from a fixed point is in a constant ratio to its perpendicular distance from a fixed straight line.

The fixed point is called the **Focus**.

The fixed straight line is called the **Directrix**.

The constant ratio is called the **Eccentricity**, and is denoted by e.

When the eccentricity e is equal to unity, the locus is called a **Parabola**.

Hence, a **parabola** is the locus of a point which moves so that its distance from a fixed point, called the focus, is equal to its perpendicular distance from a fixed straight line, called the directrix.

The name **Conic Section** is obtained from the fact that these curves (there are others besides the parabola) can be obtained by taking plane sections of a right cone.

128. *To find the equation of a parabola.*

Let **S** be the focus, **MM'** the directrix. Draw **SX** perpendicular to the directrix and bisect **SX** at **A**.

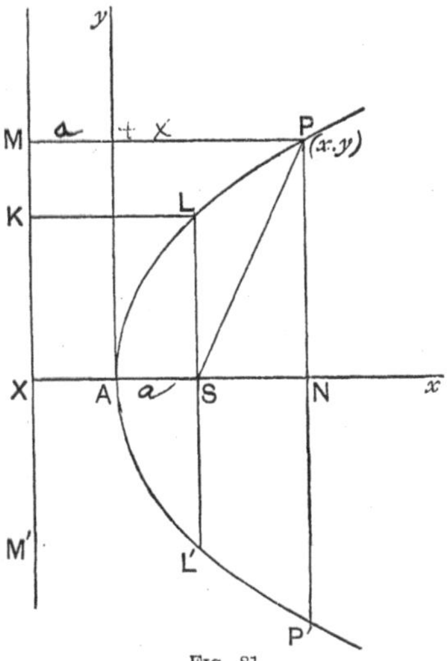

Fig. 81.

SA = **AX** ; $\therefore$ by definition, **A** is a point on the parabola.

Take **AS** for axis of x and **A**y at right angles to **AS** as axis of y.

Let **SA** $= a =$ **AX**.

Let P be any point on the curve and (x, y) its co-ordinates.

Draw PM perpendicular to the directrix and PN perpendicular to AS, produced if necessary.

$$\text{By def.,} \quad \text{SP} = \text{PM} = \text{NX} = a + x;$$
$$\therefore (a + x)^2 = \text{SP}^2 = \text{SN}^2 + \text{PN}^2 = (x - a)^2 + y^2.$$

Whence $\mathbf{y^2 = 4ax}$, the required equation.

This is the simplest form of the equation of a parabola.

N.B. The equation $y^2 = 4ax$ may be written $\mathbf{PN^2 = 4AS \cdot AN}$.

The point A is called the **vertex** of the parabola.

The line ASN　　　,,　　　**axis**　　　,,　　　,,

The point X　　　,,　　　**foot** of the directrix.

PN is called the **ordinate** of the point P and the chord PP′ a **double ordinate**.

129. *The form of the parabola.*

Consider the equation of the curve $y^2 = 4ax$.

When $y = 0$, the only value of x is zero.　$\therefore$ the curve passes through the vertex and cuts the axis at no other point.

The equation may be written $y = \pm\sqrt{4ax}$.

$\therefore$ for any value of x we obtain two equal values of y, but of opposite sign.

This shows that the curve is symmetrical about the axis of x, the axis of the curve.

If x is negative, y^2 is negative, and y is therefore imaginary.

$\therefore$ the curve lies entirely on the positive side of the axis of y.

As x increases indefinitely, the two values of y also increase indefinitely in magnitude.

$\therefore$ the two portions AP, AP′ of the curve never meet, and are of infinite length.

130. Latus Rectum. The chord LSL′ of the curve through the focus S, at right angles to the axis, is called the **latus rectum** of the parabola.

The latus rectum $= 4$AS $= 4$a.

Draw LK perpendicular to the directrix.

SL $=$ LK (by def.) $=$ SX $= 2$AS $= 2a$.

$\therefore$ LL′ $= 2$SL $= 4a$.

131. *If the point (x, y) is outside the parabola $y^2 = 4ax$, $y^2 > 4ax$; if the point (x, y) is inside the parabola $y^2 = 4ax$, $y^2 < 4ax$.*

Let P be the point (x, y), and let the ordinate PN meet the curve at Q.

When P is outside the curve

$$PN^2 > QN^2.$$

But $QN^2 = 4AS . AN = 4ax.$

$\therefore$ PN^2 or $y^2 > 4ax.$

Similarly when P is inside the curve, $PN^2 < QN^2$ and $y^2 < 4ax.$

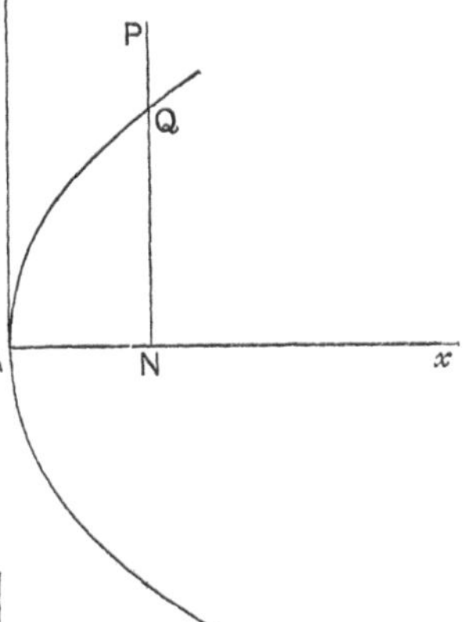

FIG. 82.

132. *To draw a parabola.* (This is most easily done on squared paper.)

Let S be the focus, MM′ the directrix. Draw SX perpendicular to the directrix and bisect SX at A.

$$SA = AX ;$$

$\therefore$ by definition, A is a point on the parabola.

In AS, or AS produced, take any point N, and draw PNP′ perpendicular to SX.

With centre S and radius XN describe a circle cutting PNP′ at P and P′.

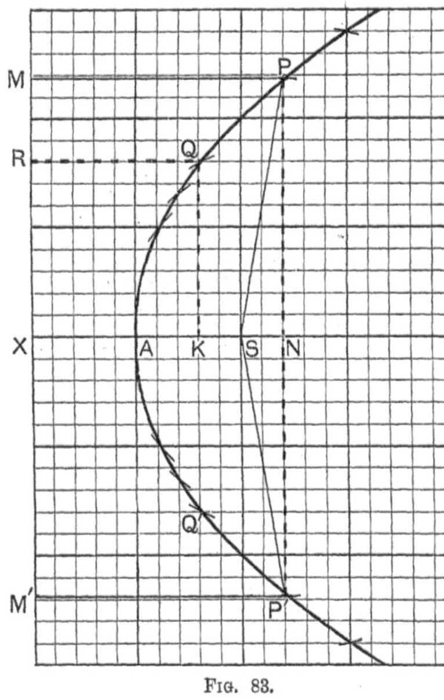

FIG. 83.

Draw PM, P′M′ perpendicular to the directrix.

Then $SP = XN = PM$; $\therefore$ by def. P is a point on the parabola.

Also $SP' = XN = P'M'$; $\therefore$ by def. P' is a point on the parabola.

We may obtain other points Q, Q' ... on the curve in the same way.

Joining them by an even curve we have the locus.

133. *Find the equation of a parabola referred to its axis as axis of x and directrix as axis of y.*

First Method. If $AS = a$, the co-ordinates of the focus are $(2a, 0)$.

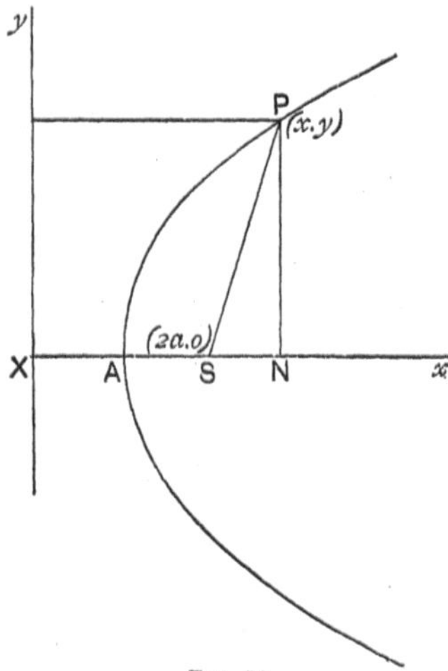

Fig. 84.

If (x, y) is any point on the curve,

the distance of (x, y) from the focus $(2a, 0) = \sqrt{(x - 2a)^2 + y^2}$.

The distance of (x, y) from the directrix $= x$;

$$\therefore \quad \sqrt{(x - 2a)^2 + y^2} = x$$

by definition.

Squaring and simplifying,

$$y^2 = 4a(x - a)$$

is the required equation.

Second Method. As in Art. 128,

$$NX^2 = SN^2 + PN^2.$$

But $NX = x$, $SN = x - 2a$, $PN = y$;

$$\therefore \quad x^2 = (x - 2a)^2 + y^2, \text{ as before.}$$

134. *Find the equation of the parabola whose focus is at the point (x_1, y_1) and whose directrix is the straight line $Ax + By + C = 0$.*

Let (x, y) be any point on the curve.

The distance of (x, y) from the focus (x_1, y_1)

$$= \sqrt{(x - x_1)^2 + (y - y_1)^2}.$$

The perpendicular distance of (x, y) from the directrix

$$Ax + By + C = 0 \text{ is } \frac{Ax + By + C}{\sqrt{A^2 + B^2}}.$$

$\therefore$ by definition of the parabola,

$$\sqrt{(x - x_1)^2 + (y - y_1)^2} = \frac{Ax + By + C}{\sqrt{A^2 + B^2}}.$$

Multiplying up and squaring, we have
$$(A^2 + B^2)[(x - x_1)^2 + (y - y_1)^2] = (Ax + By + C)^2,$$
the required equation.

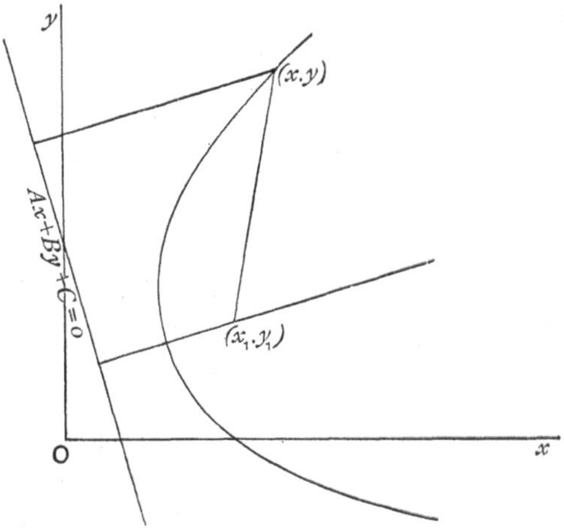

Fig. 85.

COROLLARY. Transposing all terms to the left, the terms of
the **second degree** $= x^2(A^2 + B^2 - A^2) - 2AB xy + y^2(A^2 + B^2 - B^2)$

$$= B^2 x^2 - 2AB xy + A^2 y^2$$

$$= (Bx - Ay)^2, \text{ a perfect square.}$$

Thus we see that in **the equation of any parabola, the terms
of the second degree form a perfect square.**

This is the distinguishing characteristic of the equation of a
parabola.

135. *Trace the parabola* $y^2 = 8x - 8y$, *and find* (1) *the equation of
its axis,* (2) *the equation of its directrix,* (3) *the co-ordinates of its
vertex,* (4) *the co-ordinates of its focus.*

We shall reduce the given equation to the standard form
$y^2 = 4ax$, by changing the origin.

The given equation may be written $y^2 + 8y = 8x$

$$\text{or} \quad y^2 + 8y + 16 = 8(x + 2),$$

$$(y + 4)^2 = 8(x + 2).$$

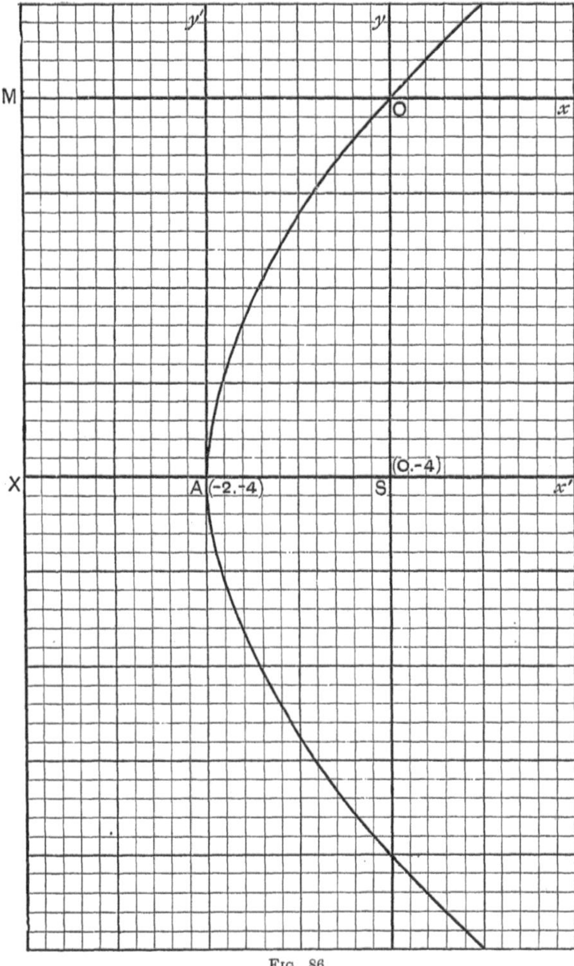

Fig. 86.

Changing the origin to the point $(-2, -4)$, *i.e.* writing $x - 2$ for x and $y - 4$ for y, the equation becomes $y^2 = 8x$.

∴ the new origin A is the vertex.

i.e. the co-ordinates of the vertex referred to the old axis are $(-2, -4)$.

Plotting this point, and noting that the curve passes through the old origin, we have the curve as shown in the figure.

Its latus-rectum $= 8$, $\therefore$ measuring $AS = 2$ along the new axis of x, S is the focus.

Its co-ordinates referred to the old axis are $(0, -4)$.

The equation of AS referred to the old axes is $y = -4$, for $OS = -4$.

Producing SA to X, and making $AX = AS = 2$, and drawing XM at rt. $\angle^s$ to XAS, XM is the directrix. Its equation is $x = -4$, for $SX = -4$.

Examples VIII. a.

Trace the curves

1. $x^2 = 4ay$.   2. $y^2 = -4ax$.   3. $x^2 = -4ay$.

4. Find the equation of the parabola taking the directrix as axis of x, and the axis of the curve as axis of y.

5. The ordinate of a point P on the parabola $y^2 = 9x$ is equal to 12. Find the length of the focal distance SP.

6. A circle passes through a fixed point and touches a fixed straight line. Find, and draw, the locus of its centre.

Find the equation of the parabola with :

[Make a freehand drawing of the curve in each case.]

7. focus $(3, -4)$, directrix $x - y + 5 = 0$.

8. focus $(a, 0)$, directrix $x + a = 0$.

9. focus $(-a, 0)$, directrix $x - a = 0$.

10. focus $(0, b)$, directrix $y + 2b = 0$.

Find (1) the co-ordinates of the vertex, (2) the co-ordinates of the focus, (3) the equation of the axis, (4) the equation of the directrix in each of the following parabolas, and draw the curve in each case :

11. $y^2 = 4(x - 2)$.   12. $x^2 = 4(y - 3)$.   13. $y^2 = 6y + 3x$.
14. $x^2 + 2ax - 4ay + a^2 + 4ab = 0$.   15. $x^2 - 6y + 4x = 0$.

136. *To find the equation of the tangent at the point* (x_1, y_1) *on the parabola* $y^2 = 4ax$.

Take a point $(x_1 + \Delta x_1, y_1 + \Delta y_1)$ on the curve and near to the point (x_1, y_1).

Since $(x_1 + \Delta x_1, y_1 + \Delta y_1)$ is on the curve,

$$(y_1 + \Delta y_1)^2 = 4a(x_1 + \Delta x_1),$$

i.e. $y_1^2 + 2y_1\Delta y_1 + (\Delta y_1)^2 = 4ax_1 + 4a\Delta x_1.$

But $y_1^2 = 4ax_1$, for $(x_1,\ y_1)$ is on the curve ;

$$\therefore \text{ by subtraction, } 2y_1\Delta y_1 + (\Delta y_1)^2 = 4a\Delta x_1.$$

And in the limit when the points $(x_1 + \Delta x_1,\ y_1 + \Delta y_1)$, $(x_1,\ y_1)$ approach one another,

$$y_1\Delta y_1 = 2a\Delta x_1$$

$$\text{and } \frac{\Delta y_1}{\Delta x_1} = \frac{2a}{y_1} ;$$

$$\therefore \frac{2a}{y_1} \text{ is the slope of the tangent.}$$

Also, the tangent passes through the point $(x_1,\ y_1)$.

$$\therefore \text{ its equation is } y - y_1 = \frac{2a}{y_1}(x - x_1), \quad \left[y - y_1 = m(x - x_1)\right]$$

$$i.e.\ yy_1 - y_1^2 = 2ax - 2ax_1,$$

$$yy_1 - 4ax_1 = 2ax - 2ax_1,$$

$$\mathbf{yy_1 = 2a\,(x + x_1).}$$

Note that the slope of the tangent at $(\mathbf{x_1},\ \mathbf{y_1})$ is $\dfrac{\mathbf{2a}}{\mathbf{y_1}}$.

CeOROLLARY 1. The equation of the tangent at the vertex $(0,\ 0)$ is $x = 0$, the axis of y.

$\therefore$ the tangent at the vertex is perpendicular to the axis.

CeOROLLARY 2. In the same way it may be proved that the equation of the tangent at $(x_1,\ y_1)$ to the parabola $x^2 = 4ay$ is

$$xx_1 = 2a\,(y + y_1).$$

The student should work this out for himself.

137. *Second Method.* By means of the Differential Calculus.

$$y^2 = 4ax.$$

Differentiating with respect to x, we have

$$2y\frac{dy}{dx} = 4a ; \quad \therefore \frac{dy}{dx} = \frac{2a}{y}.$$

$\therefore$ the slope of the tangent at the point $(x_1,\ y_1)$ is $\dfrac{2a}{y_1}$.

We now proceed as in the preceding article,

138. *To find the equation of a tangent to the parabola $y^2 = 4ax$ in terms of the slope of the tangent.*

Let m be the slope of the tangent.

We may take $y = mx + c$ to be the equation of the tangent, where c is unknown.

Where this line meets the curve, we have by substitution for x from the equation $y^2 = 4ax$,

$$y = m\frac{y^2}{4a} + c \quad \text{or} \quad my^2 - 4ay + 4ac = 0. \quad \ldots\ldots\ldots\ldots(1)$$

But by hypothesis, $y = mx + c$ is a tangent; therefore this quadratic has equal roots.

$$\therefore \ 16a^2 = 16acm; \quad (b^2 = 4ac)$$

$$\therefore \ c = \frac{a}{m}.$$

$\therefore \ y = mx + \dfrac{a}{m}$ is a tangent to the parabola $y^2 = 4ax$ for all values of m.

Corollary. The sum of the roots of the quadratic (1) is $\dfrac{4a}{m}$;

$$\therefore \ \text{each root} = \frac{2a}{m},$$

i.e. at the point of contact $y = \dfrac{2a}{m}$.

Also
$$x = \frac{y^2}{4a} = \frac{a}{m^2}.$$

$\therefore$ **the line $y = mx + \dfrac{a}{m}$ touches the parabola $y^2 = 4ax$ at the**

point
$$\left(\frac{a}{m^2} \cdot \frac{2a}{m}\right).$$

N.B. **The point $\left(\dfrac{a}{m^2} \cdot \dfrac{2a}{m}\right)$ lies on the parabola $y^2 = 4ax$ for all values of m ; and m is the slope of the tangent at that point.**

The point $(am^2, 2am)$ also lies on the parabola $y^2 = 4ax$ for all values of m, but m does not now represent the slope of the tangent,

139. *From a given point two tangents, and only two, can generally be drawn to a parabola.*

$y = mx + \dfrac{a}{m}$ is a tangent to the parabola $y^2 = 4ax$ for all values of m.

If this passes through the point (x_1, y_1),

$$y_1 = mx_1 + \frac{a}{m} \quad \text{or} \quad m^2 x_1 - m y_1 + a = 0.$$

$\therefore$ this quadratic for m gives us the slopes of the tangents which can be drawn from (x_1, y_1) to the curve.

Also, a quadratic has two roots, and not more than two.

$\therefore$ we can generally draw two tangents to a parabola from a given point.

If the roots of the quadratic are imaginary, the tangents are imaginary, and cannot be drawn.

140. *To find the condition that the straight line $lx + my + n = 0$ may touch the parabola $y^2 = 4ax$.*

Suppose that this line touches the curve at the point (x_1, y_1).

Then $yy_1 = 2a(x + x_1)$ is the equation of the tangent.

$\therefore$ this equation, *i.e.* $2ax - yy_1 + 2ax_1 = 0$ must be identical with $lx + my + n = 0$, for the two equations represent the same straight line.

$\therefore$ comparing coefficients : $\dfrac{2a}{l} = -\dfrac{y_1}{m} = \dfrac{2ax_1}{n}$;

whence $\quad y_1 = -\dfrac{2am}{l} \quad$ and $\quad x_1 = \dfrac{n}{l}.$

But $y_1^2 = 4ax$, for (x_1, y_1) is on the curve ;

$$\therefore \quad \frac{4a^2 m^2}{l^2} = \frac{4an}{l}$$

or $\quad am^2 = ln$ is the required condition.

We might have found this condition in the following manner :

From the equation $y^2 = 4ax$, substitute for x in the equation

$$lx + my + n = 0,$$

and we have at the points where the line meets the curve

$$\frac{ly^2}{4a} + my + n = 0$$

or $\quad ly^2 + 4amy + 4an = 0,$

But if the line *touches* the parabola, this quadratic has equal roots.

$$\therefore\ 16a^2m^2 = 16anl\ ;\quad (b^2 = 4ac)$$

$$\therefore\ am^2 = ln \text{ is the required condition, as before.}$$

141. *To find the equation of the normal to the parabola $y^2 = 4ax$ at the point (x_1, y_1).*

The equation of the tangent at (x_1, y_1) is $yy_1 = 2a(x + x_1)$.

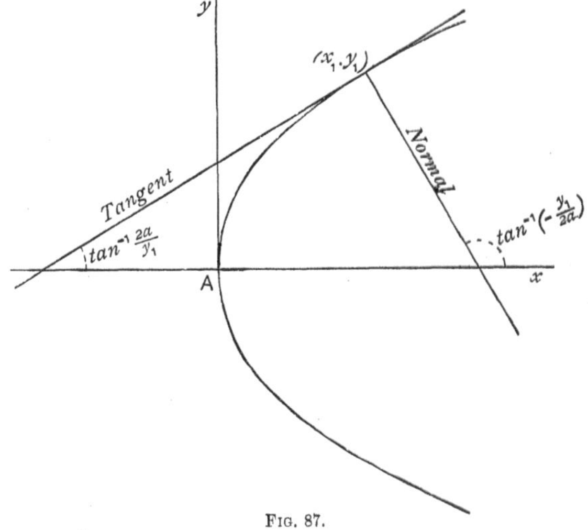

Fig. 87.

Its slope $= \dfrac{2a}{y_1}$;

$\therefore$ the slope of the normal, which is perpendicular to the tangent, is $-\dfrac{y_1}{2a}$. $(mm' = -1)$

Also, the normal passes through the point (x_1, y_1) ;

$\therefore$ its equation is $y - y_1 = -\dfrac{y_1}{2a}(x - x_1)$. $[y - y_1 = m(x - x_1)]$

142. *Tangents are drawn from the point (x_1, y_1) to the parabola $y^2 = 4ax$; to find the equation of their chord of contact.*

Let R be the point (x_1, y_1) ; RQ, RQ' the tangents.

It is required to find the equation of QQ'.

Let (h_1, k_1) be the co-ordinates of Q, (h_2, k_2) the co-ordinates of Q'.

The equation of RQ, the tangent at Q, is $yk_1 = 2a(x + h_1)$.

     ,, RQ' ,, Q', is $yk_2 = 2a(x + h_2)$.

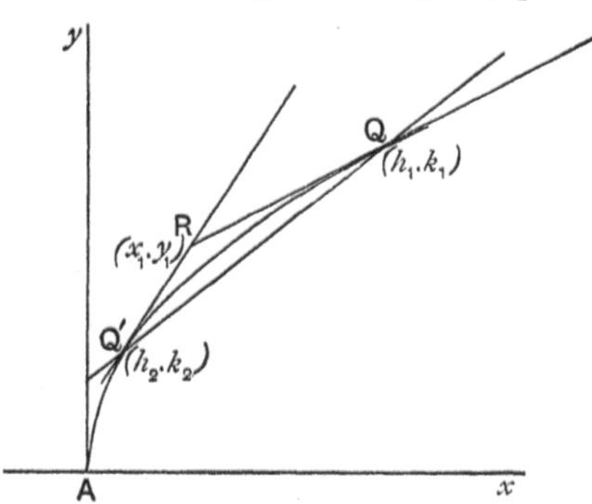

Fig. 88.

But the point $R(x_1, y_1)$ is on both these lines ;

$$\therefore \quad y_1 k_1 = 2a(x_1 + h_1) \quad\dots\dots\dots\dots\dots(1)$$

$$\text{and} \quad y_1 k_2 = 2a(x_1 + h_2) ; \quad\dots\dots\dots\dots(2)$$

$\therefore \quad y_1 y = 2a(x_1 + x)$ or $yy_1 = 2a(x + x_1)$ is the equation required.

For, firstly, it represents a straight line.

Also from (1) we see that $Q(h_1 k_1)$ lies on this line,

and from (2)   ,, $Q'(h_2 k_2)$,,

$\therefore$ it is the equation of QQ'.

143. Pole and Polar. Substitute the words "conic section" for "circle" in the definition of Art. 94.

To find the polar of the point (x_1, y_1) with respect to the parabola $y^2 = 4ax$.

Let P be the point (x_1, y_1) and PAB any chord through P; AQ, BQ the tangents at A and B.

It is required to find the locus of Q.

Let (h, k) be the co-ordinates of Q.

Then the equation of its chord of contact AB is

$$yk = 2a(x + h).$$

But $(x_1,\ y_1)$ is on this line ; $\therefore\ y_1 k = 2a(x_1 + h)$.
Also $(h,\ k)$ is *any point* on the locus ;
$\therefore$ the equation of the locus is

$$y_1 y = 2a(x_1 + x)$$

$$\text{or}\ \ yy_1 = 2a(x + x_1),\ \text{a straight line.}$$

*N.B.   When $(x_1,\ y_1)$ is outside the parabola, the polar is the same
as the chord of contact of tangents drawn from $(x_1,\ y_1)$.*

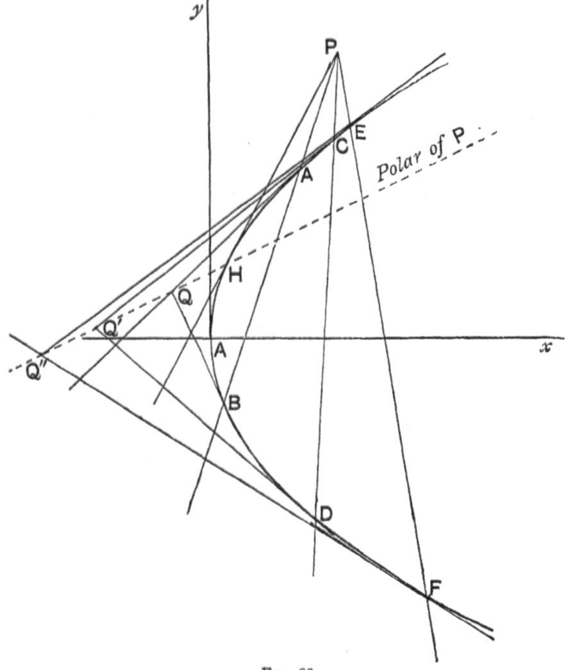

FIG. 89.

In the figure, **QA, QB** are tangents at A and B,

     Q′C, Q′D   ,,   C and D,

     Q″E, Q″F   ,,   E and F.

Q″Q′QH is the polar of P and PH is a tangent at H.

144. *If the polar of the point* P *passes through the point* Q, *the polar of the point* Q *passes through the point* P.

Let (x_1, y_1) be the co-ordinates of P, (x_2, y_2) those of Q.

The equation of the polar of P with respect to the parabola $y^2 = 4ax$ is $yy_1 = 2a(x + x_1)$. This passes through Q;

$$\therefore \ y_1 y_2 = 2a(x_1 + x_2). \ \dots\dots\dots\dots\dots\dots\dots(1)$$

But $yy_2 = 2a(x + x_2)$ is the polar of Q, and by (1) this straight line passes through (x_1, y_1) P, which proves the proposition.

145. *To find the locus of the middle points of a system of parallel chords.*

Let QQ$'$ be a chord of the system, making an angle θ with the axis of x, θ being constant.

Let (x_1, y_1) be the co-ordinates of V, its middle point.

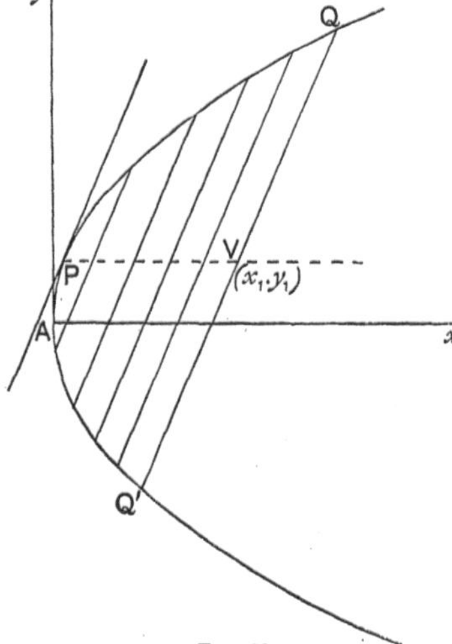

Fig. 90.

The equation of the chord may be written

$$\frac{x - x_1}{\cos \theta} = \frac{y - y_1}{\sin \theta} = r.$$

$$x = x_1 + r \cos \theta$$

and $\qquad y = y_1 + r \sin \theta$;

$\therefore$ where the chord meets the curve, we have by substitution,

$$(y_1 + r \sin \theta)^2 = 4a(x_1 + r \cos \theta)$$

or $\quad r^2 \sin^2\theta$

$$+ 2r(y_1 \sin \theta - 2a \cos \theta)$$

$$+ y_1^2 - 4ax_1 = 0.$$

The chord meets the curve at Q and Q$'$, and VQ, VQ$'$, the roots of this quadratic, are equal but of opposite sign;

$$\therefore \ y_1 \sin \theta - 2a \cos \theta = 0.$$

But (x_1, y_1) is *any* point on the locus;

$\therefore$ suppressing the suffix, $y \sin \theta = 2a \cos \theta$ is the equation of the locus of V.

This may be written $y = \dfrac{2a}{\tan \theta}$, and represents a straight line parallel to the axis.

Def. The straight line which bisects a series of parallel chords is called the **diameter** of those chords, and the semi-chords are called the **ordinates of that diameter**.

146. If the diameter PV meets the curve at P, the ordinate of P = the ordinate of $V = \dfrac{2a}{\tan\theta} = \dfrac{2a}{m}$, where m is the slope of the chord.

$\therefore$ the abscissa of $P = \dfrac{a}{m^2}$, and $y = mx + \dfrac{a}{m}$ is the tangent at P.

Therefore the tangent at the end of a diameter is parallel to the chords bisected by that diameter.

This may also be seen by letting the chord QQ′ move parallel to itself until V coincides with P.

The equal portions VQ, VQ′ vanish together when V coincides with P, and the chord becomes a tangent.

147. *To find the equation of a chord of the parabola* $y^2 = 4ax$ *in terms of the co-ordinates* $(x_1,\ y_1)$ *of its middle point.*

Let QQ′ be the chord whose middle point is $V(x_1,\ y_1)$, and let it make an angle θ with the axis of x.

Its equation is

$$\frac{x - x_1}{\cos\theta} = \frac{y - y_1}{\sin\theta} = r. \quad \ldots\ldots(1)$$

$$x = x_1 + r\cos\theta,$$

$$y = y_1 + r\sin\theta\ ;$$

$\therefore$ where it meets the curve we have by substitution,

$$(y_1 + r\sin\theta)^2 = 4a(x_1 + r\cos\theta)$$

or $r^2\sin^2\theta$

$$+ 2r(y_1\sin\theta - 2a\cos\theta)$$

$$+ y_1{}^2 - 4ax_1 = 0.$$

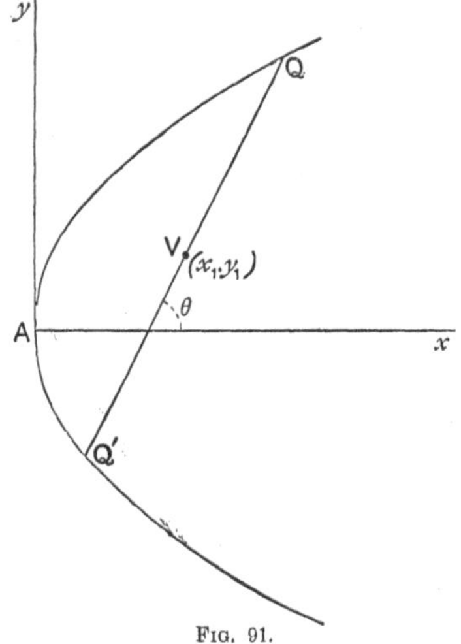

FIG. 91.

Now this chord meets the curve at Q and Q′ ;

$\therefore$ VQ and VQ′ are the roots of this quadratic.

Also VQ' and VQ are *equal but of opposite sign* ;

$$\therefore \; y_1 \sin \theta - 2a \cos \theta = 0$$

$$\text{or} \quad y_1 \sin \theta = 2a \cos \theta. \quad\dots\dots\dots\dots\dots\dots(2)$$

Multiplying (1) and (2) so as to eliminate θ,

$$(y - y_1)y_1 = 2a(x - x_1) \text{ is the equation of the chord.}$$

148. *The equation of the chord joining the two points* $(x_1,\ y_1)$, $(x_2,\ y_2)$ *on the parabola* $y^2 = 4ax$ *may be written in the form*

$$(y - y_1)(y_1 + y_2) = 4a(x - x_1).$$

The equation of the line joining the two given points is

$$\frac{y - y_1}{y_2 - y_1} = \frac{x - x_1}{x_2 - x_1}. \quad\dots\dots\dots\dots\dots\dots(1)$$

But $(x_2,\ y_2)$ is on the curve. $\quad \therefore \; y_2{}^2 = 4ax_2$.

For the same reason, $\qquad\qquad y_1{}^2 = 4ax_1$.

$\therefore$ by subtraction, $\quad (y_2 - y_1)(y_2 + y_1) = 4a(x_2 - x_1). \quad\dots\dots\dots(2)$

Multiplying (1) and (2), the equation of the chord is

$$(y - y_1)(y_2 + y_1) = 4a(x - x_1).$$

Cor. From this equation we might deduce the equation of the tangent at the point $(x_1,\ y_1)$.

Let the two points move up to one another so that x_2 becomes equal to x_1 and y_2 to y_1.

The equation becomes

$$(y - y_1)\,2y_1 = 4a(x - x_1)$$

$$\text{or} \quad yy_1 = 2ax - 2ax_1 + y_1{}^2$$

$$= 2ax - 2ax_1 + 4ax_1$$

$$\text{or} \quad yy_1 = 2a(x + x_1), \text{ the equation of the tangent.}$$

149. *To find the length of the portion of the straight line* $y = x \tan \theta + c$ *intercepted by the parabola* $y^2 = 4ax$.

Let $(x_1,\ y_1)\mathsf{P}$, $(x_2,\ y_2)\mathsf{Q}$ be the points where the straight line meets the curve.

Draw the ordinates PN, QM ; and QK parallel to the axis of x.

$$\text{The required length} = \mathsf{PQ} = \frac{\mathsf{PK}}{\sin \mathsf{PQK}} = \frac{y_1 - y_2}{\sin \theta}. \quad\dots\dots\dots\dots(1)$$

Where $y = x \tan \theta + c$ meets the curve, we have

$$y = \frac{y^2}{4a} \tan \theta + c$$

or $\quad y^2 \tan \theta - 4ay + 4ac = 0.$

Now y_1, y_2 are the roots of this quadratic.

$$\therefore \; y_1 + y_2 = \frac{4a}{\tan \theta}$$

and $\quad y_1 y_2 = \frac{4ac}{\tan \theta};$

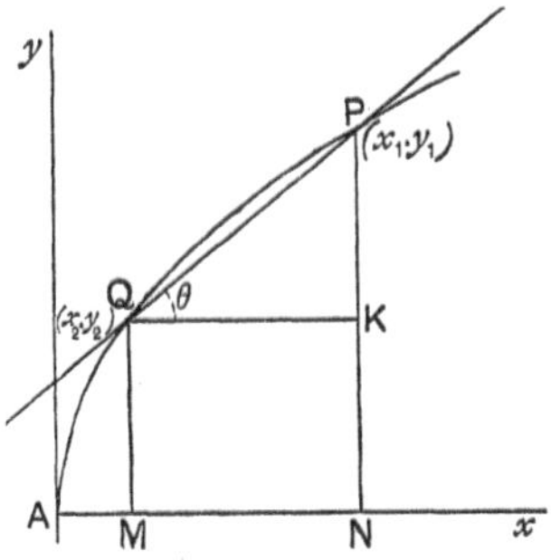

$$\therefore \; (y_1 - y_2)^2 = (y_1 + y_2)^2 - 4y_1 y_2$$

$$= \frac{16a^2}{\tan^2 \theta} - \frac{16ac}{\tan \theta} = \frac{16a}{\tan^2 \theta}(a - c \tan \theta).$$

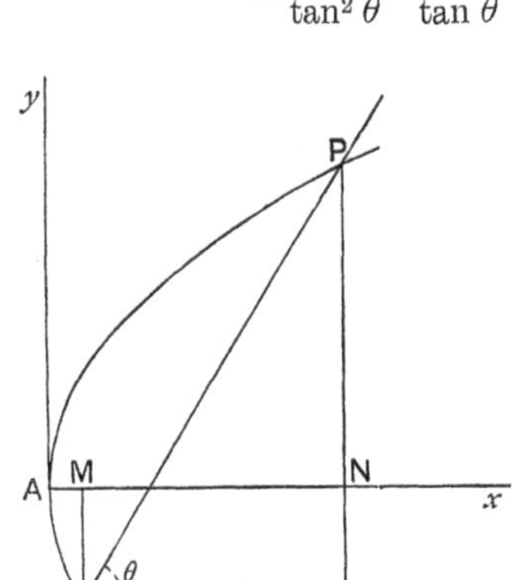

Fig. 92.

$\therefore$ from (1),

$$PQ = \frac{4\sqrt{a(a - c \tan \theta)}}{\sin \theta \cdot \tan \theta}.$$

N.B. If PQ cuts the curve as in the second figure, PK is still equal to the *algebraic difference* of the roots of the quadratic, for the ordinate of Q is negative.

150. *Infinite roots of a quadratic equation.*

Take the quadratic

$$ax^2 + bx + c = 0, \quad \dots\dots(1)$$

and let $y = \frac{1}{x}$, so that $x = \frac{1}{y}.$

On substituting in (1), we have

$$\frac{a}{y^2} + \frac{b}{y} + c = 0$$

Fig. 93.

or $\quad cy^2 + by + a = 0. \quad \dots\dots\dots\dots\dots\dots(2)$

Now if $a = 0$, one root of (2) is zero, *i.e.* one value of y is zero.

But $x = \dfrac{1}{y}$; $\quad \therefore$ when $y = 0$, $\quad x = \dfrac{1}{0} = \infty$.

$\therefore$ if $a = 0$, one root of the quadratic $ax^2 + bx + c = 0$ is infinity. In other words, we may look upon the equation

$$0 \cdot x^2 + bx + c = 0$$

as a quadratic, one of whose roots is infinity.

Again if $a = 0$ and $b = 0$, both roots of equation (2) are zero, and therefore both roots of equation (1) are infinity.

Thus we may look upon the equation

$$0 \cdot x^2 + 0 \cdot x + c = 0$$

as a quadratic, both of whose roots are infinity.

This may be seen in another manner.

When we solve equation (1),

$$x = \frac{-b \pm \sqrt{b^2 - 4ac}}{2a}$$

$$= \frac{[-b \pm \sqrt{b^2 - 4ac}][-b \mp \sqrt{b^2 - 4ac}]}{2a[-b \mp \sqrt{b^2 - 4ac}]}$$

$$= \frac{b^2 - (b^2 - 4ac)}{2a[-b \mp \sqrt{b^2 - 4ac}]}$$

$$= \frac{2c}{-b \mp \sqrt{b^2 - 4ac}}.$$

As a diminishes and approaches zero,

$$\sqrt{b^2 - 4ac} \text{ approaches } \sqrt{b^2} \text{ or } b,$$

and when $a = 0$, the two roots become

$$\frac{2c}{-2b} \text{ and } \frac{2c}{0}, \text{ or } -\frac{c}{b} \text{ and } \infty.$$

Again if $b = 0$ also, the roots are both infinity.

$\therefore$ we may look upon the equation

$$0 \cdot x^2 + 0 \cdot x + c = 0$$

as a quadratic, both of whose roots are infinity.

The proposition might be stated thus:

If in the equation $ax^2 + bx + c = 0$ the values of a and b be made smaller and smaller, both roots become larger and larger.

151. *The points of intersection of the parabola* $y^2 = 4ax$, *and the straight line* $y = mx + c$.

Where these meet we have by substitution,

$$y = m \cdot \frac{y^2}{4a} + c$$

or $my^2 - 4ay + 4ac = 0$.

This is a quadratic equation, and every quadratic has two roots, real and different, equal, or imaginary.

Therefore every straight line meets a parabola at two points, real and different, coincident, or imaginary.

When we solve this quadratic for y, the quantity under the radical sign $\qquad = 16a^2 - 16acm \quad (b^2 - 4ac)$

$$= 16a(a - cm).$$

(1) If $a = cm$, the roots are equal, and the line is a tangent as in Art. 138.

(2) If $a > cm$, the roots are real and unequal, and the line meets the parabola in two points.

(3) If $a < cm$, the roots are imaginary, and the line does not meet the parabola.

(4) When $m = 0$ (in which case $y = c$), one root of the quadratic is infinity, and the other finite.

In other words, a straight line parallel to the axis (a diameter) meets a parabola in only one point at a finite distance.

Examples VIII. b.

1. Without assuming the formula for the equation of a tangent, find the equation of the tangent to the parabola $y^2 = 8x$ at the point (2, 4). Also find the equation of the normal at the same point.

2. Prove that the straight line $6y = 4x + 39$ touches the parabola $y^2 = 12x + 36$.

3. Find the equation of the chord joining the points

$$\left(\frac{a}{m^2} \cdot \frac{2a}{m} \right), \quad \left(\frac{a}{n^2} \cdot \frac{2a}{n} \right)$$

on a parabola ; and deduce the equation of the tangent at the point

$$\left(\frac{a}{m^2} \cdot \frac{2a}{m} \right).$$

4. A tangent to the parabola $y^2 = 20x$ makes an angle of $45°$ with the axis of x. Determine its point of contact.

5. What is the equation of a tangent to the parabola $y^2 = 4ax$ which cuts off equal intercepts on the axes?

6. Write down the equation of a tangent to the parabola $y^2 = 6x$ parallel to the line $2x - 3y = 4$.

7. Find where the straight line $4y + 3x + 6 = 0$ meets the parabola

$$2y^2 = 9x.$$

8. Prove that the straight line $2x - 2y = 1$ is a tangent to the parabola $y^2 + 2x - 4y = 0$, and find its point of contact.

9. Use the method of Art. 136 or 137 to find the equation of the tangent to the parabola $x^2 = 4by$ at the point (x_1, y_1). Write down the equation of the normal at the same point.

10. Find the equation of the tangent to the parabola $y^2 + ax + by = 0$ at the origin.
Write down the equation of the normal at the same point.

11. Find the equations of the tangents drawn from the origin to the parabola $(y - k)^2 = 4a(x - h)$.

12. Prove that the latus rectum is a harmonic mean between the ordinates of the points where the straight line $4x - 9y + 18a = 0$ meets the parabola $y^2 = 4ax$.

13. Find the equation of the chord of the parabola $y^2 = 4x$ which is bisected at the point $(4, -1)$.

14. The circle described on a focal radius of a parabola as diameter touches the tangent at the vertex.

15. Find the length of the portion of the straight line $12y = 5x + 36$ intercepted by the parabola $y^2 = 8x$.

16. Find the equations of the tangents to the parabola $y^2 = 5x$ drawn from the point $(5, 13)$.

17. Find the equation of the diameter of the parabola $9y^2 = 16x$ which bisects all chords parallel to $2x + 3y + k = 0$.

18. Form the quadratic equation which will give the ordinates of the points where the normal at (x_1, y_1) meets the parabola $y^2 = 4ax$. Find the ordinate of the point, other than (x_1, y_1), where the normal cuts the curve.

19. The length of the chord of contact of tangents from (x_1, y_1) to the parabola $y^2 = 4ax$ is

$$\frac{\sqrt{(y_1^2 - 4ax_1)(y_1^2 + 4a^2)}}{a}.$$

20. The area of the triangle formed by the tangents from the point (x_1, y_1) and their chord of contact in the parabola $y^2 = 4ax$ is

$$\frac{(y_1^2 - 4ax_1)^{\frac{3}{2}}}{2a}.$$

21. The parabolas $y^2=4ax$, $x^2=4by$ cut (at the point other than the origin where they meet) at an angle whose tangent is

$$\frac{3}{2[a^{\frac{1}{3}}b^{-\frac{1}{3}}+a^{-\frac{1}{3}}b^{\frac{1}{3}}]}.$$

22. Where the parabola $(x-y)^2=3(x+2y-3)$ and the straight line $x+2y-3=0$ meet, we have by substitution $(x-y)^2=0$. What do you deduce?

23. Find the length of the chord $4y=3x-48$ intercepted by the parabola $y^2=64x$.

24. Tangents are drawn to the parabola $y^2=4x$ from the point $(4,5)$: find the angle between them.

25. Draw the parabolas $y^2=4a(x-a)$, $y^2=4a'(x-a')$. Subtracting their equations, we have $x-a-a'=0$. What can you deduce as to this straight line?

LOCUS PROBLEMS ON THE PARABOLA.

152. *To find the locus of the intersection of a tangent to a parabola and the perpendicular drawn to it from the focus.*

Let $y=mx+\dfrac{a}{m}$ ... (1) be the equation of the tangent.

The co-ordinates of the focus are $(a,0)$.

$\therefore$ the equation of the perpendicular is

$$y=-\frac{1}{m}(x-a) \ldots (2). \quad [y-y_1=m(x-x_1)]$$

To find the locus of the intersection of (1) and (2) we have to eliminate m.

By subtraction, $\qquad 0=x\left(m+\dfrac{1}{m}\right).$

Now $m+\dfrac{1}{m}$ cannot be equal to zero, for in that case m would be imaginary;

$$\therefore \ x=0 \text{ is the equation of the locus.}$$

This is the axis of y.

Hence, **a tangent to a parabola intersects the perpendicular to it from the focus on the tangent at the vertex.**

153. *Tangents are drawn to the parabola $y^2 = 4ax$ making angles θ_1, θ_2 with the axis of x: find the locus of their intersection when $\cot\theta_1 - \cot\theta_2 = c$, where c is constant.*

Let (h, k) be any point on the locus.

$y = mx + \dfrac{a}{m}$ is a tangent to the parabola $y^2 = 4ax$ for all values of m.

If this passes through the point (h, k), $k = mh + \dfrac{a}{m}$,

$$\text{or}\quad m^2 h - mk + a = 0. \quad\dotfill(1)$$

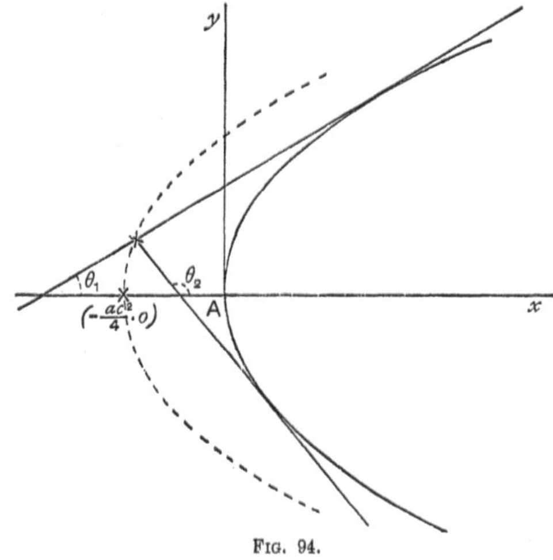

FIG. 94.

This equation therefore gives the slopes of the tangents which are drawn from the point (h, k) to the curve.

$\therefore$ $\tan\theta_1$, $\tan\theta_2$ are the roots of this quadratic ;

$$\therefore\quad \tan\theta_1 + \tan\theta_2 = \frac{k}{h} \text{ and } \tan\theta_1 \tan\theta_2 = \frac{a}{h} ;$$

$$\therefore\quad (\tan\theta_1 - \tan\theta_2)^2 = (\tan\theta_1 + \tan\theta_2)^2 - 4\tan\theta_1\tan\theta_2$$

$$= \frac{k^2}{h^2} - \frac{4a}{h} = \frac{k^2 - 4ah}{h^2}$$

$$\tan\theta_1 - \tan\theta_2 = \frac{\sqrt{k^2 - 4ah}}{h}.$$

Also $\tan \theta_1 \tan \theta_2 = \dfrac{a}{h}.$

$\therefore$ by division $\dfrac{\sqrt{k^2 - 4ah}}{a} = \cot \theta_2 - \cot \theta_1 = -c$ by hypothesis.

Squaring, $k^2 - 4ah = a^2c^2.$

But (h, k) is any point on the locus;

$\therefore y^2 - 4ax = a^2c^2$ is the equation of the locus.

This may be written $y^2 = 4a\left(x + \dfrac{ac^2}{4}\right).$ Hence the locus is a parabola, whose vertex is at the point $\left(-\dfrac{ac^2}{4},\ 0\right)$, whose latus rectum is $4a$, and the curve is co-axial with the given parabola.

154. Perpendicular tangents to a parabola intersect on the directrix, and their chord of contact passes through the focus.

Let (h, k) be any point on the locus of the intersection of perpendicular tangents.

$y = mx + \dfrac{a}{m}$ is a tangent to the parabola $y^2 = 4ax$ for all values of m.

If it passes through (h, k), $\quad k = mh + \dfrac{a}{m},$

$$\text{or } m^2h - mk + a = 0.$$

This equation therefore gives the slopes of the tangents which are drawn from (h, k) to the curve.

In this case the tangents are at right angles, therefore if m_1, m_2 are the roots of this quadratic, $m_1 m_2 = -1$;

$$\therefore \frac{a}{h} = -1, \text{ or } h + a = 0.$$

But (h, k) is any point on the locus.

$\therefore x + a = 0$ is the equation of the locus, and represents the directrix.

Again, we may take $(-a, k)$ as the co-ordinates of any point on the locus, and the equation of the polar of this point is

$$ky = 2a(x - a). \quad [yy_1 = 2a(x + x_1)]$$

This straight line passes through the focus $(a, 0)$. $\qquad$ Q.E.D.

COROLLARY.   If $\left(\dfrac{a}{m^2},\ \dfrac{2a}{m}\right)$ are the co-ordinates of one end of a focal chord, m is the slope of the tangent at this point ;

$\therefore$ $(am^2,\ -2am)$ are the co-ordinates of the other end, for the tangents at the ends are at right angles.

155. *Find the locus of the middle points of the focal radii of the parabola* $y^2 = 4ax$.

Let P be any point on the curve, and let $\left(\dfrac{a}{m^2},\ \dfrac{2a}{m}\right)$ be its co-ordinates.   Join SP, and bisect it at Q.

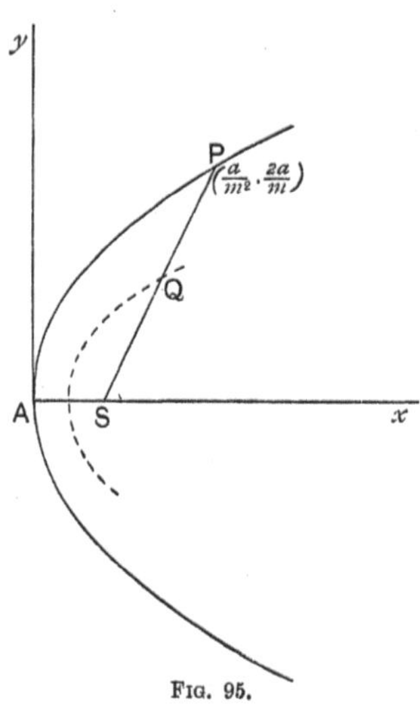

Fig. 95.

We have to find the locus of Q.
Let (x, y) be its co-ordinates.
The co-ordinates of S are $(a, 0)$,
those of $P\left(\dfrac{a}{m^2},\ \dfrac{2a}{m}\right)$;

$\therefore$ the co-ordinates of Q are

$$\left[\frac{1}{2}\left(a + \frac{a}{m^2}\right),\ \frac{a}{m}\right],$$

$$\left[\tfrac{1}{2}(x_1 + x_2),\ \tfrac{1}{2}(y_1 + y_2)\right]$$

$$i.e.\ x = \frac{a}{2}\left(1 + \frac{1}{m^2}\right),\ \ \ldots\ldots(1)$$

$$\text{and}\ y = \frac{a}{m}.\ \ \ldots\ldots\ldots\ldots(2)$$

To find the locus of Q we have to eliminate m between these equations.

From (2) $\dfrac{1}{m} = \dfrac{y}{a}$;

$\therefore$ substituting in (1) $2x = a\left(1 + \dfrac{y^2}{a^2}\right)$,

or $y^2 = 2a\left(x - \dfrac{a}{2}\right)$ is the equation of the locus.

This is a parabola, whose vertex is at the point $\left(\dfrac{a}{2},\ 0\right)$, whose latus rectum is $2a$, and it is co-axial with the given curve.

Examples VIII. c.

[It is generally advisable to use $\left(\dfrac{a}{m^2}, \dfrac{2a}{m}\right)$ as the co-ordinates of any point on the parabola $y^2 = 4ax$.]

1. Find the locus of the middle points of the ordinates of a parabola. Draw the locus.

2. Find the equation of, and draw, the locus of the middle points of chords of the parabola $y^2 = 4ax$ through the vertex.

3. Through P, any point on a parabola, PQ is drawn parallel to the axis, and SQ is drawn through the focus and parallel to the tangent at P. Find the locus of Q.

4. In any parabola, the perpendicular on the directrix from a point P and the perpendicular from the focus upon the tangent at P meet on the directrix.

5. A circle touches a given circle and a given straight line : find the locus of its centre.

6. By changing the origin, find the equation of a tangent to
$$y^2 = 4a(x + a)$$
in terms of its slope.

7. Prove that the locus of the intersection of tangents to $y^2 = 4a(x + a)$, $y^2 = 4a'(x + a')$ which are at right angles is the straight line $x + a + a' = 0$. Prove also that this straight line is the common chord of the parabolas.

8. Chords of the parabola $y^2 = 4ax$ are drawn through the point $(h, 0)$ on its axis : find the equation of, and draw, the locus of their middle points.
[Use the equation of a chord in terms of the co-ordinates of its middle point.]

9. PN is an ordinate of a parabola, and NQ is drawn perpendicular to the tangent at P to meet the diameter through P (produced) in Q. Find and draw the locus of Q.

10. If the normal at P on a parabola meets the axis at G, find the locus of the middle point of PG.

11. Pairs of tangents are drawn to a parabola so that the sum of their slopes is constant : find the locus of their intersection.

12. Find the locus of the middle points of chords of a parabola drawn through the focus.

13. Prove that the locus of the middle points of chords of the parabola $y^2 = 4ax$ which pass through a fixed point (h, k) is another parabola, and find the co-ordinates of its vertex.

14. Find the locus of the intersection of the straight lines
$$y - m(x - a) = 0, \quad y + (k - m)(x + a) = 0,$$
m being a variable quantity, k and a constants.

15. Pairs of tangents are drawn to the parabola $y^2 = 4ax$ so that the product of their slopes is constant : find the locus of their intersection.

16. Pairs of tangents are drawn to a parabola so that the sum of the angles each pair makes with the axis is constant : prove that the locus of their intersection is a straight line through the focus.

17. Through the middle point of the ordinate PN of a parabola, a straight line is drawn parallel to the axis to meet the tangent at P in Q : find the locus of Q.

18. Tangents are drawn to the parabola $y^2 = 4ax$ at the points $P(x_1,\ y_1)$, $Q(x_2,\ y_2)$: prove that the co-ordinates of their point of intersection are

$$\left(\frac{y_1 y_2}{4a},\ \frac{y_1 + y_2}{2}\right).$$

Hence find the locus of the intersection of these tangents when (1) the sum of the ordinates at P and Q is constant ; (2) the rectangle contained by the ordinates at P and Q is constant.

19. The area and base of a triangle being given, find the locus of its orthocentre.

20. Pairs of tangents to a parabola are inclined to the axis at complementary angles : find the locus of their intersection.

21. If tangents to the parabola $y^2 = 4ax$ intersect at a constant angle a, prove that the equation of the locus of their intersection is

$$y^2 - 4ax = (a + x)^2 \tan^2 a.$$

22. Find the equation of, and draw, the locus of the poles of tangents to the parabola $y^2 = 4ax$ with respect to the circle $x^2 + y^2 = 2ax$.

23. A tangent to the parabola $y^2 = 4ax$ meets the parabola $y^2 = -4ax$ at Q and R : prove that the tangents at Q and R to the second parabola intersect on the first.

24. Chords of a parabola through the vertex are divided in a constant ratio : prove that the locus of the dividing point is a parabola.

156. *To find the equation of a normal to the parabola $y^2 = 4ax$ in terms of its slope, m.*

As in article 141 the equation of the normal at the point $(x_1,\ y_1)$ is

$$y - y_1 = -\frac{y_1}{2a}(x - x_1) ; \quad \dots\dots\dots\dots\dots\dots\dots(1)$$

$$\therefore \quad -\frac{y_1}{2a} = m \quad \text{or} \quad y_1 = -2am.$$

Also $$x_1 = \frac{y_1^2}{4a} = \frac{4a^2 m^2}{4a} = am^2.$$

$\therefore$ substituting these values of x_1 and y_1 in (1), we have

$$y + 2am = m(x - am^2)$$

$$\text{or} \quad y = mx - 2am - am^3 \text{ is the equation required.}$$

157. *In general, three straight lines can be drawn from a given point which are normal to a parabola.*

$y = mx - 2am - am^3$ is a normal to the parabola $y^2 = 4ax$.

If this passes through the given point (x_1, y_1)

$$y_1 = mx_1 - 2am - am^3, \quad \text{or} \quad am^3 + 2am - mx_1 - y_1 = 0.$$

Thus we have a *cubic* equation for m, and a cubic equation has three roots.

$\therefore$ we can generally draw three normals to a parabola from a given point.

If some of the roots of the equation are imaginary, the corresponding normals are imaginary, and cannot be drawn.

Moreover imaginary roots occur in pairs, therefore we can draw three normals or one.

It must be remembered that $y = mx - 2am - am^3$ is *not* the normal at the point $\left(\dfrac{a}{m^2}, \ \dfrac{2a}{m} \right)$.

With the co-ordinates $\left(\dfrac{a}{m^2}, \ \dfrac{2a}{m} \right)$, m is the slope of the tangent.

$\therefore$ in that case $-\dfrac{1}{m}$ is the slope of the normal.

158. *Two normals are drawn to a parabola so that the product of their slopes is constant (c) : find the locus of their intersection.*

$y = mx - 2am - am^3$ is a normal to the parabola $y^2 = 4ax$ for all values of m.

If this passes through (h, k), $k = mh - 2am - am^3$

$$\text{or} \quad m^3 + m\,\frac{2a - h}{a} + \frac{k}{a} = 0.$$

If m_1, m_2, m_3 are the roots of this equation, they are the slopes of the normals drawn from the point (h, k).

By the Theory of Equations

$$m_1 + m_2 + m_3 = 0, \quad \dots\dots\dots\dots\dots\dots\dots\dots (1)$$

$$m_1 m_2 + m_2 m_3 + m_3 m_1 = \frac{2a - h}{a}, \quad \dots\dots\dots\dots\dots\dots (2)$$

$$m_1 m_2 m_3 = -\frac{k}{a}. \quad \dots\dots\dots\dots\dots\dots\dots (3)$$

Also in this case, by hypothesis,

$$m_2 m_3 = c. \quad \dots\dots\dots\dots\dots\dots\dots\dots\dots (4)$$

From (2) and (4) $\qquad m_1(m_2+m_3)=\dfrac{2a-h}{a}-c.$

$\therefore$ from (1) $\qquad\qquad -m_1{}^2=\dfrac{2a-h-ac}{a}.$ $\ \dots\dots\dots\dots\dots(5)$

Also from (3) and (4) $\qquad m_1=-\dfrac{k}{ac}.$ $\ \dots\dots\dots\dots\dots\dots(6)$

$\therefore$ from (5) and (6) $\qquad \dfrac{k^2}{a^2c^2}=\dfrac{h-2a+ac}{a}.$

$$\therefore\ \frac{y^2}{ac^2}=x-2a+ac \text{ is the equation of the locus.}$$

This may be written $y^2=ac^2(x-\overline{2a-ac})$, and represents a parabola whose vertex is at the point $(2a-ac,\ 0)$, latus rectum ac^2, and co-axial with the given parabola.

Examples VIII. d.

1. Find the equation of the normal to the parabola $y^2=4ax$ at the point $\left(\dfrac{a}{m^2},\ \dfrac{2a}{m}\right)$.

2. The straight line $y=mx-2am-am^3$ is normal to the parabola at the point $(am^2,\ -2am)$.

3. If three normals are drawn to a parabola from a given point, prove that the algebraic sum of their slopes is zero.

Hence prove that if these lines are normal to the curve at P, Q, R, the algebraic sum of the ordinates of P, Q, R is zero.

4. Find the co-ordinates of the second point where the normal at the point $(am^2,\ -2am)$ meets the parabola $y^2=4ax$.

5. The normal at $P\left(\dfrac{a}{m^2},\ \dfrac{2a}{m}\right)$ to the parabola $y^2=4ax$ meets the curve again at Q, and V is the middle point of PQ : find the co-ordinates of the point R where the diameter through V meets the curve.

6. If the normal $y=mx-2am-am^3$ meets the curve $y^2=4ax$ at P and Q, find the co-ordinates of the middle point of PQ.

Deduce the equation of the locus of the middle points of a series of normal chords.

7. Prove that the chord $y-x\sqrt{2}+4a\sqrt{2}=0$ is a normal to the parabola $y^2=4ax$, and that its length is $6\sqrt{3}\,.\,a$.

8. If the sum of the slopes of two normals to the parabola $y^2=4ax$ is constant $(=k)$, find the locus of their intersection.

9. Pairs of normals to the parabola $y^2 = 4ax$ make complementary angles with the axis : find the locus of their intersection.

10. Prove that the normal $y = mx - 2am - am^3$ to the parabola $y^2 = 4ax$ meets the perpendicular normal at a point whose abscissa is

$$a\left(m^2 + 1 + \frac{1}{m^2}\right).$$

PROPERTIES OF THE PARABOLA.

159. *If the tangent at* P, *a point on a parabola, meets the axis at* T, *and* PN *is the ordinate at* P, AN $=$ AT, *and* SP $=$ ST.

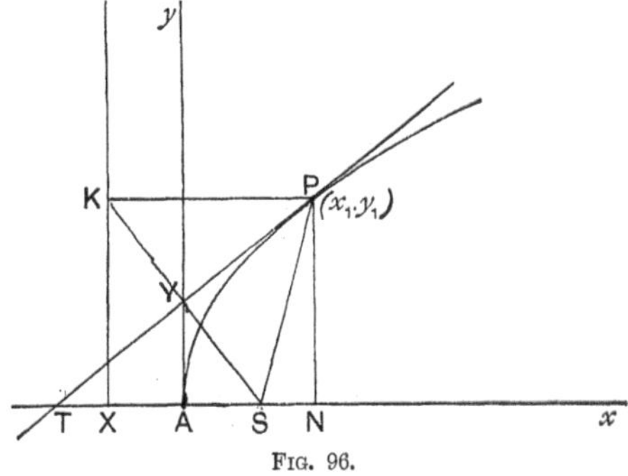

FIG. 96.

Let $(x_1,\ y_1)$ be the co-ordinates of P.
Then the equation of the tangent PT is $yy_1 = 2a(x + x_1)$.
At T, a point on this line, $y = 0$;

$\therefore\ 2a(x + x_1) = 0$; $\therefore\ x = -x_1$ (for a is not equal to zero),

or AT $=$ AN, but is of opposite sign.

The opposite sign shows that AN and AT are drawn in opposite directions.

Also SP $=$ PK $=$ NX $=$ AX $+$ AN $=$ AS $+$ AT $=$ ST.

Def. NT is called the **subtangent.**
Hence the subtangent is bisected at the vertex.

160. *The tangent at* P, *a point on a parabola, bisects the angle between* PK *the perpendicular on the directrix, and* SP *the focal distance.*

For $\angle$ KPT $= \angle$ PTS, by parallels,

$= \angle$ SPT, for ST $=$ SP.

161. *If the perpendicular from the focus upon the tangent at* P *meets it at* Y, *the point* Y *lies on the tangent at the vertex.*

This is proved in Art. 152.

162. *If* SY *is drawn from the focus* S *at right angles to the tangent at* P, SY^2 = AS . SP.

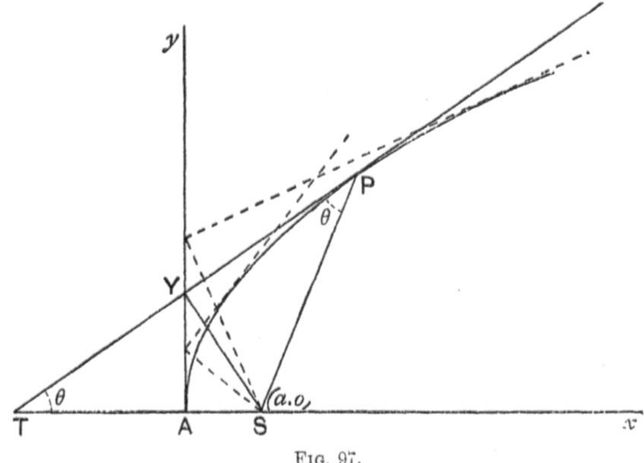

Fig. 97.

With the above figure SYT is a right angle, and YA is at right angles to ST.

∴ from the similar △^{s} SAY, SYT,

$$\frac{SA}{SY} = \frac{SY}{ST} ; \quad \therefore \ SY^2 = AS . ST = AS . SP.$$

163. **Sub-normal.** *If* PN *is the ordinate at the point* P *on a parabola, and* PG *the normal meets the axis at* G, *then* NG, *which is called the* **sub-normal,** *is equal to the semi-latus rectum* 2a, *and* SP = SG.

Let (x_1, y_1) be the co-ordinates of P.

The equation of the normal at P is

$$y - y_1 = -\frac{y_1}{2a}(x - x_1), \text{ as in Art. 141.}$$

At G, a point on this line, $y = 0$;

$$\therefore \ -y_1 = -\frac{y_1}{2a}(x - x_1),$$

$$\text{or} \quad x - x_1 = 2a,$$
$$i.e. \quad \mathsf{AG - AN} = 2a,$$
$$\mathsf{NG} = 2a.$$

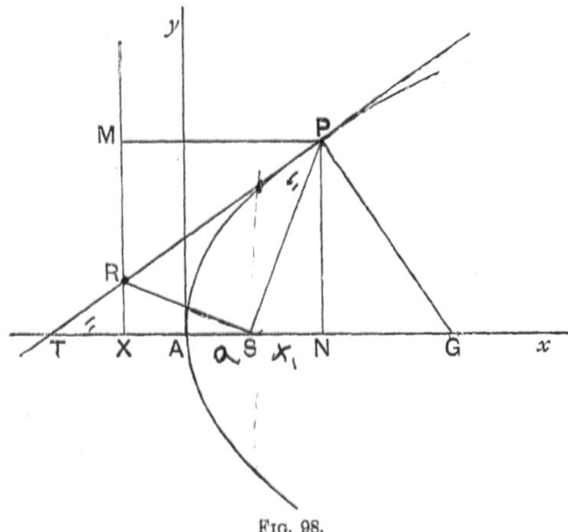

Fig. 98.

Again, $\mathsf{SG = SN + NG = AN - AS + NG}$

$$= x_1 - a + 2a = x_1 + a = \mathsf{NX = SP}.$$

164. The length of PG is generally called the **length of the normal at P.** If PG meets the curve again at Q, PQ is called the **normal chord at P.**

The length of the normal at $\mathsf{P = PG} = \sqrt{\mathsf{PN}^2 + \mathsf{NG}^2}$

$$= \sqrt{y_1^2 + 4a^2} = \sqrt{4ax_1 + 4a^2}$$
$$= \sqrt{4a(x_1 + a)} = 2\sqrt{\mathsf{AS . SP}}.$$

165. *The portion of a tangent intercepted between the point of contact and the directrix subtends a right angle at the focus.*

Let the tangent at P meet the directrix at R.   Join SR.

Draw PM perpendicular to the directrix.

In $\triangle^s \mathsf{SPR, MPR, SP = PM}$, PR is common, and $\angle \mathsf{SPR} = \angle \mathsf{MPR}$,

$$\therefore \quad \angle \mathsf{PSR} = \angle \mathsf{PMR} = \text{a right angle.} \qquad \text{Q.E.D.}$$

166. *Tangents at the ends of a focal chord intersect at right angles on the directrix.*

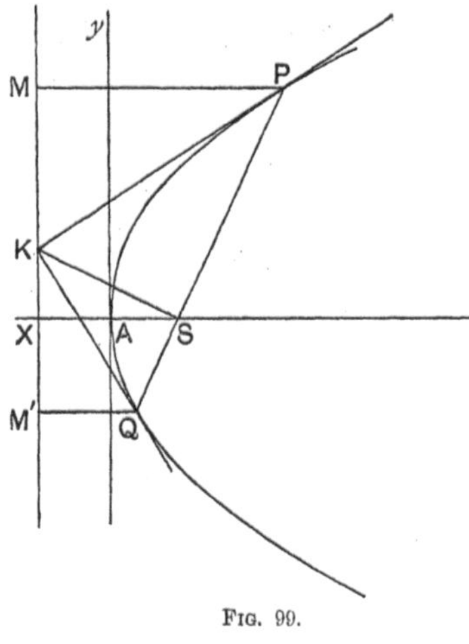

FIG. 99.

The portion of a tangent intercepted between its point of contact and the directrix subtends a right angle at the focus. (Art. 165.)

∴ if **PSQ** is a focal chord and **SK** is drawn at right angles to it, **SK** and the tangent at **P** meet on the directrix.

In the same way, **SK** and the tangent at **Q** meet on the directrix.

∴ tangents at the ends of a focal chord meet on the directrix.

Draw PM, QM′ at rt. ∠ˢ to the directrix.

In △ˢ SPK, MPK ∠ M = ∠ S (rt. ∠ˢ), ∠ MPK = ∠ SPK and KP is common ;

∴ ∠ PKM = ∠ PKS.

Similarly, from △ˢ M′QK, SQK, ∠ QKM′ = ∠ QKS.

∴ ∠ PKQ = ½ (the four angles at K) = a right angle.

167. *The semi-latus rectum is a harmonic mean between the segments of any focal chord.*

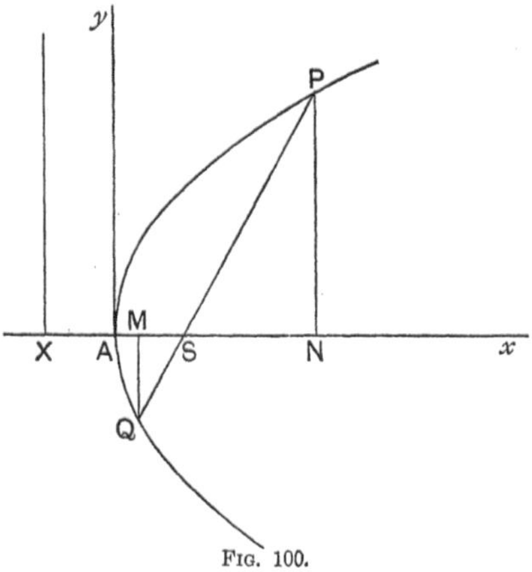

Let **PSQ** be the focal chord and $\left(\dfrac{a}{m^2}, \dfrac{2a}{m}\right)$ the co-ordinates of P.

FIG. 100.

The tangent at **Q** is at rt. ∠ˢ to that at **P**.

$\therefore$ the slope of the tangent at Q is $-\dfrac{1}{m}$ and the co-ordinates of Q are $(am^2, -2am)$.

$$\mathsf{SP} = \mathsf{NX} = a + \frac{a}{m^2}, \quad \mathsf{SQ} = \mathsf{MX} = a + am^2.$$

$\therefore$ the harmonic mean between SP and $\mathsf{SQ} = \dfrac{2\,\mathsf{SP} \cdot \mathsf{SQ}}{\mathsf{SP} + \mathsf{SQ}}$

$$= \frac{2a^2\left(1 + \dfrac{1}{m^2}\right)(1 + m^2)}{a\left(2 + m^2 + \dfrac{1}{m^2}\right)}$$

$$= \frac{2a(1 + m^2)^2}{m^4 + 2m^2 + 1} = 2a = \text{the semi-latus rectum.} \qquad \text{Q.E.D.}$$

168. *Tangents at the ends of a chord of a parabola intersect on the diameter which bisects that chord.*

Let RP, RQ be the tangents, PQ the chord, V the middle point of PQ.

Let $(x_1,\ y_1)$ be the co-ordinates of R.

The equation of PQ, the chord of contact, is

$$yy_1 = 2a(x + x_1).$$

Where this meets the curve $y^2 = 4ax$, we have by substitution,

$$yy_1 = 2a\left(\frac{y^2}{4a} + x_1\right)$$

or $\quad y^2 - 2yy_1 + 4ax_1 = 0.$

The roots of the quadratic are the ordinates of P and Q.

The ordinate of V, the middle point of PQ = half the sum of the ordinates of P and Q

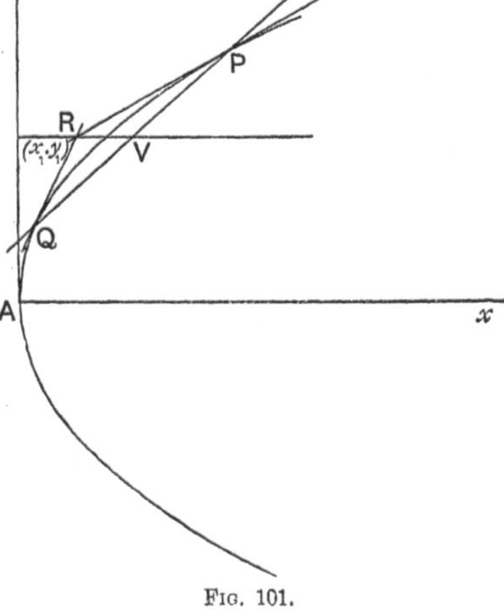

Fig. 101.

= half the sum of the roots of the quadratic

= y_1 = the ordinate of R.

$\therefore$ RV is parallel to the axis, which proves the theorem.

169. *If from* T, *any point on the tangent at* P, *perpendiculars* TR, TM *are drawn to the focal distance* SP *and the directrix,* SR = TM. (Adams' Proposition.)

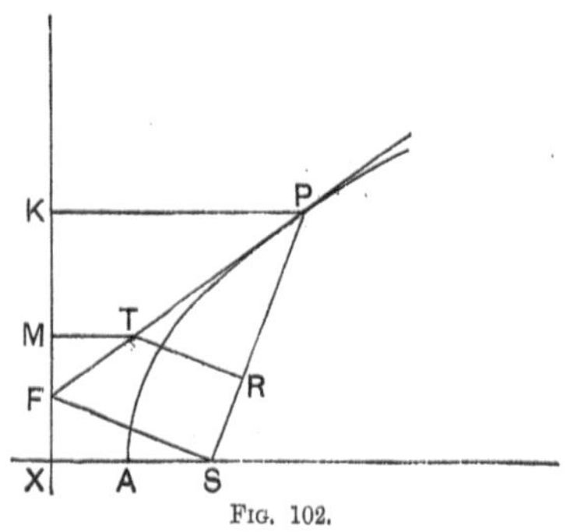

FIG. 102.

Let the tangent meet the directrix at F.   Join SF.

Draw PK perpendicular to the directrix.

$$\angle FSP = \text{a rt. } \angle \, ;$$

∴ TR and FS are parallel ;

$$\therefore \ \frac{SR}{SP} = \frac{FT}{FP} = \frac{TM}{PK}$$

from the similar $\triangle^s$ FMT, FKP.

But SP = PK ;

∴ SR = TM.   Q.E.D.

170. *Tangents to a parabola subtend equal angles at the focus.*

From the point T, where the tangents at P and Q meet, draw TR, TR′ perpendicular to SP and SQ respectively.   Also draw TM perpendicular to the directrix.

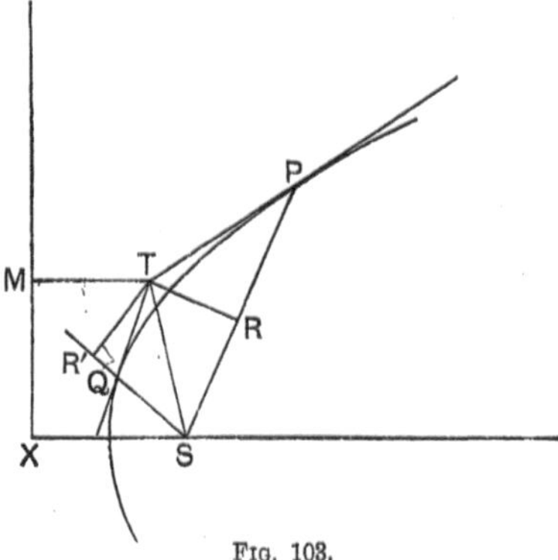

FIG. 103.

Then SR = TM by Adams' Proposition
$$= SR' \qquad \text{,,} \qquad \text{,,}$$
Also in the $\triangle^s$ TSR, TSR′, the $\angle^s$ at R and R′ are right angles.

∴ ∠TSR = ∠TSR′, which proves the proposition.

171. *Given a parabola, its focus and directrix, to draw tangents to it from an external point.*

Draw TK perpendicular to the directrix from the given point T.   With centre S and radius equal to TK describe a circle, and draw tangents TM, TM′ to it.   Join SM, SM′ and let them meet the parabola at P and Q.   TP and TQ are tangents to the parabola at P and Q.

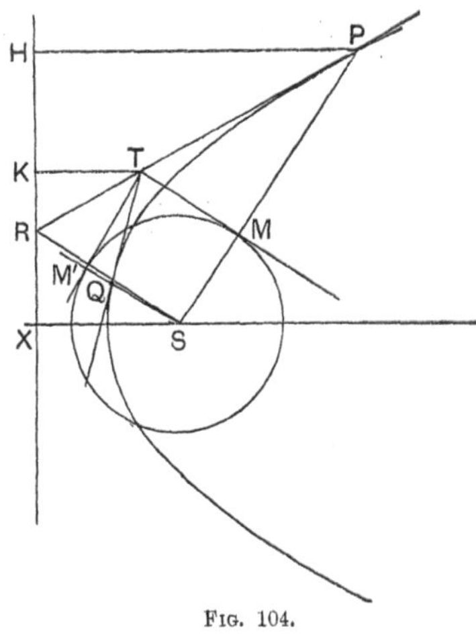

Fig. 104.

Draw PH perpendicular to the directrix, and produce PT to meet the directrix at R.   Join SR.

$$\frac{TK}{PH} = \frac{RT}{RP}.$$ But TK = SM, and PH = SP ;

$$\therefore \frac{SM}{SP} = \frac{RT}{RP} ;$$

∴ SR is parallel to TM.

∴ ∠ RSP = a right angle.

∴ RP touches the parabola at P.

Similarly TQ is a tangent.

If the point T is on the side of the directrix opposite to S, MS *produced* will meet the parabola at a point of contact.

172. *Given a curve which is known to be a parabola, find (geometrically) its axis, focus, and directrix.*

Draw any two parallel chords and bisect them at V and V'. Then since a diameter is the locus of the middle point of a series of parallel chords, VV' is a diameter.

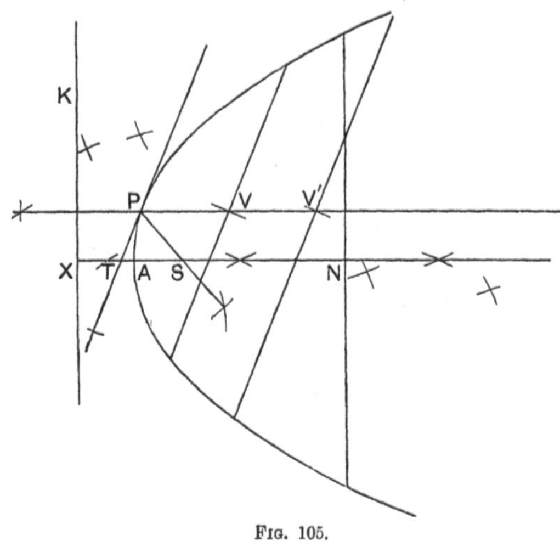

Fig. 105.

Draw any chord at rt. ∠ˢ to VV' and bisect it at N. Draw the diameter AN meeting the curve at A. AN is a diameter bisecting a chord at *right angles*;

∴ AN is the axis.

Let VV' meet the curve at P, and at P draw PT parallel to the chords through V and V' to meet the axis at T. PT is the tangent at P.

At P make the ∠ TPS = ∠ PTN, and let PS meet the axis at S.

Then SP = ST ;

∴ since PT is a tangent, S is the focus.

Produce SA to X and make AX = AS. Draw KX at rt. ∠ˢ to AN. KX is the directrix.

173. *To find the value of $y_1^2 - 4ax_1$ in connection with the parabola $y^2 = 4ax$, when (x_1, y_1), is not on the curve.*

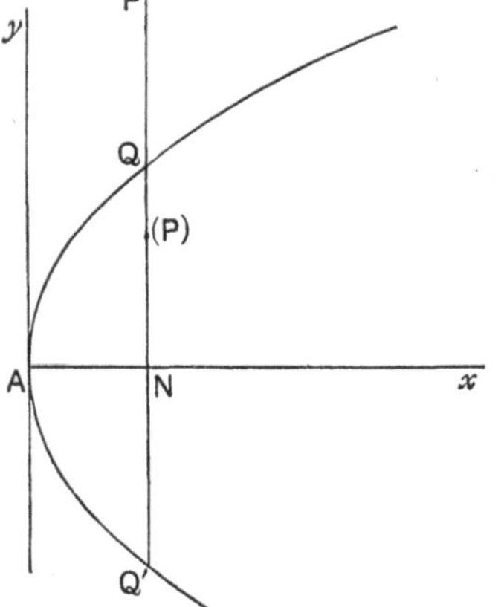

FIG. 106.

Let P be the point (x_1, y_1), and draw PN perpendicular to the axis to meet the curve in Q and Q'.

$$y_1^2 - 4ax_1 = PN^2 - 4a \cdot AN$$
$$= PN^2 - QN^2$$
$$= (PN + QN)(PN - QN)$$
$$= PQ \cdot PQ',$$

for QQ' is bisected at N.

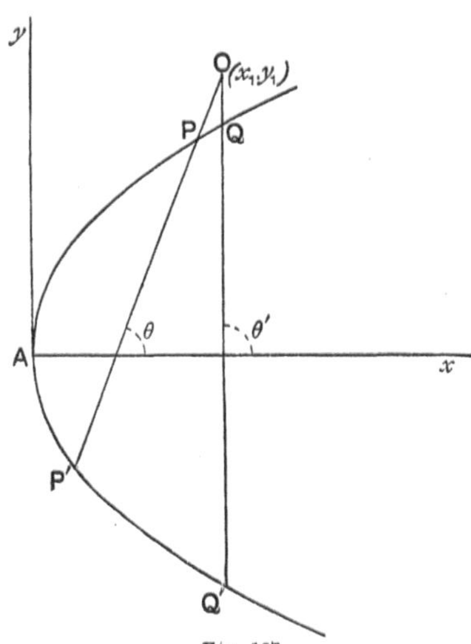

FIG. 107.

174. *If through the point O (x_1, y_1) a chord OPP' of the parabola $y^2 = 4ax$ is drawn making an angle θ with the axis of x, the rectangle*

$$OP \cdot OP' = \frac{y_1^2 - 4ax_1}{\sin^2\theta}.$$

The equation of OPP' may be taken to be

$$\frac{x - x_1}{\cos\theta} = \frac{y - y_1}{\sin\theta} = r.$$

$$x = x_1 + r\cos\theta,$$
$$y = y_1 + r\sin\theta ;$$

$\therefore$ where OPP' meets the parabola, we have, by substitution,

$$(y_1 + r\sin\theta)^2 = 4a(x_1 + r\cos\theta),$$

or

$$r^2\sin^2\theta + 2r(y_1\sin\theta - 2a\cos\theta) + y_1^2 - 4ax_1 = 0.$$

But OP, OP′ are the roots of this quadratic for r ;

$$\therefore \ \mathsf{OP}\cdot\mathsf{OP}' = \text{the product of the roots} = \frac{y_1{}^2 - 4ax_1}{\sin^2\theta}.$$

COROLLARY. If from the point O $(x_1,\ y_1)$ a tangent OP is drawn to the parabola $y^2 = 4ax$, making an angle θ with the axis of x,

$$\mathsf{OP}^2 = \frac{y_1{}^2 - 4ax_1}{\sin^2\theta}.$$

In this case $\mathsf{OP} = \mathsf{OP}'$, $\therefore$ the proposition follows from the preceding.

175. *If through any point* O, *chords* OPP′, OQQ′ *are drawn to a parabola in constant directions, the ratio of the rectangles* OP . OP′, OQ . OQ′ *is constant.*   (See fig. in preceding article.)

Let the chords make angles $\theta,\ \theta'$ with the axis of x.

As in the preceding article, $\mathsf{OP}\cdot\mathsf{OP}' = \dfrac{y_1{}^2 - 4ax_1}{\sin^2\theta}.$

Similarly, $\mathsf{OQ}\cdot\mathsf{OQ}' = \dfrac{y_1{}^2 - 4ax_1}{\sin^2\theta'}\ ;$

$$\therefore \ \frac{\mathsf{OP}\cdot\mathsf{OP}'}{\mathsf{OQ}\cdot\mathsf{OQ}'} = \frac{\sin^2\theta'}{\sin^2\theta},$$ which is a constant ratio, for the angles θ and θ' are constant.

COROLLARY. If tangents OP, OQ to a parabola make angles $\theta,\ \theta'$ with the axis of x,

$$\frac{\mathsf{OP}}{\mathsf{OQ}} = \frac{\sin\theta'}{\sin\theta}.$$

In this case $\mathsf{OP} = \mathsf{OP}'$ and $\mathsf{OQ} = \mathsf{OQ}'$.

$$\therefore \text{ by the preceding, } \frac{\mathsf{OP}^2}{\mathsf{OQ}^2} = \frac{\sin^2\theta'}{\sin\theta} \text{ and } \frac{\mathsf{OP}}{\mathsf{OQ}} = \frac{\sin\theta'}{\sin\theta}.$$

176. *If* PV *is the diameter through the point* P *on a parabola, and a chord* QVQ′ *is drawn parallel to the tangent at* P, *to prove that* $\mathsf{QV}^2 = 4\mathsf{SP}\cdot\mathsf{PV}.$

Let the chord QVQ′ and the tangent PT make an angle θ with the axis.

Let $(x_1,\ y_1)$ be the co-ordinates of V, the middle point of QQ′.

Since PV is parallel to Ox, we may take $(x_2,\ y_1)$ for the co-ordinates of P.

Also $x_2 = \dfrac{a}{m^2}$ where $m = \tan\theta$. $\left(\dfrac{a}{m^2},\ \dfrac{2a}{m}\right)$

The equation of QVQ$'$ is $\dfrac{x - x_1}{\cos\theta} = \dfrac{y - y_1}{\sin\theta} = r.$

$$\therefore\quad x = x_1 + r\cos\theta,\quad y = y_1 + r\sin\theta.$$

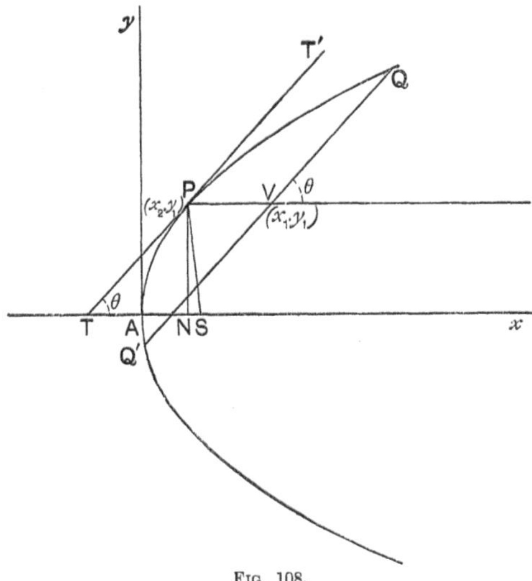

Fig. 108.

$\therefore$ where this line meets the curve, $y^2 = 4ax$,

$$(y_1 + r\sin\theta)^2 = 4a(x_1 + r\cos\theta)$$

or $\qquad r^2\sin^2\theta + 2r(y_1\sin\theta - 2a\cos\theta) + y_1{}^2 - 4ax_1 = 0.$

Now the roots of this equation are QV, Q$'$V, which are equal *but of opposite sign,*

$$\therefore\quad -\,\text{QV}^2 = \text{the product of the roots} = \frac{y_1{}^2 - 4ax_1}{\sin^2\theta}.$$

But $y_1{}^2 = 4ax_2$ for $(x_2,\ y_1)$ is on the curve,

$$\therefore\quad \text{QV}^2 = \frac{4a(x_1 - x_2)}{\sin^2\theta} = \frac{4a\,\text{PV}}{\sin^2\theta}. \qquad\ldots\ldots\ldots\ldots\ldots(1)$$

Also $SP = NX = a + x_2 = a + \dfrac{a}{\tan^2\theta} = \dfrac{a(1 + \tan^2\theta)}{\tan^2\theta} = \dfrac{a}{\sin^2\theta}.$

$$\left[\text{Or,} \quad SP = \dfrac{PN}{\sin 2\theta} \ (\text{for } SP = ST) = \dfrac{2a}{\tan\theta \,.\, \sin 2\theta} = \dfrac{a}{\sin^2\theta}. \right]$$

$$\therefore \text{ from (1),} \quad QV^2 = 4SP\,.\,PV.$$

***177.** *To find the equation of a parabola referred to a diameter as axis of x and the tangent at the extremity of the diameter as axis of y.*

With the figure of the preceding article, PV and PT' are the axes of x and y.

First, prove as in that article that

$$QV^2 = 4SP\,.\,PV. \ \dotfill (1)$$

Now, if (x, y) are the co-ordinates of Q,

$$QV = y, \quad PV = x.$$

SP is now constant and equal to the focal distance of the origin.

Let $SP = a'$.

Then from (1) $y^2 = 4a'x$ is the equation of the curve, where $a' = a \cosec^2\theta$, θ being the angle between the axes of co-ordinates.

***178.** We see that the equation of a parabola referred to a diameter and tangent at its extremity as axes is of the *same form* as when referred to its axis and tangent at the vertex.

$$yy_1 = 2a'(x + x_1)$$

is the equation of the tangent at the point (x_1, y_1).

The method of Art. 136 may be used to prove this, remembering that $y - y_1 = m(x - x_1)$ is the equation of the tangent, where m is the limiting value of $\dfrac{\Delta y_1}{\Delta x_1}$.   (See Art. 115.)

$y = mx + \dfrac{a'}{m}$ touches the parabola $y^2 = 4a'x$, for all values of m,

at the point $\left(\dfrac{a'}{m^2}, \dfrac{2a'}{m} \right)$.   The proof of Art. 138 holds.

In this equation $m = \dfrac{\sin a}{\sin(\theta - a)}$, where θ is the angle between the axes of co-ordinates, and a the angle which the tangent makes with the axis of the parabola.   (Art. 66.)

$yy_1 = 2a'(x + x_1)$ is the chord of contact of tangents drawn from (x_1, y_1).   The proof of Art. 142 holds.

$yy_1 = 2a'(x + x_1)$ is the polar of the point (x_1, y_1).

The equation of a normal to the curve $y^2 = 4a'x$ will not be the same as with rectangular axes, for the condition of perpendicularity of straight lines is not the same.   (Art. 69.)

At the point (x_1, y_1) the equation of the normal is

$$y - y_1 = -\left(\frac{y_1 + 2a' \cos \theta}{2a' + y_1 \cos \theta}\right)(x - x_1),$$

θ being the angle between the axes.

At the point $\left(\dfrac{a'}{m^2}, \dfrac{2a'}{m}\right)$, its equation is

$$y - \frac{2a'}{m} = -\left(\frac{1 + m \cos \theta}{m + \cos \theta}\right)\left(x - \frac{a'}{m^2}\right).$$

The proofs are left as an exercise for the student.

*179. *If* OQ, OQ' *are tangents to a parabola, and the diameter through* O *meets the curve at* P *and the chord of contact* QQ' *at* V, OP = PV.

OPV bisects QQ', and the tangent PT at P is parallel to QQ'.

Taking PV and PT as axes of co-ordinates,

$y^2 = 4a'x$ is the equation of the parabola, where SP $= a'$.

Let (x_1, y_1) be the co-ordinates of Q.

The equation of the tangent at Q is

$$yy_1 = 2a'(x + x_1).$$

At O, a point on this line, $y = 0$,

$$\therefore x = -x_1,$$

i.e. PO = PV.

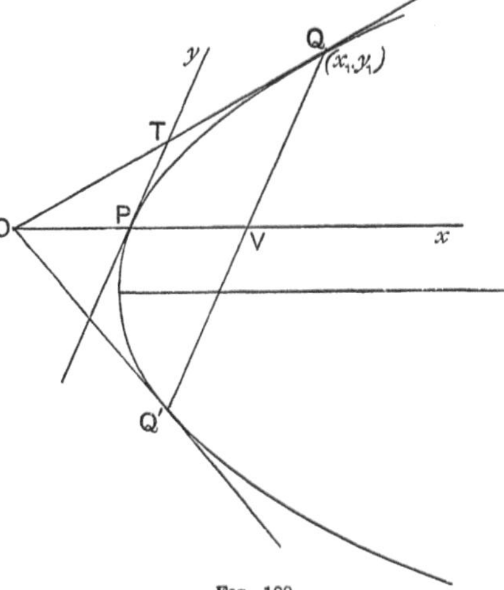

Fig. 109.

The opposite signs of PO and PV show that these lines are drawn in opposite directions,

*180. *Second method of drawing tangents to a parabola from an external point.* (See Art. 171 for first method.)

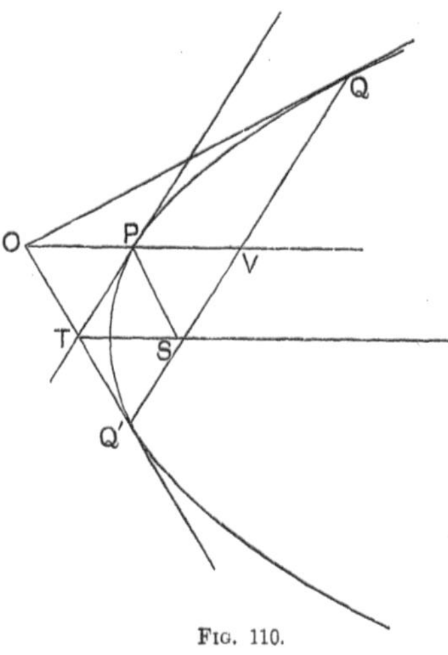

Fig. 110.

Let O be the external point and draw OPV parallel to the axis to meet the curve at P, and make PV = OP. From SA produced cut off ST = SP.

PT is the tangent at P.

Draw QVQ′ parallel to PT to meet the parabola at Q and Q′.

OQ, OQ′ are tangents at Q and Q′.

For QQ′ is parallel to the tangent PT,

$$\therefore \quad QV = Q'V.$$

Also   OP = PV.

$\therefore$ OQ and OQ′ are tangents at Q and Q′.

*181. *To find the equation of a pair of tangents drawn to the parabola* $y^2 = 4ax$ *from the point* (x_1, y_1).

Let OD, OE be the tangents, P (x, y) any point on either tangent.

Draw OL, PM perpendicular to DE the chord of contact. Let OPE make an angle θ with the axis of x.

From the similar

$\triangle^s$ EPM, EOL,

$$\frac{OL}{PM} = \frac{OE}{PE}. \quad ......(1)$$

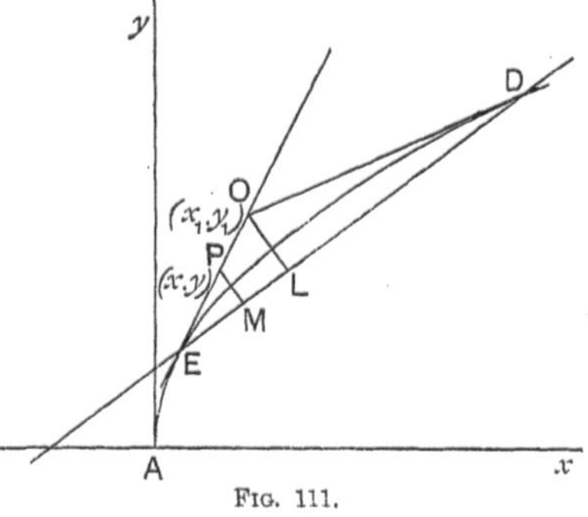

Fig. 111.

As in Art. 174, Cor.,

$$OE^2 = \frac{y_1^2 - 4ax_1}{\sin^2\theta}, \quad \text{and} \quad PE^2 = \frac{y^2 - 4ax}{\sin^2\theta}. \quad ....(2)$$

The equation of DE is $yy_1 = 2a(x + x_1)$;

$$\therefore\ \mathsf{OL} = \frac{y_1{}^2 - 4ax_1}{\sqrt{y_1{}^2 + 4a^2}}. \qquad \left(p = \frac{\mathsf{A}x_1 + \mathsf{B}y_1 + \mathsf{C}}{\sqrt{\mathsf{A}^2 + \mathsf{B}^2}}\right)$$

Also $\qquad \mathsf{PM} = \dfrac{yy_1 - 2a(x + x_1)}{\sqrt{y_1{}^2 + 4a^2}}.\quad (\quad ,, \qquad ,, \quad)$

$\therefore$ from (1) and (2) $\qquad \dfrac{\dfrac{y_1{}^2 - 4ax_1}{\sqrt{y_1{}^2 + 4a^2}}}{\dfrac{yy_1 - 2a(x + x_1)}{\sqrt{y_1{}^2 + 4a^2}}} = \dfrac{\sqrt{y_1{}^2 - 4ax_1}}{\sqrt{y^2 - 4ax}}.$

Squaring and multiplying up,

$$(y^2 - 4ax)(y_1{}^2 - 4ax_1) = [yy_1 - 2a(x + x_1)]^2.$$

Also P is *any point on either tangent*, therefore this is the equation required.

Examples VIII. e.

1. Prove that $y = mx + c$ touches the parabola

$$y^2 = 4a(x + a) \text{ if } c = am + \frac{a}{m}.$$

2. Through the vertex of a parabola a perpendicular is drawn to any tangent meeting it at P and the curve at Q. Show that AP . AQ is constant.

3. If two tangents be drawn to a parabola making complementary angles with the axis, their chord of contact must pass through the foot of the directrix.

4. Find the equations of the tangents drawn from the point (1, 3) to the parabola $y^2 = 8x$.

5. A parabola touches two given straight lines at two given points. Find its focus and axis by geometrical construction.

6. Find the equation of the parabola whose vertex and focus are each on the axis of x at positive distances a and a' from the origin.

7. If on a given base triangles be described such that the sum of the tangents of the base angles is constant, prove that the locus of the vertices is a parabola.

8. The tangents at the ends of the latus rectum of a parabola meet an ordinate PN produced at R and R': prove that $\mathsf{RP} . \mathsf{R'P} = \mathsf{SN}^2$, where S is the focus.

9. The equation of a focal chord PSQ of the parabola may be written

$$\frac{x - a}{\cos \theta} = \frac{y}{\sin \theta} = r.$$

Use this to find the value of (i) SP + SQ ; (ii) the rectangle SP . SQ.
(*N.B.* SP, SQ will be drawn in *opposite* directions.)

10. If SY is drawn perpendicular to the tangent at P the point (x_1, y_1) on the parabola $y^2 = 4ax$, $AY^2 = ax_1$.

11. Given a curve which is known to be a parabola, draw a tangent to it which is parallel to a given straight line.

12. If PSQ is a focal chord of the parabola $y^2 = 4ax$, prove that the product of the abscissae of P and Q is equal to a^2.

13. If a circle cuts the parabola $y^2 = 4ax$ in four points, the algebraic sum of their ordinates is zero.

14. Tangents are drawn to a given parabola from a given point : draw a third tangent making equal angles with them.

15. Find the equation of a parabola when the axes of co-ordinates are two tangents, drawn from the foot of the directrix.

16. The chord PQ is normal to a parabola at P, and is bisected at V.   If the diameter through V meets the curve at R, prove that $RP = RV$.
Prove also that the tangents at R and P meet on the directrix.

17. Two equal parabolas are co-axial but have different vertices ; show that the portion of a tangent to the inner one intercepted by the other is bisected at the point of contact.

18. Tangents are drawn to the parabola $y^2 = 4ax$ from the point (x_1, y_1) : find the equation of the lines joining the vertex to their points of contact.

19. Chords of a parabola subtend a right angle at the vertex : find the locus of their middle points.
[Use the equation of a chord in terms of the co-ordinates at its middle point, and the method of Art. 52.]

20. Tangents OP, OQ, are drawn to the parabola $y^2 = 4ax$.   Prove that the orthocentre of the triangle POQ will be at the vertex if $2mm' + 1 = 0$, where m, m' are the slopes of the tangents.

21. If a normal to a parabola make an acute angle ϕ with the axis, show that it will cut the curve again at an angle $\tan^{-1}(\tfrac{1}{2}\tan\phi)$.

22. Prove that two parabolas, having the same focus and their axes in opposite directions, cut at right angles.

23. The triangle which has any chord of a parabola for base, and its vertex at the vertex of the parabola $y^2 = 4ax$, has an area $h\sqrt{k^2 - 4ah}$, (h, k) being the pole of the chord.

24. Tangents to the parabola $y^2 = 4ax$ are drawn at points whose abscissae are in the ratio $\mu : 1$: prove that they intersect on the curve

$$y^2 = (\mu^{\frac{1}{4}} + \mu^{-\frac{1}{4}})^2 ax.$$

25. If P, Q, R be three points on a parabola whose ordinates are in geometrical progression, prove that the tangents at P and R meet on the ordinate of Q.

26. A point V is taken on an ordinate PN produced, of a parabola, and NE is taken on NP a mean proportional between NP and NV ; if the diameters through E and V meet the curve in R and Q, prove that PQ meets the axis in the foot of the ordinate of R.

27. The normal at P meets the parabola $y^2=4ax$ again at Q ; if x_1 is the abscissa of P, prove that $PQ^2=\dfrac{16a}{x_1{}^2}(x_1+a)^3$.

28. In the parabola $y^2=4ax$, show that $\sqrt{y_1{}^2-4ax_1}$ is the projection of the tangent from $(x_1,\ y_1)$ to the curve upon the tangent at the vertex.

29. Perpendicular lines AP, AQ are drawn from the vertex of a parabola to meet it in P, Q : prove that the normals at P and Q intersect on a parabola.

30. TPZ is a tangent at P to a parabola cutting the axis of the curve at T, and PZ is taken in a constant ratio to TP ; prove that the locus of Z is a parabola, and find its parameter (latus rectum).

Revision Questions on the Parabola.

[These may be taken orally, or the answers may be written down without any working.]

What are the equations of the directrices of the following parabolas :

1. $y^2=4ax.$ **2.** $y^2=-4ax.$ **3.** $x^2=4ay.$
4. $x^2=-4ay.$ **5.** $ax^2+by=0.$ **6.** $ax^2=by.$

Give the co-ordinates of the vertices of the parabolas :

7. $y^2=4a(x+a).$ **8.** $x^2=8a(y-2a).$ **9.** $(y-k)^2=4a(x-h).$
10. $(x+h)^2=a(y+k).$ **11.** $(y+k)^2=2a(h-x).$ **12.** $(x-3)^2+2y=0.$

What is the equation of the latus rectum of the parabola :

13. $y^2=4ax.$ **14.** $y^2=2ax.$ **15.** $y^2=-2ax.$
16. $x^2=4ay.$ **17.** $x^2=-4ay.$ **18.** $x^2=8ay.$

What are the co-ordinates of the focus of the parabola :

19. $y^2=-ax.$ **20.** $x^2=ay.$ **21.** $x^2=-ay.$

Give the equation of the axis of the parabola :

22. $x^2=by.$ **23.** $x^2=-4y.$ **24.** $(y-k)^2=4a(x-h).$
25. $(x+h)^2=4a(y-k).$ **26.** $(x-2)^2=4y.$ **27.** $y^2=4(x-3).$

28. If $y^2=4ax$, $y=\pm2\sqrt{ax}$; what do you deduce as to the shape of the curve ?

Give the equation of the tangent at the point $(x_1,\ y_1)$ to the following curves :

29. $x^2=4ay.$ **30.** $y^2=-4ax.$
31. $ax^2+by=0.$ **32.** $ax^2=4by.$

In each of the following curves, give the equation of a tangent in terms of its slope :

33. $y^2=8ax.$ **34.** $y^2=-4ax.$
35. $ay^2=bx.$ **36.** $x^2=4ay.$

37. What is the equation of the tangent at the origin to the parabola $y^2 = ax + by$?

38. If a tangent to the curve $y^2 = 4ax$ makes an angle of 45° with the axis of the curve, what are the co-ordinates of its point of contact?

39. The point $(an^2,\ 2an)$ is on the parabola $y^2 = 4ax$: what is the slope of the tangent at this point?

40. Give the co-ordinates of any point on the parabola $y^2 = -4ax$ in terms of the slope (m) of the tangent at that point.

41. If $y = mx + c$ is the equation of the tangent to a curve at the point $(h,\ k)$, what is the equation of the normal at that point?

42. What is the equation of the normal to $y^2 = 4ax$ at the point $\left(\dfrac{a}{m^2},\ \dfrac{2a}{m}\right)$?

43. What is the equation of the polar of the point $(4,\ 6)$ with respect to the parabola $y^2 = 2x$?

44. What is the equation of a parabola whose latus rectum is 8, vertex at the point $(h,\ k)$, and whose axis is parallel to the axis of x? Two parabolas can be thus drawn.

45. What would be the equations of the curve in the preceding question if its axis were parallel to the axis of y?

46. Where the straight line $x = h$ meets the parabola $x^2 = ay + b$, we have $h^2 = ay + b$. What can you deduce?

47. What is the equation of the tangent to the parabola $y^2 = 8x$ which is parallel to $y = 3x + 4$?

48. What is the equation of the normal to the parabola $y^2 = 4x$ which is parallel to $y = 2x$?

49. $y = mx + \dfrac{a}{m}$, $y = -\dfrac{1}{m}(x - a)$. Subtracting, we have $x = 0$. Give a geometrical interpretation to this.

50. A point moves so that the square of its perpendicular distance from one fixed straight line varies as its perpendicular distance from another fixed straight line at right angles to the first: what is its locus?

51. Prove that $ax + by = 0$ is a tangent to the parabola $(y - k)^2 = ax + by$.

52. If $Ax + By + C$ is a tangent to a parabola at the point $(h,\ k)$, what is the equation of the normal at the same point?

53. Prove that $lx + my + n = 0$ is a tangent to the parabola
$$(Ax + By + C)^2 = lx + my + n.$$
How would you find the point of contact?

54. What is the equation of a normal to the parabola $y^2 = 4ax$ in terms of its slope? What can you deduce as to the number of normals which can be drawn to the parabola from a given point?

CHAPTER IX.

REVISION PAPERS.

Revision Paper IX. a.

1. Show in a diagram the lines $x=3$, $x+y=2$, $x=y$.
Find the co-ordinates of their intersections, and the area of the triangle contained by them.

2. Determine the equation of the bisector of the acute angle between the lines $12x+5y=4$, and $24y-7x=9$.

3. In the equation $x^2+y^2-2ax\tan a-2ay+a^2=0$, interpret the result (i) when $x=0$; (ii) when $y=a(1+\tan a)$.
Draw the circle.

4. Find the equation of the circle described on the line joining the points $(4, 6)$, $(1, -2)$ as diameter.

5. Find the equation of the parabola whose focus is at the point $(2, 0)$ and whose directrix is the line $3x-4y=0$. Draw a rough sketch of the curve.

6. On squared paper draw the parabola $y^2=2(x+3)$ and find the equation of its directrix.

7. Find the equation of the chord of the parabola $y^2=2x$ which is bisected at the point $(3, -1)$.

Revision Paper IX. b.

1. Find the equation and length of the perpendicular from the intersection of $bx+ay=ab$, and $ax-by=ab$ upon $ax-by=0$.

2. Taking an inch as the unit of length, represent in a figure the three lines $2x+3y-6=0$, $4x-2y+3=0$, $6x+y+1=0$; and show that the area of the triangle they form is half a square inch.

3. A circle of radius a, with both co-ordinates of its centre positive, touches the axis of x and the straight line $3y=4x$. Find its equation.

4. Draw the circle $x^2+y^2-2ax-2ay\tan a+a^2=0$, and find the equations of the tangents drawn to it from the origin. $(a<90°.)$

5. Find the value of c if the straight line $y=x+c$ touches the parabola
$$y^2=3x-4y.$$

6. Double ordinates of the parabola $y^2=4ax$ are drawn. Find the equation of, and draw, the locus of their points of trisection.

7. Prove that $y+2x=12a$ is a normal to the parabola $y^2=4ax$, and find the co-ordinates of the point where it cuts the curve at right angles.

Revision Paper IX. c.

1. Prove that if the line joining the origin and the point (x_1, y_1) is at right angles to the line joining the points $(a, 0)(x_2, y_2)$, then

$$x_1 x_2 + y_1 y_2 = a x_1.$$

2. Find the area of the triangle included by the three points $(1, 2)$, $(-3, 1)$, $(4, -5)$, and the co-ordinates of a point equidistant from them.

3. Find the equation of the tangent to the circle $x^2 + y^2 + 11x - 7y = 0$ at the origin.

4. Find the locus of the intersection of the straight lines $y - mx = \beta - ma$, and $my + x = m\beta' + a'$, where m is a variable quantity.

5. Show that the roots of the equation $y^4 - 4y^3 + 5y^2 - 9 = 0$ are the ordinates of the points of intersection of the circle $x^2 + y^2 = 9$ with the parabola $(y - 1)^2 = x + 1$.

Find two of the roots by drawing the curves on squared paper.

What do you deduce as to the other roots?

6. Trace the parabolas $x^2 = 8ay$, $y^2 = 27ax$, and show that they intersect at right angles and also at an angle $\tan^{-1} 1\frac{9}{13}$.

7. Find the locus of the vertex of a parabola having a given point as focus and touching a given straight line.

Revision Paper IX. d.

1. Find the equation of the straight line joining the origin to the intersection of the straight lines $x - y - 4 = 0$, $7x + y + 20 = 0$; and show that it bisects one of the angles between them.

2. If lines are drawn parallel to the axes of co-ordinates from the points where the line $x \cos a + y \sin a - p = 0$ meets the axes, so as to meet the perpendicular on the line from the origin in P and Q; prove that

$$\mathsf{PQ} = 4p \cot 2a \,.\, \operatorname{cosec} 2a.$$

3. Write down the equation of the tangent to the circle $x^2 + y^2 = 169$ at the point $(5, 12)$.

Deduce the equation of the tangent to the circle $(x - 2)^2 + (y + 3)^2 = 169$ at the point $(7, 9)$.

4. Find the equation of the straight lines joining the origin to the points of intersection of $x^2 + y^2 - 2x + 3y + 4 = 0$ and $2x + 3y = k$.

Find the values of k for which the lines are at right angles to one another.

5. Show that the equations $y^2 = 4a(x + a)$, $y^2 = -4b(x - b)$ represent a pair of parabolas having the same focus and axis. Find the co-ordinates of their points of intersection, and prove that at each of these points the tangents to the two parabolas are at right angles.

6. Draw the curve $x^2 = 4(1 - y)$ and find the equation of its directrix.

7. The normal to the parabola $y^2 = 4ax$, at any point P, meets the axis in G, and GP is produced outwards to Q so that $\mathsf{PQ} = \mathsf{PG}$: find the locus of Q.

Revision Paper IX. e.

1. Find the equation of the straight line which passes through the point $(2, -6)$ and is perpendicular to the straight line $3x + 4y = 7$, and determine the area of the triangle contained by these lines and the axis of x.

2. Find the angle between the straight lines represented by

$$2x^2 + 5xy - 3y^2 = 0.$$

3. Find the condition or conditions that the circle

$$x^2 + y^2 + Ax + By + C = 0$$

(i) should touch the axis of x; (ii) should touch the axis of x at the origin.

4. Find the equation of the two straight lines joining the origin to the intersections of the circle $x^2 + y^2 - 3x - 5y = 0$, and the straight line

$$3x - 5y = 1.$$

Find the tangent of the angle between the lines.

5. Trace the curve $y = 2x - x^2$, taking an inch as unit. Find, approximately, the co-ordinates of the point where it cuts the line $y + 4 = 0$.

6. Prove that if two tangents are drawn to the parabola $y^2 = 4ax$ from a point on the line $x = -4a$, their chord of contact subtends a right angle at the vertex.

7. Assuming that $y = mx + \dfrac{a}{m}$ is a tangent to the parabola $y^2 = 4ax$, show that the equation of the tangent at the other extremity of the focal chord through the point of contact is $x + my + m^2a = 0$.

Revision Paper IX. f.

1. Find the equations of the two straight lines joining the point $(a, 2a)$ with the points in which the straight line $x + 2y = 2a$ cuts the axes of co-ordinates; and find the angle between the two lines.

2. Find the equation of the two straight lines drawn through the point $(1, 1)$ which are parallel respectively to the straight lines given by the equation $x^2 + 5xy + 2y^2 = 0$.

Find also the equations of the two diagonals of the parallelogram formed by the four lines.

3. Prove that $y(a^2 - c^2) = (a\beta \pm c\sqrt{a^2 + \beta^2 - c^2})x$ are the equations of the tangents drawn from the origin to the circle $(x - a)^2 + (y - \beta)^2 = c^2$.

4. Find the equation which represents the two straight lines joining the origin to the points of intersection of the line $x + 4y = 2a$, and the circle $x^2 + y^2 = a^2$.

Find the angle between the lines.

5. A circle is described with its centre on the axis, and to pass through the vertex of a parabola: show that it will not cut the parabola in any other point unless the centre is at a distance from the vertex greater than half the latus rectum.

6. The two parabolas $y^2 = 4ax$, $x^2 = 4by$ intersect at the origin and at P. Find the equations of OP, and of the tangents to the curves at P.

7. In a parabola the diameter bisecting the normal chord meets in Q the parallel to the normal through the focus : prove that the locus of Q is the parabola $y^2 = 2a(x - a)$.

Revision Paper IX. g.

1. Solve the following quadratic for y, and draw the locus given by the equation : $4x^2 - 9y^2 - 2x + 27y - 20 = 0$.

2. Prove that the length of the portion of the line $2y = x + 2$ intercepted by the circle $x^2 + y^2 = 4$ is $\dfrac{4}{\sqrt{5}}$.

3. Prove that the line $12x + 5y - 240 = 0$ is a tangent to the circle
$$x^2 + y^2 - 6x - 14y - 111 = 0.$$

4. If the polar of the point (2, 3) with respect to the circle $x^2 + y^2 = 17$ touches the circle $(x - 3)^2 + (y - 5)^2 = r^2$, prove that $r = \dfrac{4}{\sqrt{13}}$.

5. From first principles find the equations of the tangent and normal to the parabola $x^2 = 2ay$ at the point (x_1, y_1).

6. From the point (6a, 0) a normal is drawn to the parabola $y^2 = 4ax$; prove that its length is $2\sqrt{5} \, . \, a$.

7. Find the equation of the parabola whose vertex is at the point (0, 2), whose axis is parallel to the axis of y, and which passes through the point (2, 5).

Revision Paper IX. h.

1. Write down the general equation of a straight line through the intersection of $3x - 2y = 4$, and $5x + 3y = 7$.

Deduce the equation of the straight line through their point of intersection, and having a slope equal to -2.

2. Find the equations of the tangents to the circle
$$x^2 + y^2 - 6x + 8y + 21 = 0$$
which are parallel to the axis of x.

Verify by a figure.

3. Determine the co-ordinates of the centre of the circle circumscribing the triangle formed by joining the points (0, 0), (4, −6), (1, −3).

4. Find the equations of the tangents drawn to the circle $x^2 + y^2 = 36$ from the point (10, 0).

5. Find the equation of the diameter of the parabola $y^2 = 8x$ which bisects chords parallel to $y = 2x$.

6. Find the equations of the tangents drawn from the origin to the parabola $(y - 3)^2 = 4(x - 4)$.

7. If $y = mx + c$ is a chord of the parabola $y^2 = 4ax$, find the co-ordinates of its middle point.

Revision Paper IX. k.

1. Use the formula for the equation of the straight line joining two given points to determine the equation of the straight line joining the points $(2, 3)$, $(2, -5)$.

2. Find the equations of the tangents to the circle $x^2 + y^2 + 2x - 2y + 1 = 0$ which are parallel to the axis of y.

Verify your results by means of a figure.

3. Determine the co-ordinates of the centre of the circle inscribed in the triangle formed by the lines $x = 8$, $y = 0$, $3y = 4x$.

4. Find the equation of the circle passing through the points $(2, 4)$, $(2, 6)$, $(4, 10)$.

5. Prove that $ny = x + an^2$ is a tangent to the parabola $y^2 = 4ax$ and find the co-ordinates of its point of contact.

6. Given a parabola and its axis, draw a tangent to the curve at a given point.

7. Prove that the straight line $x + 2y + \lambda = 0$ cuts the parabola

$$(x + 2y)^2 = 4 (x - y)$$

at one point only (not two coincident points). What do you deduce?

Revision Paper IX. l.

1. Two straight lines are drawn through the origin, each at a perpendicular distance p from the point (h, k). Prove that the equation of the lines is $(kx - hy)^2 = p^2(x^2 + y^2)$.

2. Tangents are drawn to the circle $x^2 + y^2 - 6x - 2y + 9 = 0$ from the origin : find their equations.

3. If θ is the acute angle between the tangents at a common point of the circles $(x - 2)^2 + (y - 1)^2 = 16$, and $(x - 1)^2 + (y - 3)^2 = 4$, prove that $\cos \theta = \frac{15}{16}$.

4. Under what condition will the straight line $y = mx + c$ touch the circle $x^2 + y^2 = 2ay$?

5. Prove that $4y = 2x - 9$ is a normal to the parabola $y^2 = 8x$. Find the equation of the tangent at the same point.

6. Write down the equation of a tangent to the parabola $y^2 = 9x$ in terms of its slope.

Deduce the equation of a tangent to the parabola $(y - 2)^2 = 9(x + 2)$.

7. Use the property $PN^2 = 4AS \cdot AN$ to determine the equation of the parabola which has its vertex at the point $(4, 0)$, $3x + 4y = 12$ for its axis, and which passes through the point $(12, 4)$.

Draw a rough sketch of the curve.

CHAPTER X.

THE ELLIPSE.

182. A conic section whose eccentricity is less than unity is called an ellipse. We may therefore define an ellipse thus :

Def. : An **ellipse** is a curve traced out by a point which moves so that its distance from a fixed point, called the focus, is in a constant ratio, which is less than unity, to its perpendicular distance from a fixed straight line which is called the directrix. The constant ratio is called the **eccentricity** of the curve.

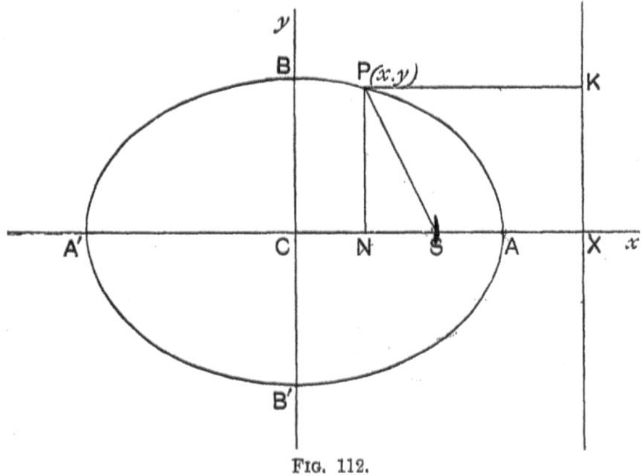

Fig. 112.

To find the equation of an ellipse.

Let S be the focus, KX the directrix, e the eccentricity. Draw SX perpendicular to the directrix, and in it take a point A such that $\dfrac{SA}{AX} = e$. By def., A is a point on the curve.

Since a line can be divided internally and externally in the same ratio, we can find a point A′ in XS produced so that $\dfrac{SA'}{A'X} = e$. This point A′ is also on the curve, by definition.

Bisect AA′ at C, and let $CA = a = CA'$.

$$SA' = e \cdot A'X, \text{ and } SA = e \cdot AX \ ;$$

$$\therefore \ SA' + SA = e(A'X + AX),$$

or $\qquad 2a = e(\overline{CX + CA'} + \overline{CX - CA}) = 2e \cdot CX \ ;$

$$\therefore \ \mathbf{CX} = \frac{\mathbf{a}}{\mathbf{e}}. \dots\dots\dots\dots\dots\dots\dots(1)$$

Also $SA' - SA = e(A'X - AX)$, or $(\overline{CA' + CS} - \overline{CA - CS}) = 2ae$;

$$\therefore \ 2CS = 2ae,$$

$$\therefore \ \mathbf{CS} = \mathbf{ae}. \ \dots\dots\dots\dots\dots\dots\dots(2)$$

Now take CX as axis of x, and Cy at rt. $\angle^s$ to it as axis of y.

Let P(x, y) be any point on the curve, and draw PK perpendicular to the directrix, and PN perpendicular to CX.

By definition, $\qquad SP = e \cdot PK = e \cdot NX \ ;$

$$\therefore \ SP^2 = e^2 \cdot NX^2,$$

$$SN^2 + PN^2 = e^2 \cdot NX^2.$$

But $\qquad SN = CS - CN = ae - x, \ \ PN = y,$

and $\qquad NX = CX - CN = \dfrac{a}{e} - x \ ;$

$$\therefore \ (ae - x)^2 + y^2 = e^2\left(\frac{a}{e} - x\right)^2,$$

Rearranging, $\qquad x^2(1 - e^2) + y^2 = a^2(1 - e^2),$

or $\quad \dfrac{x^2}{a^2} + \dfrac{y^2}{a^2(1 - e^2)} = 1. \ \dots\dots\dots\dots\dots\dots(3)$

Let $y = b$, when $x = 0$.

Then $\qquad \dfrac{b^2}{a^2(1 - e^2)} = 1 \ \text{ or } \ \mathbf{b}^2 = \mathbf{a}^2(1 - \mathbf{e}^2), \ \dots\dots\dots(4)$

and equation (3) becomes $\dfrac{\mathbf{x}^2}{\mathbf{a}^2} + \dfrac{\mathbf{y}^2}{\mathbf{b}^2} = \mathbf{1}.$

This is the simplest form of the equation of an ellipse.

Important. $\qquad \mathbf{CS} = \mathbf{ae}, \ \mathbf{CX} = \dfrac{\mathbf{a}}{\mathbf{e}}.$

The different forms of (4)

$$b^2 = a^2(1 - e^2), \quad a^2 - b^2 = a^2 e^2, \quad \frac{b^2}{a^2} = 1 - e^2.$$

We notice that the **terms of the second degree**, $\frac{x^2}{a^2} + \frac{y^2}{b^2}$, are in the form of the sum of two squares, and therefore **have no real factors**. This will be found to be the distinguishing characteristic of the equation of any ellipse.

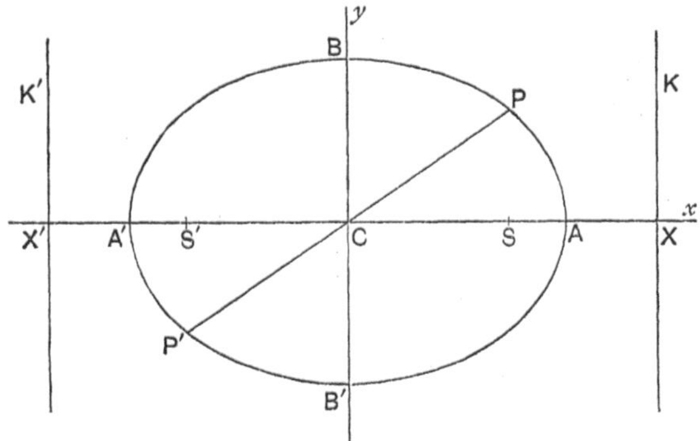

Fig. 113.

183. *Form of the ellipse.*

The equation may be written

$$\frac{y^2}{b^2} = 1 - \frac{x^2}{a^2} = \frac{a^2 - x^2}{a^2}, \quad \text{or} \quad \frac{y}{b} = \pm\sqrt{\frac{a^2 - x^2}{a^2}} ;$$

$\therefore$ for any value of x, we have two values of y, equal but opposite in sign. This proves that **the curve is symmetrical about the axis of x**.

Also a is the maximum value of x, and $-a$ its minimum value, for if $x > a$ or $< -a$, $\dfrac{a^2 - x^2}{a^2}$ is negative, and the values of y are imaginary.

The equation may also be written, $\dfrac{x}{a} = \pm\sqrt{\dfrac{b^2 - y^2}{b^2}}.$

$\therefore$ for any value of y, we have two values of x, equal but opposite in sign. This proves that **the curve is symmetrical about the axis of y**.

Also, as above, we see that b is the maximum value of y, and $-b$ its minimum value.

AA$'$ ($2a$) is called the **major axis**.

If the curve meets the axis of y at B and B$'$

BB$'$ ($2b$) is called the **minor axis**.

The axes AA$'$, BB$'$ are called the **principal axes** of the ellipse.

Observe that if the curve is folded about AA$'$, the upper and lower parts then coincide. Also if it is folded about BB$'$, the right and left hand parts then coincide.

The symmetry of the curve shows that it has two foci, and two directrices.

In XC produced take CS$'$ = CS, and CX$'$ = CX. Draw K$'$X$'$ parallel to KX. Then it is evident, by symmetry, that the *same curve* might be drawn with S$'$ as focus and X$'$K$'$ as directrix.

Since x cannot be greater than a, and y cannot be greater than b, we see that the ellipse is a *closed curve*, lying entirely within a rectangle whose sides are $2a$, and $2b$.

Again if (x_1, y_1) satisfy the equation $\dfrac{x^2}{a^2}+\dfrac{y^2}{b^2}=1$, ($-x_1$, $-y_1$) also satisfy the equation. If P and P$'$ are these points, we see from a figure that PCP$'$ is a straight line, and CP = CP$'$.

Hence every chord through C is bisected at C. C is therefore called the **centre**.

Def. **A centre of a conic is a point which bisects all chords of the conic drawn through it.**

Def. Any chord of an ellipse drawn through the centre is called a **diameter**.

184. *The point* (x_1, y_1) *is outside or inside the ellipse* $\dfrac{x^2}{a^2}+\dfrac{y^2}{b^2}=1$, *according as* $\dfrac{x_1^{\,2}}{a^2}+\dfrac{y_1^{\,2}}{b^2}-1$ *is greater or less than zero.*

Let Q be the point (x_1, y_1) and let the ordinate QN meet the ellipse at P.

If P is outside the curve,

$$\frac{x_1^{\,2}}{a^2}+\frac{y_1^{\,2}}{b^2}-1=\frac{\mathsf{CN}^2}{a^2}+\frac{\mathsf{QN}^2}{b^2}-1>\frac{\mathsf{CN}^2}{a^2}+\frac{\mathsf{PN}^2}{b^2}-1$$

which is equal to zero.

For CN . PN are the co-ordinates of P, and P is on the ellipse

$$\frac{x^2}{a^2} + \frac{y^2}{b^2} = 1 \; ; \quad \therefore \; \frac{x_1^2}{a^2} + \frac{y_1^2}{b^2} - 1 > 0.$$

Similarly $\dfrac{x_1^2}{a^2} + \dfrac{y_1^2}{b^2} - 1 < 0$ if (x_1, y_1) is inside the curve.

185. *To find the equation of the ellipse whose focus is at the point (h, k), directrix the straight line* $\mathsf{A}x + \mathsf{B}y + \mathsf{C} = 0$, *and eccentricity e.*

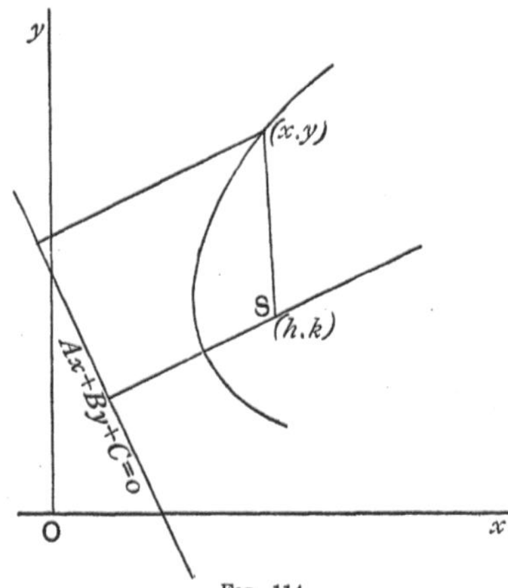

Fig. 114.

Let (x, y) be any point on the curve.

The distance of (x, y) from the focus (h, k)

$$= \sqrt{(x - h)^2 + (y - k)^2}.$$

The distance of (x, y) from the directrix

$$= \frac{\mathsf{A}x + \mathsf{B}y + \mathsf{C}}{\sqrt{\mathsf{A}^2 + \mathsf{B}^2}} \; ;$$

$\therefore$ by definition,

$$\sqrt{(x - h)^2 + (y - k)^2}$$
$$= \frac{e(\mathsf{A}x + \mathsf{B}y + \mathsf{C})}{\sqrt{\mathsf{A}^2 + \mathsf{B}^2}} \; ;$$

$\therefore$ squaring, $(x - h)^2 + (y - k)^2 = \dfrac{e^2(\mathsf{A}x + \mathsf{B}y + \mathsf{C})^2}{\mathsf{A}^2 + \mathsf{B}^2}$

is the required equation.

186. When we have multiplied both sides of the above equation by $\mathsf{A}^2 + \mathsf{B}^2$, and transposed all terms to the left-hand side, the terms of the second degree

$$= x^2[\mathsf{A}^2(1 - e^2) + \mathsf{B}^2] - 2xye^2 \,.\, \mathsf{AB} + y^2[\mathsf{A}^2 + \mathsf{B}^2(1 - e^2)].$$

This expression will have no real factors if

$$4e^4\mathsf{A}^2\mathsf{B}^2 - 4[\mathsf{A}^2(1 - e^2) + \mathsf{B}^2][\mathsf{A}^2 + \mathsf{B}^2(1 - e^2)] \text{ is negative.} \quad (b^2 - 4ac)$$

Let us see if this is the case.

This last expression

$$= 4\left[e^4 \mathsf{A}^2\mathsf{B}^2 - \mathsf{A}^4(1-e^2) - \mathsf{B}^4(1-e^2) - \mathsf{A}^2\mathsf{B}^2\left\{(1-e^2)^2+1\right\}\right]$$
$$= -4\left[\mathsf{A}^4(1-e^2) + \mathsf{B}^4(1-e^2) + \mathsf{A}^2\mathsf{B}^2\left\{(1-e^2)^2+(1-e^4)\right\}\right]$$

which is negative for $e^2 < 1$.

Thus in the equation of any ellipse, the terms of the second degree have no real factors.

187. *If* S, S′ *are the foci of the ellipse* $\dfrac{x^2}{a^2}+\dfrac{y^2}{b^2}=1$, *and* P$(x, y)$ *is any point on the curve*

$$\mathsf{SP} = a - ex, \quad \mathsf{S'P} = a + ex, \quad and \quad \mathsf{SP} + \mathsf{S'P} = 2a,$$

the major axis.

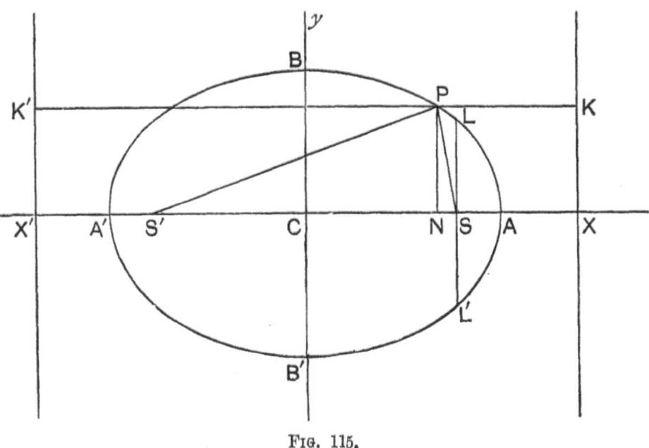

Fɪɢ. 115.

From the figure,

$$\mathsf{SP} = e \,.\, \mathsf{PK} = e \,.\, \mathsf{NX} = e(\mathsf{CX} - \mathsf{CN}) = e\left(\frac{a}{e} - x\right) = a - ex.$$

$$\mathsf{S'P} = e \,.\, \mathsf{PK'} = e \,.\, \mathsf{NX'} = e(\mathsf{CX'} + \mathsf{CN}) = e\left(\frac{a}{e} + x\right) = a + ex.$$

Adding, $\qquad\qquad\qquad \mathsf{SP} + \mathsf{S'P} = 2a.$

Note. If P lies in the second or third quadrant, SP is still equal to $a - ex$, for x is then negative.

188. *The latus rectum* $= \dfrac{2b^2}{a}$.

If LSL' is the latus rectum, the co-ordinates of L are (CS, SL) or (*ae*, SL), and this point is on the ellipse $\dfrac{x^2}{a^2} + \dfrac{y^2}{b^2} = 1$;

$$\therefore \quad \frac{a^2 e^2}{a^2} + \frac{SL^2}{b^2} = 1.$$

$$SL^2 = b^2(1 - e^2) = \frac{b^4}{a^2} ;$$

$$\therefore \quad SL = \frac{b^2}{a} \text{ and } LL' = \frac{2b^2}{a}.$$

Example. *To draw the ellipse* $\dfrac{(x-3)^2}{16} + \dfrac{(y-2)^2}{4} = 1.$

If the origin is changed to the point (3, 2), the given equation becomes

$$\frac{x^2}{16} + \frac{y^2}{4} = 1.$$

Thus we see that the given equation represents an ellipse whose centre is at the point (3, 2) and whose semi-axes are 4 and 2 units in length. Also the axes of the curve are parallel to the axes of co-ordinates.

The student should draw the curve.

Examples X. a.

Find the eccentricity, the distance between the foci, and the latus rectum in each of the following curves :

1. $\dfrac{x^2}{25} + \dfrac{y^2}{16} = 1.$ 2. $3x^2 + 4y^2 = 12.$ 3. $4x^2 + 3y^2 = 24.$ Draw the curve.

Find the equation of the ellipse referred to its principal axes as axes of co-ordinates :

4. Which passes through the points (1, 4), (−6, 1).

5. Whose latus rectum is $\frac{5}{2}$, and eccentricity $\frac{1}{2}$.

6. Whose foci are at the points (3, 0), (−3, 0) and eccentricity $\frac{1}{3}$.

7. Find the lengths of the focal radii of the ellipse $3x^2 + 4y^2 = 48$ drawn to the point whose abscissa is 2.

8. Find the eccentricity of the ellipse whose latus rectum is one-half its minor axis.

9. Find the equation of the ellipse whose eccentricity is e, whose focus is at the point $(c, 0)$, and whose directrix is the axis of y.

10. Find the equation of the ellipse whose focus is at the point (3, 0), whose directrix is the line $x = 5$, and whose eccentricity is $\frac{1}{2}$.

11. Draw the ellipse $\dfrac{(x-2)^2}{4}+y^2=1$.

[Change the origin to the point (2, 0).]

12. Draw the ellipse $\dfrac{(x-3)^2}{4}+\dfrac{(y+1)^2}{9}=1$.

13. Find the co-ordinates of the foci of the ellipse whose equation is
$$4x^2+3y^2=27.$$

14. Find the equation of the ellipse whose focus is at the point (3, 4), whose directrix is the line $x+y=2$, and whose eccentricity is $\frac{1}{3}$.
Draw a rough sketch of the curve.

189. *To find the equation of the tangent to the ellipse* $\dfrac{x^2}{a^2}+\dfrac{y^2}{b^2}=1$ *at the point* (x_1, y_1).

Take a point $(x_1+\Delta x_1,\, y_1+\Delta y_1)$ on the curve, and near to the point (x_1, y_1).

The point (x_1, y_1) is on the curve ; $\therefore\ \dfrac{x_1^{\,2}}{a^2}+\dfrac{y_1^{\,2}}{b^2}=1.$

For the same reason, $\dfrac{(x_1+\Delta x_1)^2}{a^2}+\dfrac{(y_1+\Delta y_1)^2}{b^2}=1\ ;$

$\therefore$ by subtraction, $\dfrac{2x_1\Delta x_1+(\Delta x_1)^2}{a^2}+\dfrac{2y_1\Delta y_1+(\Delta y_1)^2}{b^2}=0\ ;$

$\therefore$ in the limit, when the points approach one another,
$$\frac{x_1\Delta x_1}{a^2}+\frac{y_1\Delta y_1}{b^2}=0\ ;$$

$i.e.\ \dfrac{\Delta y_1}{\Delta x_1}=-\dfrac{b^2x_1}{a^2y_1}\ ;$

$\therefore$ the slope of the tangent $=-\dfrac{b^2x_1}{a^2y_1}.$

Also, the tangent passes through the point (x_1, y_1),

$\therefore$ its equation is $\quad y-y_1=-\dfrac{b^2x_1}{a^2y_1}\,(x-x_1)\ ;$

$$\text{or}\quad \frac{yy_1}{b^2}-\frac{y_1^{\,2}}{b^2}=-\frac{xx_1}{a^2}+\frac{x_1^{\,2}}{a^2},$$

$$\frac{xx_1}{a^2}+\frac{yy_1}{b^2}=\frac{x_1^{\,2}}{a^2}+\frac{y_1^{\,2}}{b^2}=1\ ;$$

$\therefore\ \dfrac{xx_1}{a^2}+\dfrac{yy_1}{b^2}=1$ is the required equation.

B.A.G. M

190. *Second method of finding the equation of a tangent by means of the Differential Calculus.*

$$\frac{x^2}{a^2} + \frac{y^2}{b^2} = 1 \; ;$$

$\therefore$ differentiating with respect to x, $\dfrac{2x}{a^2} + \dfrac{2y}{b^2}\dfrac{dy}{dx} = 0 \; ;$

$$\therefore \frac{dy}{dx} = -\frac{b^2x}{a^2y} \; ;$$

$\therefore$ the slope of the tangent at the point (x_1, y_1) is $-\dfrac{b^2x_1}{a^2y_1}$. Now proceed as in the first method.

191. *To find the equation of the normal to the ellipse* $\dfrac{x^2}{a^2} + \dfrac{y^2}{b^2} = 1$ *at the point* (x_1, y_1).

The equation of the tangent at (x_1, y_1) is $\dfrac{xx_1}{a^2} + \dfrac{yy_1}{b^2} = 1$.

Its slope $= -\dfrac{b^2x_1}{a^2y_1} \; ;$

$\therefore$ the slope of the normal $= \dfrac{a^2y_1}{b^2x_1} \; ; \quad (mm' = -1)$

$$\therefore \; y - y_1 = \frac{a^2y_1}{b^2x_1}(x - x_1) \quad [y - y_1 = m(x - x_1)]$$

is the equation of the normal.

The equation may be written $\dfrac{y - y_1}{\frac{y_1}{b^2}} = \dfrac{x - x_1}{\frac{x_1}{a^2}}$.

192. *To find the equation of a tangent to the ellipse* $\dfrac{x^2}{a^2} + \dfrac{y^2}{b^2} = 1$ *in terms of its slope,* m.

Let $y = mx + c$ be the equation of the tangent, where c is unknown. Where it meets the curve, we have by substitution,

$$\frac{x^2}{a^2} + \frac{(mx + c)^2}{b^2} = 1$$

$$x^2\left(\frac{m^2}{b^2} + \frac{1}{a^2}\right) + \frac{2mcx}{b^2} + \frac{c^2}{b^2} - 1 = 0.$$

But, by hypothesis, $y = mx + c$ is a tangent,

$\therefore$ this quadratic has equal roots ;

$$\therefore\ \frac{m^2c^2}{b^4} = \left(\frac{m^2}{b^2} + \frac{1}{a^2}\right)\left(\frac{c^2}{b^2} - 1\right) \quad (b^2 = 4ac)$$

$$= \frac{m^2c^2}{b^4} + \frac{c^2}{a^2b^2} - \frac{m^2}{b^2} - \frac{1}{a^2},$$

whence
$$\frac{c^2}{a^2b^2} = \frac{m^2}{b^2} + \frac{1}{a^2}$$

$$c^2 = a^2m^2 + b^2$$

$$c = \pm\sqrt{a^2m^2 + b^2}.$$

$\therefore\ \mathbf{y = mx \pm \sqrt{a^2m^2 + b^2}}$ is a tangent to the ellipse $\dfrac{x^2}{a^2} + \dfrac{y^2}{b^2} = 1$ for all values of m.

193. *Def.* The circle described on the major axis, AA', of an ellipse as diameter is called the **auxiliary circle**.

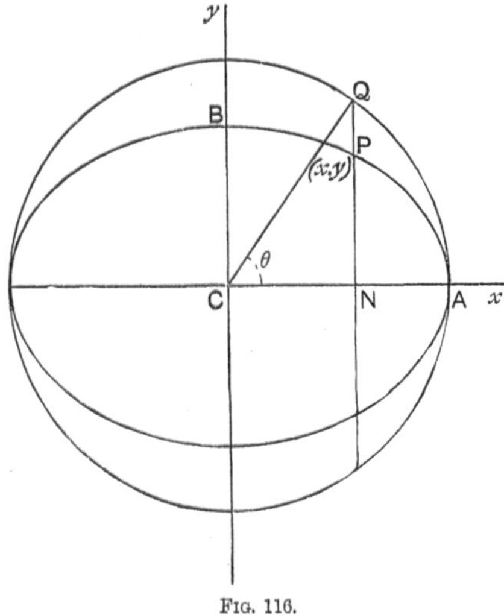

FIG. 116.

Eccentric Angle.

Take Q any point on the auxiliary circle, and draw the ordinate QN, meeting the ellipse at P.

Join CQ, and let $\angle$ QCN $= \theta$.

If (x, y) are the co-ordinates of P,

$$x = \text{CN} = \text{CQ} \cos \theta = a \cos \theta, \text{ for } \text{CQ} = \text{CA}.$$

Also $\dfrac{x^2}{a^2} + \dfrac{y^2}{b^2} = 1$; $\quad \therefore \dfrac{y^2}{b^2} = 1 - \dfrac{x^2}{a^2} = 1 - \cos^2\theta = \sin^2\theta$;

$$\therefore \ y = b \sin \theta.$$

$\therefore$ **the co-ordinates of P, any point on the ellipse** $\dfrac{x^2}{a^2} + \dfrac{y^2}{b^2} = 1$, **may be taken to be (a cos θ, b sin θ), where** $\theta = \angle \text{QCN}$.

The $\angle$ QCN is called the **eccentric angle** of the point P.

194. *To find the equation of the tangent to the ellipse* $\dfrac{x^2}{a^2} + \dfrac{y^2}{b^2} = 1$ *at the point* $(a \cos \theta, \ b \sin \theta)$.

Take the point $[a \cos(\theta + a), \ b \sin(\theta + a)]$ near to the given point.

The equation of the chord joining the two points is

$$\frac{y - b \sin \theta}{b[\sin(\theta + a) - \sin \theta]} = \frac{x - a \cos \theta}{a[\cos(\theta + a) - \cos \theta]}$$

or

$$\frac{y - b \sin \theta}{2b \cos\left(\theta + \frac{a}{2}\right) \sin\frac{a}{2}} = - \frac{x - a \cos \theta}{2a \sin\left(\theta + \frac{a}{2}\right) \sin\frac{a}{2}}$$

Now let the two points move up to one another, so that a becomes equal to zero.

Then the required equation of the tangent is

$$\frac{y - b \sin \theta}{b \cos \theta} = - \frac{x - a \cos \theta}{a \sin \theta} \ ;$$

$$\therefore \frac{y \sin \theta}{b} - \sin^2\theta = - \frac{x \cos \theta}{a} + \cos^2\theta,$$

or

$$\frac{\mathbf{x} \cos \theta}{\mathbf{a}} + \frac{\mathbf{y} \sin \theta}{\mathbf{b}} = 1.$$

Note. This equation might be deduced from $\dfrac{xx_1}{a^2} + \dfrac{yy_1}{b^2} = 1$, by writing $a \cos \theta$ for x_1 and $b \sin \theta$ for y_1.

195. *To find the equation of the normal at the point* $(a \cos \theta, \ b \sin \theta)$ *to the ellipse* $\dfrac{x^2}{a^2} + \dfrac{y^2}{b^2} = 1$.

The equation of the tangent at the point $(a \cos \theta,\ b \sin \theta)$ is

$$\frac{x \cos \theta}{a} + \frac{y \sin \theta}{b} = 1.$$

The normal passes through the point $(a \cos \theta,\ b \sin \theta)$ and is perpendicular to the tangent, therefore its equation is

$$(x - a \cos \theta)\,\frac{a}{\cos \theta} - (y - b \sin \theta)\,\frac{b}{\sin \theta} = 0. \qquad [aa' + bb' = 0.]$$

This may be written $\dfrac{\mathbf{a x}}{\cos \theta} - \dfrac{\mathbf{b y}}{\sin \theta} = \mathbf{a}^2 - \mathbf{b}^2.$

Note. **If a point is on an ellipse, the eccentric angle co-ordinates should generally be used** in preference to x, y co-ordinates, for we shall then have but one variable, θ, to deal with instead of two, x and y.

196. *To find the condition that the straight line* $lx + my + n = 0$ *may touch the ellipse* $\dfrac{x^2}{a^2} + \dfrac{y^2}{b^2} = 1.$

Suppose that it touches the curve at the point $(a \cos \theta,\ b \sin \theta)$. The equation of the tangent at that point is

$$\frac{x \cos \theta}{a} + \frac{y \sin \theta}{b} = 1\ ;$$

$\therefore$ this equation is identical with

$$lx + my = -n,$$

for the two equations represent the same straight line.

$\therefore$ comparing coefficients,

$$\frac{\cos \theta}{al} = \frac{\sin \theta}{bm} = -\frac{1}{n}.$$

$$\cos \theta = -\frac{al}{n}, \quad \sin \theta = -\frac{bm}{n}.$$

But $\cos^2\theta + \sin^2\theta = 1$;

$$\therefore\ \frac{a^2l^2}{n^2} + \frac{b^2m^2}{n^2} = 1,$$

or $\ a^2l^2 + b^2m^2 = n^2$ is the required condition.

This condition might also be obtained by using the second method of Art. 140 on the parabola.

197. *Given that $lx + my + n = 0$ is a tangent to the ellipse $\dfrac{x^2}{a^2} + \dfrac{y^2}{b^2} = 1$, to find the co-ordinates of its point of contact.*

The equation of the tangent at the point (x_1, y_1) is

$$\frac{xx_1}{a^2} + \frac{yy_1}{b^2} = 1 ; \qquad \dots\dots\dots\dots\dots\dots(1)$$

$\therefore$ if (x_1, y_1) are the required co-ordinates, equation (1) and $lx + my = -n$ are identical, for they represent the same straight line. $\therefore$ comparing coefficients,

$$\frac{x_1}{a^2 l} = \frac{y_1}{b^2 m} = -\frac{1}{n} ;$$

$$\therefore x_1 = -\frac{a^2 l}{n}, \ y_1 = -\frac{b^2 m}{n}.$$

198. *The equation of the chord joining the two points (x_1, y_1), (x_2, y_2) on the ellipse $\dfrac{x^2}{a^2} + \dfrac{y^2}{b^2} = 1$ may be written in the form*

$$\frac{(x - x_1)(x_1 + x_2)}{a^2} + \frac{(y - y_1)(y_1 + y_2)}{b^2} = 0.$$

The equation of the chord is $\dfrac{y - y_1}{y_2 - y_1} = \dfrac{x - x_1}{x_2 - x_1}.$ $\qquad\dots\dots\dots\dots\dots(1)$

(x_1, y_1) is on the ellipse, $\qquad \therefore \dfrac{x_1^{\,2}}{a^2} + \dfrac{y_1^{\,2}}{b^2} = 1 ;$

(x_2, y_2) ,, ,, $\qquad \therefore \dfrac{x_2^{\,2}}{a^2} + \dfrac{y_2^{\,2}}{b^2} = 1.$

By subtraction, and transposition,

$$\frac{(y_2 - y_1)(y_2 + y_1)}{b^2} = -\frac{(x_2 - x_1)(x_2 + x_1)}{a^2}. \qquad\dots\dots\dots\dots(2)$$

Multiplying equations (1) and (2)

$$\frac{(y - y_1)(y_1 + y_2)}{b^2} = -\frac{(x - x_1)(x_1 + x_2)}{a^2},$$

$$\text{or} \ \frac{(x - x_1)(x_1 + x_2)}{a^2} + \frac{(y - y_1)(y_1 + y_2)}{b^2} = 0,$$

the required equation.

Note. We might deduce the equation of the tangent at (x_1, y_1), by letting the points (x_1, y_1), (x_2, y_2) move up to one another and coincide.

199. *To find the equation of the chord joining the two points*
$(a \cos \theta,\ b \sin \theta),\ (a \cos \phi,\ b \sin \phi).$

The equation of the chord is

$$\frac{y - b \sin \theta}{b(\sin \phi - \sin \theta)} = \frac{x - a \cos \theta}{a(\cos \phi - \cos \theta)},$$

or

$$\frac{y - b \sin \theta}{2b \cos \dfrac{\phi + \theta}{2} \sin \dfrac{\phi - \theta}{2}} = -\frac{x - a \cos \theta}{2a \sin \dfrac{\phi + \theta}{2} \sin \dfrac{\phi - \theta}{2}}.$$

This may be written

$$\frac{x}{a} \cos \frac{\phi + \theta}{2} + \frac{y}{b} \sin \frac{\phi + \theta}{2} = \cos \theta \,.\, \cos \frac{\phi + \theta}{2} + \sin \theta \,.\, \sin \frac{\phi + \theta}{2},$$

$$\text{or } \frac{x}{a} \cos \frac{\phi + \theta}{2} + \frac{y}{b} \sin \frac{\phi + \theta}{2} = \cos \frac{\phi - \theta}{2}.$$

Note. The equation of the tangent at $(a \cos \theta,\ b \sin \theta)$ may be deduced from this by letting ϕ become equal to θ.

Example. *Find the equations of the tangents drawn to the ellipse* $2x^2 + 7y^2 = 14$ *from the point* (5, 2).

The equation of the ellipse may be written $\dfrac{x^2}{7} + \dfrac{y^2}{2} = 1.$

$\therefore\ y = mx \pm \sqrt{7m^2 + 2}$ is a tangent to the curve for all values of m. If this tangent passes through the point (5, 2),

$$2 = 5m \pm \sqrt{7m^2 + 2},$$

$$\therefore\ 2 - 5m = \pm \sqrt{7m^2 + 2}.$$

Squaring, $4 - 20m + 25m^2 = 7m^2 + 2\ ;$

$$\therefore\ 18m^2 - 20m + 2 = 0,$$

$$9m^2 - 10m + 1 = 0,$$

$$(9m - 1)(m - 1) = 0\ ;$$

$$\therefore\ m = \tfrac{1}{9} \text{ or } 1.$$

The tangents pass through the point (5, 2).

$\therefore\ $ the equations required are $y - 2 = \tfrac{1}{9}(x - 5)$ and $y - 2 = x - 5$;

$$\textit{i.e. } 9y - x = 13 \text{ and } x - y = 3.$$

Note. When the slopes of the tangents are found, it is better to use the formula $y - 2 = m(x - 5)$, rather than $y = mx \pm \sqrt{7m^2 + 2}$ in order to avoid the ambiguity of sign in the square root.

Examples X. b.

1. Find the equations of the tangents to the ellipse $3x^2 + 4y^2 = 12$ which are parallel to the straight line $y + 2x = 4$.

2. Prove that the tangents at the ends of the latera recta of an ellipse pass through the feet of the corresponding directrices.

3. A tangent to the ellipse $4x^2 + 9y^2 = 36$ makes an angle $\tan^{-1}\frac{2}{3}$ with the major axis : find the co-ordinates of its point of contact.

4. A straight line drawn through the point $P(x_1, y_1)$ on the ellipse $\frac{x^2}{a^2} + \frac{y^2}{b^2} = 1$, and through its centre, meets the auxiliary circle at Q and Q'. Prove that the rectangle $PQ . PQ' = e^2(a^2 - x_1^2)$. Show also that this rectangle has the same value when the straight line QPQ' does not pass through the centre.

5. Prove that the equation of the tangent to the ellipse $Ax^2 + By^2 = C$ at the point (x_1, y_1) is $Axx_1 + Byy_1 = C$.

6. Show that the line $y = x + \sqrt{\frac{5}{6}}$ touches the ellipse $2x^2 + 3y^2 = 1$.

7. If PN is the ordinate at the point $P(x_1, y_1)$ on the ellipse $\frac{x^2}{a^2} + \frac{y^2}{b^2} = 1$, and the tangent at P meets the major axis produced at T, find the value of CT, and prove that $CN . CT$ is constant for all positions of P.

8. If PN is the ordinate at the point (x_1, y_1) on the ellipse $\frac{x^2}{a^2} + \frac{y^2}{b^2} = 1$, and the normal at P meets the major axis at G, prove that the ratio $\dfrac{CG}{CN}$ is constant for all positions of P.

9. If from any point $P(x_1, y_1)$ on the ellipse $\frac{x^2}{a^2} + \frac{y^2}{b^2} = 1$, a perpendicular Pn is drawn to the minor axis, and the tangent at P meets the minor axis produced at t, find the value of Ct and prove that $Cn . Ct$ is constant for all positions of P.

10. Find the equations of the tangents drawn from the point $(2, 3)$ to the ellipse $9x^2 + 16y^2 = 144$. Make a figure.

11. Prove that the normal to the ellipse $\frac{x^2}{a^2} + \frac{y^2}{b^2} = 1$ divides the major axis into segments whose product is $a^2 - e^4x_1^2$.

12. Find the equations of the tangents drawn from the point $(12, 25)$ to the ellipse $25x^2 + 144y^2 = 3600$. Make a figure.

13. Find the length of the line $x = y$ intercepted by the ellipse

$$3x^2 + 7y^2 = 500.$$

14. If the line $x\cos a + y\sin a = p$ is a tangent to the ellipse $\frac{x^2}{a^2} + \frac{y^2}{b^2} = 1$, prove that $p = \sqrt{a^2\cos^2 a + b^2\sin^2 a}$.

15. Draw a number of ordinates of the circle $x^2 + y^2 = a^2$. Bisect them, and join these middle points by an even curve. The curve is an ellipse. Prove that its equation is $x^2 + 4y^2 = a^2$.

16. Prove that $x + y = \sqrt{a^2 + b^2}$ is a tangent to the ellipse $\dfrac{x^2}{a^2} + \dfrac{y^2}{b^2} = 1$, and find the co-ordinates of its point of contact.

17. A tangent to the ellipse $\dfrac{x^2}{a^2} + \dfrac{y^2}{b^2} = 1$ makes equal intercepts of positive sign on the axes : find its point of contact.

18. If P and Q are corresponding points on an ellipse and its auxiliary circle, prove that the tangents at P and Q meet on the major axis produced.

19. Tangents are drawn from the point $(h,\ k)$ to the ellipse $\dfrac{x^2}{a^2} + \dfrac{y^2}{b^2} = 1$. Prove that if they make angles θ, θ' with the major axis

$$\tan \theta \tan \theta' = \frac{k^2 - b^2}{h^2 - a^2}.$$

20. Find the equations of the tangents to the ellipse $16x^2 + 25y^2 = 400$ drawn from the point $(5,\ 3)$. Draw a figure.

21. Find the condition that the straight line $\dfrac{x}{m} + \dfrac{y}{n} = 1$ may touch the ellipse $\dfrac{x^2}{a^2} + \dfrac{y^2}{b^2} = 1$.

22. Find the equation of the tangent at the origin to the ellipse

$$ax^2 + by^2 + gx + fy = 0.$$

23. If the normal at the point P on an ellipse cuts the major and minor axes at G and g, prove that

$$a^2 \cdot \mathsf{CG}^2 + b^2 \cdot \mathsf{C}g^2 = (a^2 - b^2)^2.$$

24. Prove that the slopes of the tangents, drawn from the origin to the ellipse $ax^2 + by^2 + 2gx + 2fy + c = 0$, are given by the quadratic equation

$$m^2(f^2 - bc) + 2fgm + g^2 - ac = 0.$$

What do you deduce if $f^2 = bc$?

25. If a tangent to an ellipse meets the major and minor axes at T and t respectively, prove that $\dfrac{a^2}{\mathsf{CT}^2} + \dfrac{b^2}{\mathsf{C}t^2} = 1$.

26. $\dfrac{x^2 - ax}{p^2} + \dfrac{y^2 - by}{q^2} = 0$ represents an ellipse, and $\dfrac{ax - a^2}{p^2} + \dfrac{by - b^2}{q^2} = 0$ represents a straight line. By subtraction, we have

$$\frac{(x - a)^2}{p^2} + \frac{(y - b)^2}{q^2} = 0.$$

What do you deduce ?

27. Find the equations of the tangents drawn from the point $(3, 4)$ to the ellipse $4x^2 + 9y^2 = 36$. Draw a figure.

DRAWING ELLIPSES.

200. *To draw an ellipse having given its focus, directrix, and eccentricity.*

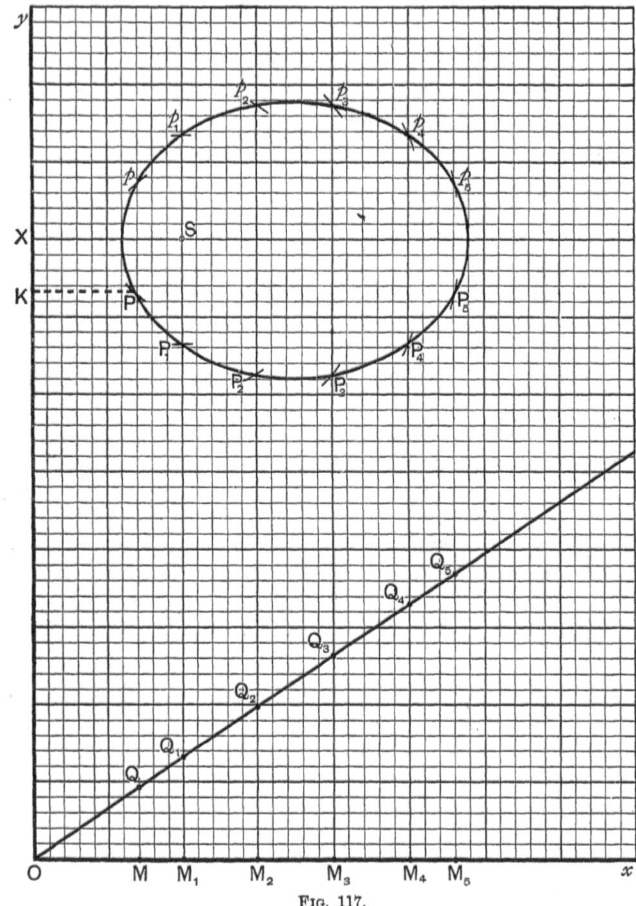

FIG. 117.

Using squared paper, let **S** be the focus, **O**Xy the directrix, $\dfrac{25}{38}$ the eccentricity.

Taking O, the origin, anywhere in the directrix, OXy as axis of y, and Ox at right angles to Oy as axis of x, draw OQ$_5$ the graph of $y = \dfrac{25}{38} \cdot x$.

Take any point Q on this line and its ordinate QM.

With centre S and radius QM, describe a circle cutting MQ produced in P, p.

P and p are points on the ellipse.

For $\dfrac{SP}{PK} = \dfrac{QM}{OM} = \dfrac{25}{38}$, by construction.

In the same way we obtain the points P$_1$, p_1, P$_2$, p_2, P$_3$, p_3. ⋯
Joining them by an even curve we have the ellipse required.

201. *To trace an ellipse by means of the property* SP + S′P = 2a.

Take a thread of length 2a, and attach its ends to two points S, S′ by means of drawing pins.

Placing a pencil-point in the angle SPS′, move it, keeping the thread taut. The pencil-point traces out an ellipse, whose foci are S and S′, and whose major axis = 2a.

Or we may use the method of the following Example :

Describe an ellipse whose foci are 10 cms. apart and whose major axis is 15 cms. in length.

Take SS′ = 10 cms., and C its middle point.

C is the centre of the ellipse.

Produce SS′ to A and A′ making CA = CA′ = 7·5 cms.

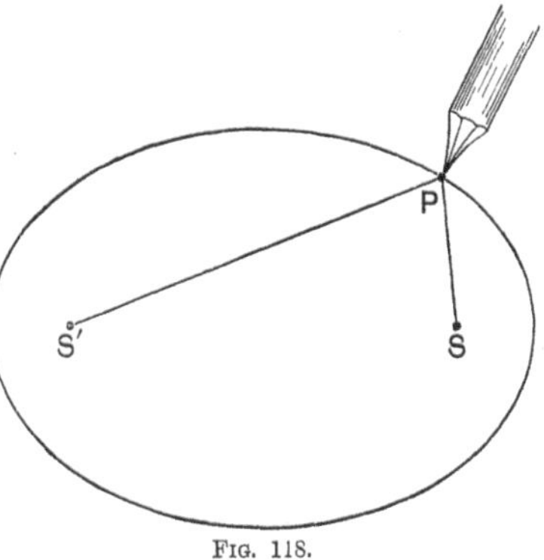

Fig. 118.

A and A′ are the ends of the major axis.

With centres S, S′ and radii 7·5 cms. describe arcs cutting at B and B′.

BS = BS′ = 7·5 cms. ∴ B and B′ are the ends of the minor axis.

With centres S, S′ and radii 5 and 10 cms. describe arcs cutting at P and Q.

$$SP + S'P = 5 + 10 = 15 \text{ cms.}$$

∴ P is a point on the ellipse.

Similarly Q is a point on the ellipse.

In this way, by describing arcs the sum of whose radii is always 15 cms., we can obtain any number of points on the curve.   Joining them by an even curve, we have the ellipse.

202. *Given the major and minor axes of an ellipse, to draw the curve.*

Draw the axes AA′, BB′ bisecting one another at right angles.

On AA′, and BB′ as diameters describe circles, and draw a number of radii of these circles,

CQP, CQ$_1$P$_1$, CQ$_2$P$_2$, etc., as shown in the figure.

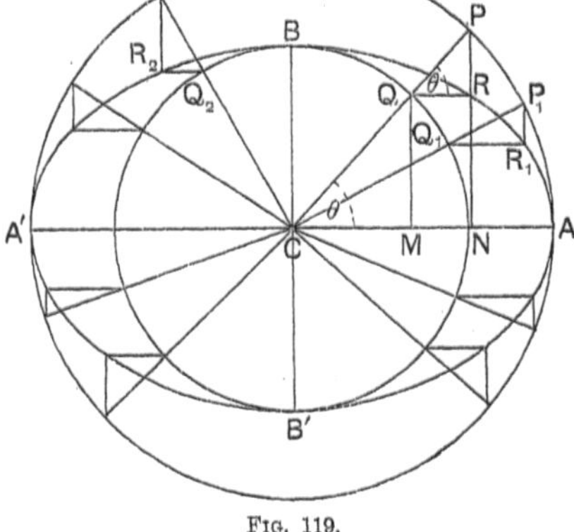

Through Q, Q$_1$, Q$_2$, etc., where these radii cut the smaller circle, draw straight lines parallel to the major axis.

Through P, P$_1$, P$_2$, etc., where the radii cut the larger circle, draw straight lines parallel to the minor axis, and meeting the lines parallel to the major axis at R, R$_1$, R$_2$, etc.

Fɪɢ. 119.

The curve joining the points R$_1$, R$_1$, R$_2$, etc., is the ellipse required.

Proof.   Let ∠ PCA = θ = ∠ PQR.

Let PR produced meet the major axis at N, and draw QM perpendicular to the major axis.

$$CN = CP \cos \theta = a \cos \theta \text{ if } AA' = 2a.$$

$$RN = QM = CQ \sin \theta = b \sin \theta \text{ if } BB' = 2b.$$

∴ R is a point on the ellipse whose axes are AA′ and BB′.

Similarly the points R$_1$, R$_2$, etc. lie on the curve.

203. When there is no term involving the product xy in the equation of an ellipse, the curve can easily be drawn.

To draw the ellipse $9x^2 + 4y^2 - 36x + 8y + 4 = 0.$

The equation may be written $9(x^2 - 4x) + 4(y^2 + 2y) = -4.$

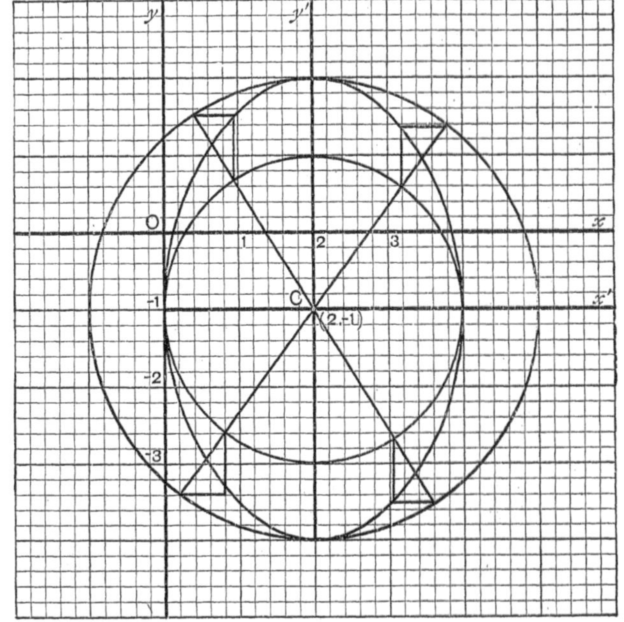

Fig. 120.

Completing the squares inside the brackets, the equation becomes

$$9(x-2)^2 + 4(y+1)^2 = 36 + 4 - 4 = 36,$$

$$\text{or } \frac{(x-2)^2}{4} + \frac{(y+1)^2}{9} = 1.$$

If the origin is transferred to the point $(2, -1)$, this becomes

$$\frac{x^2}{4} + \frac{y^2}{9} = 1.$$

$\therefore$ the given equation represents an ellipse whose centre is at the point $(2, -1)$, and whose semi-axes are 3 and 2 in length. The major axis is parallel to the axis of y, and the minor axis is parallel to the axis of x. It will be seen that the curve touches the axis of y.

Examples X. c.

[Squared paper should be used in most of the following.]

If S is a focus of an ellipse, SX the perpendicular on the corresponding directrix, e the eccentricity, draw the curve, when :

1. SX $=2$ inches, $e=\frac{2}{3}$.
        2. SX $=2$ inches, $e=\frac{3}{5}$.

3. SX $=1$ inch, $e=\frac{7}{10}$.

Draw the following curves :

4. $\dfrac{(x-2)^2}{4}+y^2=1.$
        5. $9x^2+4y^2+18x-8y=23.$

6. $4x^2+25y^2-100y=0.$
        7. $x^2+4y^2+4x=0.$

8. $4x^2+y^2+8x-4y+4=0.$

9. Given that the distance between the foci of an ellipse is $2\frac{1}{2}$ inches, and that its major axis is 3 inches long, find a number of points on the curve and draw it.

10. Find a number of points on, and draw, an ellipse whose axes are 4 and 2 inches long respectively.

S, S′ being the foci and AA′ the major axis of an ellipse, draw the curve in each of the following cases :

11. SS′ $=8$ cms., AA′ $=13$ cms.
        12. SS′ $=4$ in., AA′ $=6$ in.

13. SS′ $=6$ cms., AA′ $=9$ cms.
        14. SS′ $=8$ cms., AA′ $=16$ cms.

204. *Tangents are drawn to the ellipse* $\dfrac{x^2}{a^2}+\dfrac{y^2}{b^2}=1$ *from the point* $(x_1,\ y_1)$: *to find the equation of their chord of contact.*

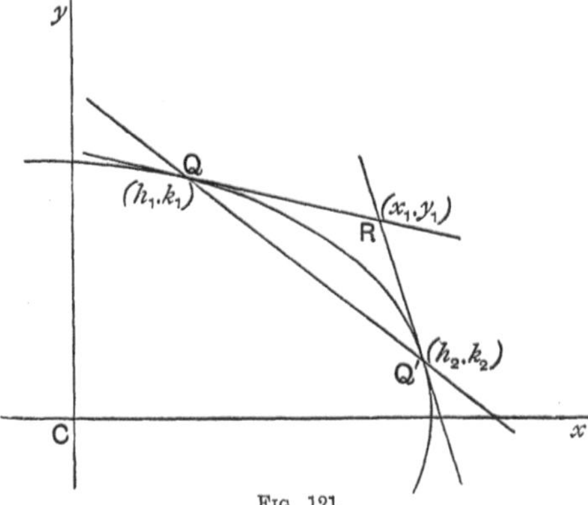

Fig. 121.

Let R be the point (x_1, y_1); RQ, RQ′ the tangents.

Let (h_1, k_1) be the co-ordinates of Q, (h_2, k_2) the co-ordinates of Q′.

The equation of RQ, the tangent at Q, is

$$\frac{xh_1}{a^2}+\frac{yk_1}{b^2}=1.$$

The equation of RQ′, the tangent at Q′, is $\dfrac{xh_2}{a^2}+\dfrac{yk_2}{b^2}=1.$

But the point R (x_1, y_1) is on both these lines,

$$\therefore \quad \frac{x_1 h_1}{a^2} + \frac{y_1 k_1}{b^2} = 1, \dots\dots\dots\dots\dots\dots\dots(1)$$

and

$$\frac{x_1 h_2}{a^2} + \frac{y_1 k_2}{b^2} = 1, \dots\dots\dots\dots\dots\dots(2)$$

$$\therefore \ \frac{x_1 x}{a^2} + \frac{y_1 y}{b^2} = 1, \ \text{ or } \ \frac{x x_1}{a^2} + \frac{y y_1}{b^2} = 1 \text{ is the equation reqd.}$$

For firstly, it represents a straight line.

Also from (1) we see that $Q(h_1, k_1)$ lies on this line,

and   ,,   (2)   ,, $Q'(h_2, k_2)$,,

$\qquad \therefore$ it is the equation of QQ', the chord of contact.

205. *To find the polar of the point (x_1, y_1) with respect to the ellipse*
$\dfrac{x^2}{a^2} + \dfrac{y^2}{b^2} = 1.$

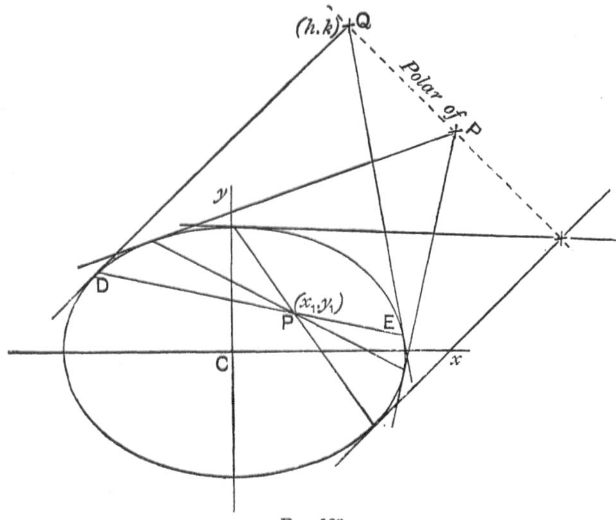

Fig. 122.

Let P be the point (x_1, y_1) and DPE any chord through P; DQ,
EQ the tangents at D and E.

It is required to find the locus of Q.

Let (h, k) be the co-ordinates of Q.

Then the equation of its chord of contact DE is

$$\frac{xh}{a^2} + \frac{yk}{b^2} = 1.$$

But (x_1, y_1) is on this line,

$$\therefore \ \frac{x_1 h}{a^2} + \frac{y_1 k}{b^2} = 1.$$

Also (h, k) is *any point* on the locus,

$\therefore$ the equation of the locus is

$$\frac{x_1 x}{a^2} + \frac{y_1 y}{b^2} = 1, \ \text{ or } \ \frac{\mathbf{x x_1}}{\mathbf{a^2}} + \frac{\mathbf{y y_1}}{\mathbf{b^2}} = 1,$$

a straight line.

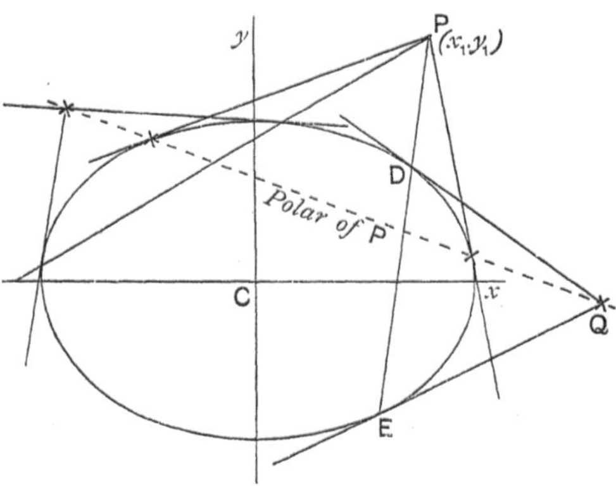

Fig. 123.

When (x_1, y_1) is outside the ellipse, the polar is the same as the chord of contact of tangents drawn from (x_1, y_1).

If the polar of the point P passes through the point Q, the polar of the point Q passes through the point P.

This may be proved as in Art. 144.

206. *Find the co-ordinates of the pole of the straight line*
$$lx + my + n = 0$$
with respect to the ellipse $\dfrac{x^2}{a^2} + \dfrac{y^2}{b^2} = 1.$

If (x_1, y_1) are the co-ordinates of the pole, the equations
$$\frac{xx_1}{a^2} + \frac{yy_1}{b^2} = 1 \quad \text{and} \quad lx + my = -n$$
are identical, for they represent the same straight line.

Now proceed as in Art. 197.

207. *To find the equation of a chord of the ellipse* $\dfrac{x^2}{a^2} + \dfrac{y^2}{b^2} = 1,$ *in terms of the co-ordinates of its middle point* $(x_1, y_1).$

We may take as the equation of the chord PQ, whose middle point is O (x_1, y_1),

$$\frac{x - x_1}{\cos \theta} = \frac{y - y_1}{\sin \theta} = r. \quad \ldots\ldots\ldots(1)$$

$$x = x_1 + r \cos \theta, \quad y = y_1 + r \sin \theta,$$

$\therefore$ where the chord meets the ellipse, we have by substitution,

$$\frac{(x_1 + r \cos \theta)^2}{a^2} + \frac{(y_1 + r \sin \theta)^2}{b^2} = 1,$$

or

$$r^2 \left(\frac{\cos^2 \theta}{a^2} + \frac{\sin^2 \theta}{b^2} \right)$$
$$+ 2r \left(\frac{x_1 \cos \theta}{a^2} + \frac{y_1 \sin \theta}{b^2} \right)$$
$$+ \frac{x_1^2}{a^2} + \frac{y_1^2}{b^2} - 1 = 0.$$

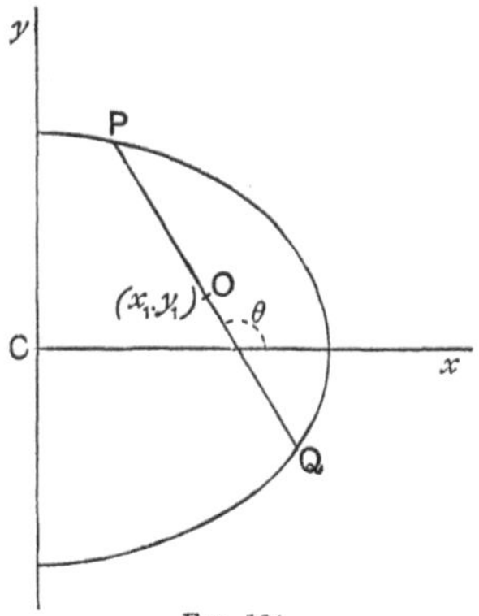

FIG. 124.

Now the roots of this equation OP, OQ are equal but of opposite sign, $\therefore$ the coefficient of $r = 0$,

$$\therefore \quad \frac{x_1 \cos \theta}{a^2} = -\frac{y_1 \sin \theta}{b^2}. \quad \ldots\ldots\ldots\ldots\ldots\ldots\ldots(2)$$

Multiplying (1) and (2), $\quad \dfrac{(x - x_1)x_1}{a^2} = -\dfrac{(y - y_1)y_1}{b^2}$

is the required equation.

It may be written $\quad \dfrac{xx_1}{a^2} + \dfrac{yy_1}{b^2} = \dfrac{x_1^2}{a^2} + \dfrac{y_1^2}{b^2}.$

B.A.G.   N

Examples X. d.

1. Tangents are drawn to an ellipse from a point on the directrix : prove that their chord of contact passes through the corresponding focus.

2. If R is the point on the directrix in the above example, prove that RS is perpendicular to the chord of contact.

Find with respect to the ellipse $\dfrac{x^2}{a^2}+\dfrac{y^2}{b^2}=1$ the pole of the line :

3. $y=mx+c.$ **4.** $xx_1+yy_1=ab.$ **5.** $x=c.$

6. Find the poles of the directrices of an ellipse.

7. Find the equation of the chord of the ellipse $3x^2+7y^2=84$ which is bisected at the point $(1, -2)$.

8. Tangents are drawn to an ellipse from a point Q on the auxiliary circle ; prove that their chord of contact and the tangent to the ellipse at the point corresponding to Q meet on the major axis.

9. PQN is drawn perpendicular to the major axis of an ellipse, and meets the ellipse at Q. Prove that the polar of P and the tangent at Q meet on the major axis.

10. $y=m(x-ae)$ is a straight line through a focus of the ellipse $\dfrac{x^2}{a^2}+\dfrac{y^2}{b^2}=1$. Find the co-ordinates of its pole with respect to the ellipse.

Deduce a geometrical property of the ellipse.

11. Tangents OP, OQ are drawn to the ellipse $\dfrac{x^2}{a^2}+\dfrac{y^2}{b^2}=1$, centre C ; prove that the product of the slopes of PQ and OC is $-\dfrac{b^2}{a^2}$.

12. Find the condition that the pole of $lx+my=1$ with respect to the ellipse $\dfrac{x^2}{a^2}+\dfrac{y^2}{b^2}=1$ may lie on the ellipse $\dfrac{x^2}{4a^2}+\dfrac{y^2}{4b^2}=1$.

13. CP and CQ are at right angles, P and Q lying on an ellipse whose centre is C : prove that $\dfrac{1}{CP^2}+\dfrac{1}{CQ^2}=\dfrac{1}{a^2}+\dfrac{1}{b^2}$, where a and b are the semi-axes of the ellipse.

[With the usual axes, let $\angle PCx=\alpha$, so that $(CP\cos\alpha,\ CP\sin\alpha)$ are the co-ordinates of P.]

14. If the chord of contact of tangents drawn to the ellipse $\dfrac{x^2}{a^2}+\dfrac{y^2}{b^2}=1$ from the point (x_1, y_1) subtends a right angle at the centre, prove that

$$\frac{x_1^{2}}{a^4}+\frac{y_1^{2}}{b^4}=\frac{1}{a^2}+\frac{1}{b^2}.$$

15. If O is the point (x_1, y_1) and the straight line $\dfrac{x-x_1}{\cos\theta}=\dfrac{y-y_1}{\sin\theta}=r$ meets the ellipse $\dfrac{x^2}{a^2}+\dfrac{y^2}{b^2}=1$ at the points P and Q, prove that the rectangle $OP\cdot OQ=\dfrac{b^2x_1^{2}+a^2y_1^{2}-a^2b^2}{b^2\cos^2\theta+a^2\sin^2\theta}.$

208. *To find the locus of the middle points of a series of parallel chords of an ellipse.*

Let $O(x_1, y_1)$ be the mid-point of any chord QQ' of the series.

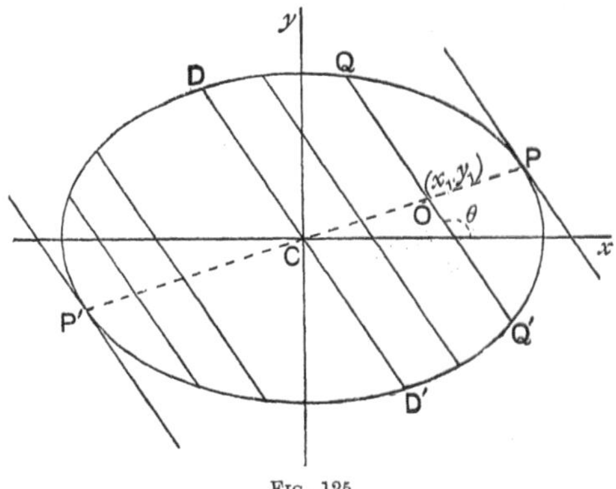

Fig. 125.

We may take for the equation of QQ'

$$\frac{x - x_1}{\cos \theta} = \frac{y - y_1}{\sin \theta} = r,$$

$$x = x_1 + r \cos \theta, \quad y = y_1 + r \sin \theta.$$

$\therefore$ where the chord meets the ellipse $\dfrac{x^2}{a^2} + \dfrac{y^2}{b^2} = 1$ we have, by substitution,

$$\frac{(x_1 + r \cos \theta)^2}{a^2} + \frac{(y_1 + r \sin \theta)^2}{b^2} = 1,$$

or $\quad r^2 \left(\dfrac{\cos^2\theta}{a^2} + \dfrac{\sin^2\theta}{b^2} \right) + 2r \left(\dfrac{x_1 \cos \theta}{a^2} + \dfrac{y_1 \sin \theta}{b^2} \right) + \dfrac{x_1^2}{a^2} + \dfrac{y_1^2}{b^2} - 1 = 0.$

Now O is the mid-point of QQ'; $\therefore$ the roots OQ, OQ', of this equation are equal but of opposite sign;

$$\therefore \frac{x_1 \cos \theta}{a^2} + \frac{y_1 \sin \theta}{b^2} = 0.$$

But (x_1, y_1) is *any point* on the locus; $\therefore$ suppressing suffixes,

$$\frac{x \cos \theta}{a^2} + \frac{y \sin \theta}{b^2} = 0 \quad \dotfill (1)$$

is the equation of the locus, for θ is a constant angle,

This is a straight line through the centre, a **diameter**. (See Def. Art. 183.)

If m is the slope of QQ′, $\tan \theta = m$, and the equation of CO may be written $y = -\dfrac{b^2}{a^2 m}\,x.$

$\therefore$ if $y = mx$ bisects all chords parallel to $y = m'x$,

$$m' = -\frac{b^2}{a^2 m} \text{ or } \mathbf{mm'} = -\frac{\mathbf{b^2}}{\mathbf{a^2}}.$$

Vice versâ, if $y = m'x$ bisects all chords parallel to $y = mx$, we have in the same way, $mm' = -\dfrac{b^2}{a^2}.$

$\therefore$ if the diameter PCP′ bisects chords parallel to the diameter DCD′,

then the diameter DCD′ bisects chords parallel to the diameter PCP′.

Def. **Conjugate** diameters are such that each bisects chords parallel to the other.

209. *The tangents at the extremities of a diameter are parallel to the chords bisected by that diameter.*

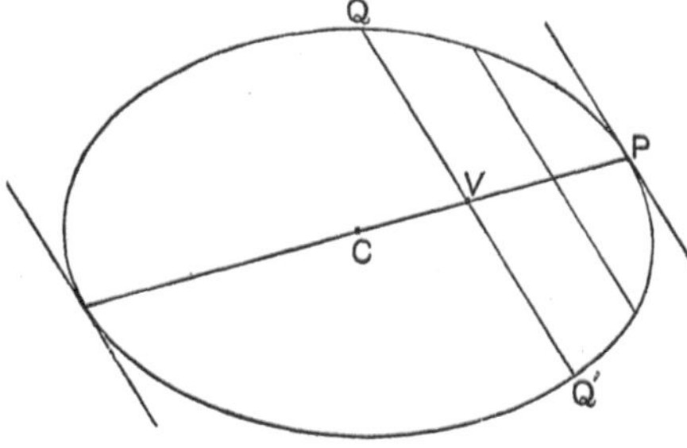

Fig. 126.

Let the diameter CP bisect the chord QQ′ at V, and let the chord move parallel to itself, the point V approaching P. The diameter always bisects the chord. Therefore when V coincides with P, the equal portions QV, Q′V *vanish together*, and the chord becomes the tangent at P. This proves the proposition.

210. Let PCP′, DCD′ be two conjugate diameters, and let θ and ϕ be the eccentric angles of the points P and D.

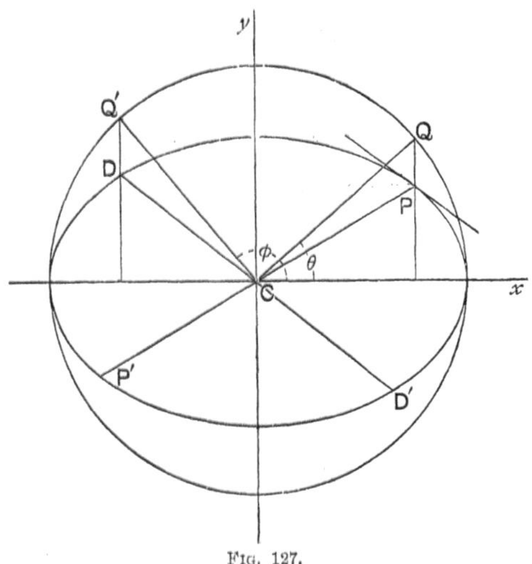

FIG. 127.

The equation of CP is $\dfrac{y}{b\sin\theta}=\dfrac{x}{a\cos\theta}$.

   ,,   ,,   CD ,, $\dfrac{y}{b\sin\phi}=\dfrac{x}{a\cos\phi}$.

But CP, CD are conjugate ;

$$\therefore\ \frac{b\sin\theta}{a\cos\theta}\times\frac{b\sin\phi}{a\cos\phi}=-\frac{b^2}{a^2}\qquad\left(mm'=-\frac{b^2}{a^2}\right),$$

i.e. $\cos\theta\cos\phi+\sin\theta\sin\phi=0$, or $\cos(\phi-\theta)=0$.

$\therefore$ if P be taken in the first quadrant, and D in the second,

$$\phi-\theta=\frac{\pi}{2}.$$

Hence the co-ordinates of D are

$$\left[a\cos\left(\theta+\frac{\pi}{2}\right),\ b\sin\left(\theta+\frac{\pi}{2}\right)\right]\ \text{or}\ (-a\sin\theta,\ b\cos\theta).$$

N.B. If Q and Q′ are the points on the auxiliary circle corresponding to P and D respectively, $\angle\,QCQ'=\phi-\theta=$ a right angle.

211. The equi-conjugate diameters.

The equations of the diameters of the ellipse $\dfrac{x^2}{a^2}+\dfrac{y^2}{b^2}=1$ which are conjugate and equal are $\dfrac{y}{b}=\pm\dfrac{x}{a}$.

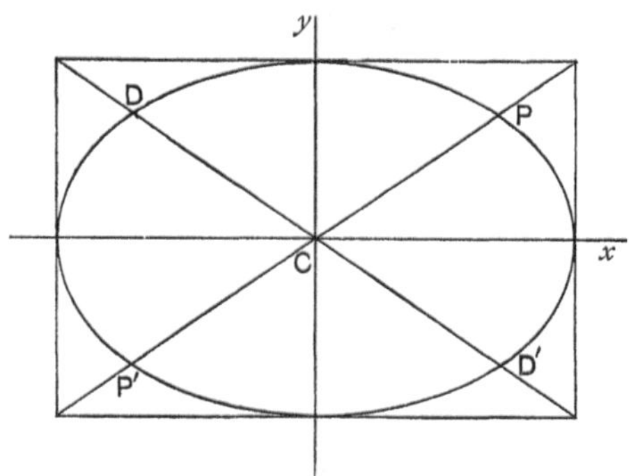

Fig. 128.

Let PCP', DCD' be the diameters.

Then if $(a\cos\theta,\ b\sin\theta)$ are the co-ordinates of P,

$$\left[a\cos\left(\theta+\frac{\pi}{2}\right),\ b\sin\left(\theta+\frac{\pi}{2}\right)\right]\quad\text{or}\quad(-a\sin\theta,\ b\cos\theta)$$

are the co-ordinates of D ;

$$\therefore\ \ \text{CP}^2=a^2\cos^2\theta+b^2\sin^2\theta\quad\text{and}\quad \text{CD}^2=a^2\sin^2\theta+b^2\cos^2\theta.$$

$\therefore$ by hypothesis,

$$a^2\cos^2\theta+b^2\sin^2\theta=a^2\sin^2\theta+b^2\cos^2\theta,$$

$$\text{and}\quad a^2\cos 2\theta=b^2\cos 2\theta,$$

$$\therefore\ \cos 2\theta=0,\ 2\theta=\frac{\pi}{2},\ \text{and}\ \theta=\frac{\pi}{4}\ ;$$

$\therefore$ the equation of CP is

$$\cfrac{y}{b\sin\dfrac{\pi}{4}}=\cfrac{x}{a\cos\dfrac{\pi}{4}}\qquad\left(\frac{y}{y_1}=\frac{x}{x_1}\right)$$

$$\text{or}\quad \frac{y}{b}=\frac{x}{a},$$

and the equation of CD is

$$\frac{y}{b\cos\dfrac{\pi}{4}} = \frac{x}{-a\sin\dfrac{\pi}{4}}$$

or $\dfrac{y}{b} = -\dfrac{x}{a}.$

These diameters are the diagonals of the rectangle formed by drawing tangents at the extremities of the principal axes.

212. Supplemental chords.

Def. If QCQ′ is a diameter, and P any point on an ellipse, the chords PQ, PQ′ are called **supplemental chords**.

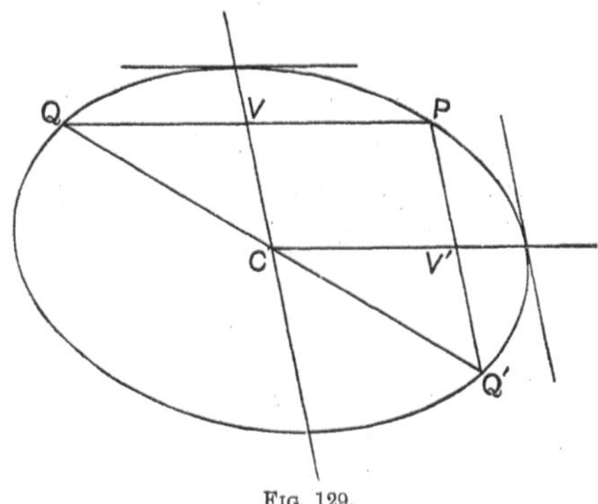

Fig. 129.

Supplemental chords of an ellipse are parallel to a pair of conjugate diameters.

Take V, V′ the mid-points of the supplemental chords PQ, PQ′. Join CV, CV′.

QC = CQ′ and QV = VP ; ∴ CV is parallel to PQ′.

Similarly CV′ ,, PQ.

∴ CV is a diameter bisecting chords parallel to PQ or CV′,

and CV′ ,, ,, ,, PQ′ or CV.

∴ CV, CV′ are conjugate diameters, and are parallel to the supplemental chords PQ′, PQ.

213. *If an ellipse has its centre at the origin, its equation contains no terms of the first degree.*

If possible let $ax^2 + 2hxy + by^2 + 2gx + 2fy + c = 0$ be the equation of the ellipse.

Let (x, y) be *any point* on the curve. Then since the origin is at the centre, $(-x, -y)$ is also on the curve.

$$\therefore \quad ax^2 + 2hxy + by^2 + 2gx + 2fy + c = 0$$
$$\text{and} \quad ax^2 + 2hxy + by^2 - 2gx - 2fy + c = 0 ;$$

$\therefore$ by subtraction, $\qquad 4gx + 4fy = 0,$

$$\text{or} \quad gx + fy = 0.$$

But (x, y) is *any point* on the curve,

$\therefore$ $gx + fy = 0$ for *every point* on the curve.

$\therefore$ we must have $f = 0$ and $g = 0$.

Hence the equation of an ellipse referred to any axes through the centre is of the form

$$ax^2 + 2hxy + by^2 + c = 0.$$

A more useful form is $ax^2 + 2hxy + by^2 = 1$.

***214.** *To find the equation of an ellipse referred to two conjugate diameters as axes of co-ordinates.* (Oblique.)

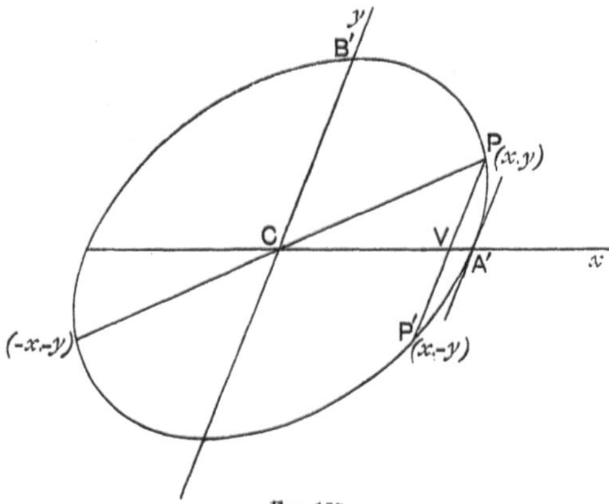

Fɪɢ. 180.

The origin being at the centre, if (x, y) lies on the curve, $(-x, -y)$ also lies on the curve ;

$\therefore$ its equation can contain no terms of the first degree.

We may therefore take

$$Ax^2 + 2Hxy + By^2 = 1 \quad\dots\dots\dots\dots\dots\dots\dots\dots (1)$$

to be the equation of the ellipse.

Again, if P is the point $(x,\ y)$ and the chord PVP′ is parallel to CB′ the axis of y, P′V = PV, or the co-ordinates of P′, a point on the curve, are $(x,\ -y)$;

$\therefore$ $Ax^2 - 2Hxy + By^2 = 1$ by substitution ;

$\therefore$ from (1) by subtraction, $H = 0$;

$\therefore$ the equation of the curve reduces to

$$Ax^2 + By^2 = 1.$$

Again, let a', b' be the lengths of the given conjugate semi-diameters.

When $y = 0,\ \ x = a'$; $\therefore$ $Aa'^2 = 1$ and $A = \dfrac{1}{a'^2}$.

,, $x = 0,\ \ y = b'$; $\therefore$ $Bb'^2 = 1$ and $B = \dfrac{1}{b'^2}$.

$\therefore$ $\dfrac{x^2}{a'^2} + \dfrac{y^2}{b'^2} = 1$ is the required equation.

***215.** We see that the equation of an ellipse referred to a pair of conjugate diameters is of the *same form* as when referred to its principal axes.

$$\frac{xx_1}{a'^2} + \frac{yy_1}{b'^2} = 1 \text{ is a tangent to the ellipse } \frac{x^2}{a'^2} + \frac{y^2}{b'^2} = 1.$$

The proof in Art. 189 holds, remembering that the limiting value of $\dfrac{\Delta y_1}{\Delta x_1}$ is not the slope of the tangent, the axes being oblique.

$$\frac{xx_1}{a'^2} + \frac{yy_1}{b'^2} = 1$$

is the chord of contact of tangents drawn from $(x_1,\ y_1)$.

The proof of Art. 204 holds.

$$\frac{xx_1}{a'^2} + \frac{yy_1}{b'^2} = 1 \text{ is the polar of } (x_1,\ y_1).$$

The proof of Art. 205 holds.

$$y = mx \pm \sqrt{a^2m^2 + b^2} \text{ touches the curve for all values of } m.$$

The proof of Art. 192 holds.

Also $(a'\cos\theta,\ b'\sin\theta)$ is a point on the curve for all values of θ.

It must be remembered here that θ is no longer the eccentric angle.

Examples X. e.

1. Chords of the ellipse $3x^2+5y^2=45$ are drawn parallel to the line $y=2x$. Find the locus of their middle points.

2. Prove that the polar of a point P with respect to an ellipse is parallel to the diameter which is conjugate to the diameter through P.

3. Find the equation of the chord of the ellipse $2x^2+5y^2=20$ which is bisected at the point $(2,\ 1)$.

4. Find the equations of the diameters of the ellipse $\dfrac{x^2}{a^2}+\dfrac{y^2}{b^2}=1$ which are conjugate to ;

$$\text{(i)}\ x=y, \qquad \text{(ii)}\ lx+my=0, \qquad \text{(iii)}\ ay=bx.$$

5. In the ellipse $4x^2+5y^2=100$, find the equation of the diameter conjugate to $y=2x$.

6. Find the equations of the tangents to the ellipse $3x^2+4y^2=48$ at the extremities of the diameter $y=3x$.

7. Find the equation of the diameter of the ellipse $3x^2+4y^2=48$ which bisects the chord $4x+y=8$.

8. Find the angle between the equi-conjugate diameters of the ellipse $x^2+4y^2=100$.

9. Prove that the tangents to the ellipse $\dfrac{x^2}{a^2}+\dfrac{y^2}{b^2}=1$ at the points whose eccentric angles are θ, and $\dfrac{\pi}{2}-\theta$ meet on one of the equi-conjugate diameters.

10. If θ is the eccentric angle of a point P on an ellipse, and SP is bisected at Q, prove that $\mathsf{CQ}=\dfrac{a}{2}(1-e\cos\theta)$.

11. If with centre, C, of an ellipse a circle is described cutting the curve at P, Q, P', Q', prove that PQ, $\mathsf{P}'\mathsf{Q}$ are parallel to the axes of the ellipse.

12. Given that $(x_1,\ y_1)$ is the extremity of a semi-diameter of the ellipse $ax^2+by^2+c=0$, find the co-ordinates of the extremities of the conjugate diameter.

13. Find the equation of the tangent at the point $(x_1,\ y_1)$ to the ellipse $ax^2+2hxy+by^2=1$.

14. In the ellipse $ax^2+2hxy+by^2=1$, prove that the equation of a chord in terms of the co-ordinates of its middle point $(x_1,\ y_1)$ is

$$(x-x_1)(ax_1+hy_1)+(y-y_1)(hx_1+by_1)=0.$$

15. If the two diameters $y=mx$, $y=m'x$ of the ellipse $ax^2+2hxy+by^2=1$ are conjugate, prove that $a+h(m+m')+bmm'=0$.

16. Find the general equation of the system of parallel chords of the ellipse $3x^2 + 5y^2 = 27$ which are bisected by the diameter $y = x$.

17. P being any point on an ellipse, S a focus, Q the middle point of SP, prove that CQ + SQ is constant.

What is the locus of Q?

18. On the tangent at P to an ellipse equal lengths PR, PR′ are taken. CR, CR′ cut the curve at Q and Q′, C being the centre. Prove that QQ′ is parallel to the tangent at P and is bisected by CP.

19. If CP, CD are conjugate semi-diameters of the ellipse $\dfrac{x^2}{a^2} + \dfrac{y^2}{b^2} = 1$, and $(a\cos\theta,\ b\sin\theta)$ are the co-ordinates of P, prove that the tangents at P and D meet at the point $\left[a\cos\left(\theta + \dfrac{\pi}{4}\right),\ b\cos\left(\theta - \dfrac{\pi}{4}\right) \right]$.

20. In the ellipse $3x^2 + 7y^2 = 21$, find the equations of the equi-conjugate diameters, and their lengths.

21. In the ellipse $\dfrac{x^2}{a^2} + \dfrac{y^2}{b^2} = 1$, CP, CD are conjugate semi-diameters. If $(x_1,\ y_1)$ are the co-ordinates of P, find those of D.

22. If CP, CD are two conjugate semi-diameters of an ellipse, p, d the corresponding points on the auxiliary circle to P and D, and pPN, dDM the ordinates at P and D, prove that pN = CM, and dM = CN.

***23.** What does the equation $x^2 + y^2 = a^2$ represent when the axes are oblique?

***24.** If a be the acute angle between the axes of co-ordinates, the semi-axes of the ellipse $x^2 + y^2 = c^2$ are $\sqrt{2}\,c\cos\dfrac{a}{2}$ and $\sqrt{2}\,c\sin\dfrac{a}{2}$.

***25.** If e be the eccentricity of the ellipse in the previous question,

$$e = \left(\frac{2\cos a}{1 + \cos a}\right)^{\frac{1}{2}}.$$

LOCUS PROBLEMS ON THE ELLIPSE.

216. *To find the locus of the intersection of perpendicular tangents to an ellipse.*

The straight line $y = mx + \sqrt{a^2m^2 + b^2}$ is a tangent to the ellipse $\dfrac{x^2}{a^2} + \dfrac{y^2}{b^2} = 1$ for all values of m. If this passes through $(x_1,\ y_1)$,

$$y_1 = mx_1 + \sqrt{a^2m^2 + b^2},$$
$$(y_1 - mx_1)^2 = a^2m^2 + b^2,$$
$$m^2(a^2 - x_1{}^2) + 2mx_1y_1 + b^2 - y_1{}^2 = 0.$$

Hence this quadratic for m gives the slopes of the tangents which can be drawn from the point (x_1, y_1) to the ellipse.

In this case the tangents are at rt. $\angle^s$.

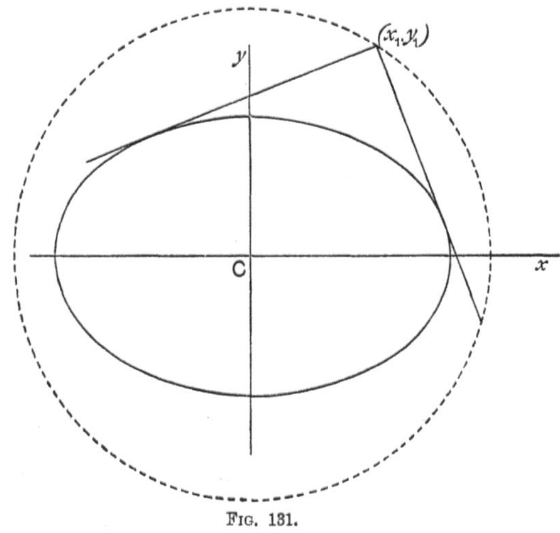

Fig. 131.

$\therefore$ if $m_1 m_2$ are the roots of this quadratic,

$$m_1 m_2 = -1 ;$$

$$\therefore \frac{b^2 - y_1^2}{a^2 - x_1^2} = -1 \quad \text{or} \quad x_1^2 + y_1^2 = a^2 + b^2.$$

$\therefore x^2 + y^2 = a^2 + b^2$ is the equation of the locus.

A circle, whose centre is at the centre of the ellipse, and whose radius is $\sqrt{a^2 + b^2}$. This is called the *Director Circle*.

217. *Chords of an ellipse are drawn from one end of the major axis ; find the locus of their middle points.*

Let $(a \cos \theta, \ b \sin \theta)$ be any point P on the ellipse, A$(a, 0)$ the end of the major axis from which the chords are drawn.

Let (x, y) be the co-ordinates of the middle point of PA.

$$x = \tfrac{1}{2}(a + a \cos \theta) \quad \text{and} \quad y = \frac{b \sin \theta}{2} ;$$

$$\therefore \frac{2x - a}{a} = \cos \theta \quad \text{and} \quad \frac{2y}{b} = \sin \theta.$$

Squaring and adding, $\left(\dfrac{2x-a}{a}\right)^2 + \dfrac{4y^2}{b^2} = 1$ is the equation of the locus.

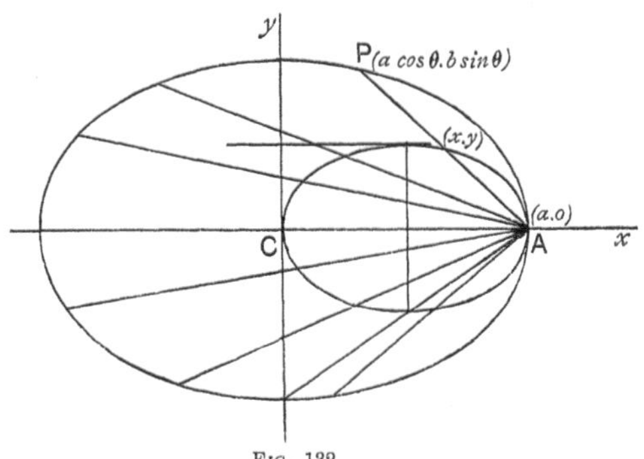

Fig. 132.

This may be written $\dfrac{\left(x-\dfrac{a}{2}\right)^2}{\dfrac{a^2}{4}} + \dfrac{y^2}{\dfrac{b^2}{4}} = 1.$

Hence we see that the locus is an ellipse whose centre is at the point $\left(\dfrac{a}{2},\ 0\right)$, whose semi-axes are $\dfrac{a}{2}$ and $\dfrac{b}{2}$, and whose axes are parallel to those of the given ellipse.

218. *Tangents to an ellipse make angles θ_1, θ_2 with the major axis; find the equation of the locus of their intersection when*

$$\cot\theta_1 + \cot\theta_2 = k^2,$$

a constant.

$y = mx + \sqrt{a^2m^2 + b^2}$ is a tangent to the ellipse

$$\dfrac{x^2}{a^2} + \dfrac{y^2}{b^2} = 1 \text{ for all values of } m.$$

Now if the values of x and y, the co-ordinates of a point on this line, are known, m is the only unknown quantity in the equation.

$\therefore$ if we solve this equation for m, the roots obtained give the slopes of the tangents which can be drawn from (x, y) to the curve.

$$y - mx = \sqrt{a^2m^2 + b^2}.$$

Squaring and transposing, $m^2(x^2 - a^2) - 2mxy + y^2 - b^2 = 0.$

Also if (x, y) is a point on the locus, $\tan \theta_1$ and $\tan \theta_2$ are the roots of this quadratic.

$$\therefore \ \tan \theta_1 + \tan \theta_2 = \frac{2xy}{x^2 - a^2},$$

$$\text{and} \quad \tan \theta_1 \tan \theta_2 = \frac{y^2 - b^2}{x^2 - a^2} \ ;$$

$$\therefore \ \text{by division,} \quad \cot \theta_2 + \cot \theta_1 = \frac{2xy}{y^2 - b^2} \ ;$$

$$\therefore \ \frac{2xy}{y^2 - b^2} = k^2 \text{ is the equation of the locus.}$$

This may be written, $k^2 y^2 - 2xy = b^2 k^2$.

Examples X. f.

LOCUS PROBLEMS ON THE ELLIPSE.

1. Find the locus of the vertex of a triangle whose base, and sum of the other sides are given.

2. The ordinates of the circle $x^2 + y^2 = a^2$ are bisected : find the locus of the points of bisection.   Draw the curve.

3. If CP, CD are two conjugate semi-diameters of an ellipse, find the locus of the middle point of PD.

[Let $(a \cos \theta \, . \, b \sin \theta)$ be the co-ordinates of P, so that

$$\left[a \cos \left(\theta + \frac{\pi}{2} \right), \ b \sin \left(\theta + \frac{\pi}{2} \right) \right]$$

or $(-a \sin \theta, \ b \cos \theta)$ are the co-ordinates of D.

If (x, y) is the middle point of PD,

$$x = \frac{a}{2} (\cos \theta - \sin \theta) \dots\dots\dots\dots\dots\dots\dots\dots\dots\dots\dots\dots(1)$$

$$y = \frac{b}{2} (\sin \theta + \cos \theta). \ \dots\dots\dots\dots\dots\dots\dots\dots\dots\dots(2)$$

The equation of the locus is found by eliminating θ from (1) and (2).]

4. Find the locus of the vertex of a triangle having given the base $2c$, and the product k^2 of the tangents of the angles at the base.

[Take the mid-point of the base as origin, and the axis of y at right angles to the base.]

5. Tangents to the ellipse $\frac{x^2}{a^2} + \frac{y^2}{b^2} = 1$ are drawn at points whose eccentric angles differ by a constant $(2a)$: find the locus of their intersection.

[Let $\theta + a, \ \theta - a$ be the eccentric angles of the points, so that

$$\frac{x}{a} \cos (\theta + a) + \frac{y}{b} \sin (\theta + a) = 1 \ \dots\dots\dots\dots\dots\dots\dots(1)$$

$$\text{and} \quad \frac{x}{a} \cos (\theta - a) + \frac{y}{b} \sin (\theta - a) = 1 \ \dots\dots\dots\dots\dots\dots\dots(2)$$

are the equations of the tangents.   To find the equation of the locus we must eliminate θ.

Adding (1) and (2) and dividing by $2\cos\alpha$

$$\frac{x\cos\theta}{a}+\frac{y\sin\theta}{b}=\sec\alpha. \quad\dots\dots\dots\dots\dots\dots\dots\dots\dots\dots(3)$$

Subtracting (1) from (2) and dividing by $2\sin\alpha$

$$-\frac{x\sin\theta}{a}+\frac{y\cos\theta}{b}=0. \quad\dots\quad\dots\dots\dots\dots\dots\dots\dots\dots\dots(4)$$

Now square (3) and (4), and add.]

6. Chords of the ellipse $\dfrac{x^2}{a^2}+\dfrac{y^2}{b^2}=1$ are drawn through the point (h, k) : find the equation of, and draw the locus of their middle points.

7. CP, CD are conjugate semi-diameters of the ellipse $\dfrac{x^2}{a^2}+\dfrac{y^2}{b^2}=1$ whose axes are AA′, BB′ ; prove that the locus of the point of intersection of AP, BD is the ellipse $\dfrac{x\,(x-a)}{a^2}+\dfrac{y\,(y-b)}{b^2}=0.$ Draw the locus.

8. Find the locus of the intersection of tangents to the ellipse $\dfrac{x^2}{a^2}+\dfrac{y^2}{b^2}=1$, when the sum of the eccentric angles of their points of contact is constant and equal to $2a$.

9. Tangents to the ellipse $\dfrac{x^2}{a^2}+\dfrac{y^2}{b^2}=1$ make angles θ_1, θ_2 with the major axis.   Find the equation of the locus of their intersection when $\tan\theta_1+\tan\theta_2$ is constant, and equal to k.

10. Tangents to the ellipse $\dfrac{x^2}{a^2}+\dfrac{y^2}{b^2}=1$ make angles θ_1, θ_2 with the major axis.   Find the equation of the locus of their intersection when $\tan(\theta_1+\theta_2)=k$, where k is constant.

11. Prove that the locus of a pole with respect to the ellipse $\dfrac{x^2}{a^2}+\dfrac{y^2}{b^2}=1$, when the perpendicular on its polar from the centre is constant, is a concentric and co-axial ellipse.

12. Find, and draw, the locus of the middle point of a perpendicular drawn from a point on the circle $(x-a)^2+y^2=r^2$ to the axis of y.

13. Find the equation of the locus of the middle points of the portions of tangents intercepted by the axes (produced) of the ellipse $\dfrac{x^2}{a^2}+\dfrac{y^2}{b^2}=1.$

14. Ellipses are drawn on the same major axis.   Prove that the locus of the two ends of the latera recta which are on the same side of the major axis is a parabola.

15. A straight line of given length moves so that its extremities are always on two perpendicular straight lines : prove that any point on it describes an ellipse whose semi-axes are the segments of the line.

16. A circle is described which passes through the point $(0, 2)$ and touches the circle $x^2 + y^2 = 16$ internally. Prove that the locus of its centre is an ellipse whose foci are the origin, and the point $(0, 2)$.

Also find the equation of the locus.

17. Prove that the equation of a tangent to the ellipse $\dfrac{x^2}{a^2} + \dfrac{y^2}{b^2} = 1$ may be written in the form $x \cos a + y \sin a = \sqrt{a^2 \cos^2 a + b^2 \sin^2 a}$. Hence find the locus of the intersection of perpendicular tangents.

18. Through the foci of an ellipse, perpendiculars are drawn to a pair of conjugate diameters : prove that they meet on a fixed concentric ellipse.

19. Radii vectores, *i.e.* focal radii, are drawn from one of the foci of an ellipse : find the locus of their middle points.

20. The perpendicular from the centre of an ellipse upon the tangent at P meets the focal distances of P in Q and Q' : find the loci of these points.

PROPERTIES OF THE ELLIPSE.

219. *To prove that* $\qquad \dfrac{PN^2}{AN \cdot A'N} = \dfrac{BC^2}{AC^2}.$

Let $(a \cos \theta,\ b \sin \theta)$ be the co-ordinates of P.

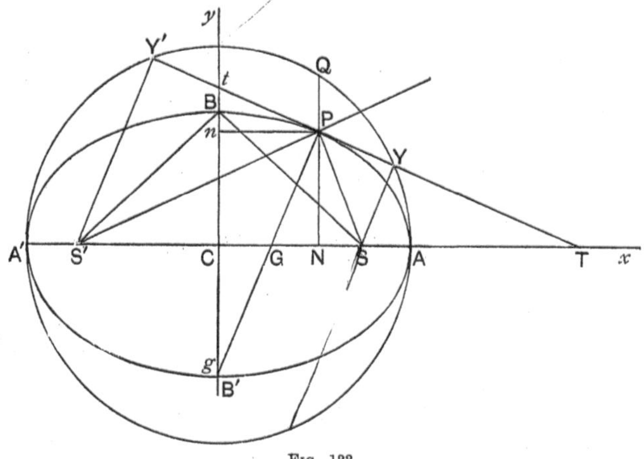

Fig. 133.

$$\frac{PN^2}{AN \cdot A'N} = \frac{PN^2}{CA^2 - CN^2} = \frac{b^2 \sin^2\theta}{a^2 - a^2 \cos^2\theta} = \frac{b^2}{a^2}. \qquad \text{Q.E.D.}$$

220. $SP + S'P =$ *the major axis.*

This is proved in Art. 187.

221. *To prove that* $BC^2 = AS \cdot A'S$.

$AS \cdot A'S = (CA - CS)(CA + CS) = CA^2 - CS^2 = a^2 - a^2e^2 = b^2$. Q.E.D.

COROLLARY. $BS = CA$ for $BS = \frac{1}{2}(BS + BS') = \frac{1}{2}$ the major axis.

222. *If the ordinate* NP, *produced, meets the auxiliary circle at* Q,

$$\frac{PN}{QN} = \frac{BC}{AC}.$$

If θ is the eccentric angle of P,

$$\frac{PN}{QN} = \frac{b \sin \theta}{a \sin \theta} = \frac{b}{a}. \qquad \text{Q.E.D.}$$

223. *If* PN *is the ordinate of* P, *and the tangent at* P *meets the major axis produced at* T, $CN \cdot CT = CA^2$.   (See fig., Art. 219.)

Let $(a \cos \theta,\ b \sin \theta)$ be the co-ordinates of P.

The equation of PT is $\dfrac{x \cos \theta}{a} + \dfrac{y \sin \theta}{b} = 1$.

At T, a point on this line, $y = 0$; $\therefore\ x = \dfrac{a}{\cos \theta}$ or $CT = \dfrac{a}{\cos \theta}$.

$$\therefore\ CN \cdot CT = a \cos \theta \times \frac{a}{\cos \theta} = a^2. \qquad \text{Q.E.D.}$$

224. *If* Pn *is drawn perpendicular to the minor axis and the tangent at* P *meets the minor axis produced at* t, $Cn \cdot Ct = BC^2$.   (See fig., Art. 219.)

Let $(a \cos \theta,\ b \sin \theta)$ be the co-ordinates of P.

The equation of Pt is $\dfrac{x \cos \theta}{a} + \dfrac{y \sin \theta}{b} = 1$.

At t, a point on this line, $x = 0$; $\therefore\ y = \dfrac{b}{\sin \theta}$, or $Ct = \dfrac{b}{\sin \theta}$;

$$\therefore\ Cn \cdot Ct = PN \cdot Ct = b \sin \theta \times \frac{b}{\sin \theta} = b^2. \qquad \text{Q.E.D.}$$

225. *If the normal at* P *meets the major axis at* G, $CG = e^2 \cdot CN$.
(See fig., Art. 219).

Let $(a \cos \theta,\ b \sin \theta)$ be the co-ordinates of P.

The equation of the normal at P is $\dfrac{ax}{\cos \theta} - \dfrac{by}{\sin \theta} = a^2 - b^2$.

At G, a point on this line, $y = 0$;

$$\therefore\ x = \frac{(a^2 - b^2)\cos\theta}{a} = ae^2\cos\theta,$$

$$i.e.\ \ \mathsf{CG} = e^2 \cdot \mathsf{CN}.$$

226. *If the normal meets the major axis at* G, SG $= e$. SP. *Also the normal bisects the interior angle, and the tangent bisects the exterior angle between the focal distances of* P. (See fig., Art. 219.)

We may take as the equation of the normal at P,

$$\frac{ax}{\cos\theta} - \frac{by}{\sin\theta} = a^2 - b^2.$$

At G (as above) $y = 0$, $x = \dfrac{(a^2 - b^2)\cos\theta}{a}$, and CG $= ae^2\cos\theta$.

$$\therefore\ \ \mathsf{SG} = \mathsf{CS} - \mathsf{CG} = ae - ae^2\cos\theta = e(a - ae\cos\theta) = e\cdot\mathsf{SP}$$

(Art. 187.)

In the same way, $\mathsf{S'G} = e \cdot \mathsf{S'P}$;

$$\therefore\ \frac{\mathsf{SG}}{\mathsf{S'G}} = \frac{\mathsf{SP}}{\mathsf{S'P}}\ ;\ \ \therefore\ \ \mathsf{PG}\ \text{bisects the angle}\ \mathsf{SPS'}.$$

Again, the normal bisects the interior angle between SP and S'P; $\therefore$ the tangent, which is at right angles to it, bisects the exterior angle between SP and S'P.

227. *If perpendiculars* SY, S'Y' *are drawn to any tangent,* Y *and* Y' *lie on the auxiliary circle.* (See fig., Art. 219.)

We may take as the equation of the tangent

$$y = mx + \sqrt{a^2m^2 + b^2}. \quad \dots\dots\dots\dots\dots(1)$$

SY passes through the focus $(ae, 0)$ and is at rt. $\angle^s$ to the tangent,

$$\therefore\ \text{its equation is} \qquad y = -\frac{1}{m}(x - ae). \quad \dots\dots\dots\dots(2)$$

To find the locus of Y we must eliminate m between (1) and (2).

From (1) $(y - mx)^2 = a^2m^2 + b^2.$

„   (2) $(my + x)^2 = a^2e^2.$

Adding, $y^2(1 + m^2) + x^2(1 + m^2) = a^2m^2 + b^2 + a^2 - b^2$

$$= a^2(1 + m^2)\ ;$$

$\therefore\ x^2 + y^2 = a^2$ is the equation of the locus of Y.

$$\therefore\ \ \mathsf{Y}\ \text{lies on the auxiliary circle.}$$

In the same way we can prove that Y' lies on the auxiliary circle.

N.B.   In dealing with tangents and perpendiculars to them, it is advisable to use the 'm' equation of a tangent.

228. *If perpendiculars* SY, $S'Y'$ *are drawn to any tangent,*

$$SY \cdot S'Y' = BC^2. \quad \text{(See fig., Art. 219.)}$$

Taking $y = mx + \sqrt{a^2m^2 + b^2}$ as the equation of the tangent,

$SY =$ the length of the perpendicular from the point $(ae, 0)$ upon this line

$$= \frac{-aem - \sqrt{a^2m^2 + b^2}}{\sqrt{1 + m^2}}, \quad \left[p = \frac{Ax_1 + By_1 + C}{\sqrt{A^2 + B^2}} \right]$$

$SY' =$ the length of the perpendicular from the point $(-ae, 0)$ upon the tangent

$$= \frac{aem - \sqrt{a^2m^2 + b^2}}{\sqrt{1 + m^2}}.$$

$$\therefore \ SY \cdot S'Y' = \frac{a^2m^2 + b^2 - a^2e^2m^2}{1 + m^2} = \frac{m^2a^2(1 - e^2) + b^2}{1 + m^2}$$

$$= \frac{m^2b^2 + b^2}{1 + m^2} = b^2. \qquad \text{Q.E.D.}$$

229. *Tangents at the extremities of a diameter are parallel to one another.*

Let $\dfrac{x^2}{a^2} + \dfrac{y^2}{b^2} = 1$ be the equation of the ellipse.

If PCP' is a diameter, and (x_1, y_1) the co-ordinates of P, then $(-x_1, -y_1)$ are the co-ordinates of P'.

The equation of the tangent at P is $\dfrac{xx_1}{a^2} + \dfrac{yy_1}{b^2} = 1$.

$$\quad\quad\quad\quad\quad\quad\quad P' \text{ is } -\frac{xx_1}{a^2} - \frac{yy_1}{b^2} = 1.$$

The slopes of these lines are equal.

$$\therefore \text{ the tangents are parallel.} \quad [\text{See also Art. 209.}]$$

230. *Tangents at the ends of a chord intersect on the diameter which bisects that chord.*

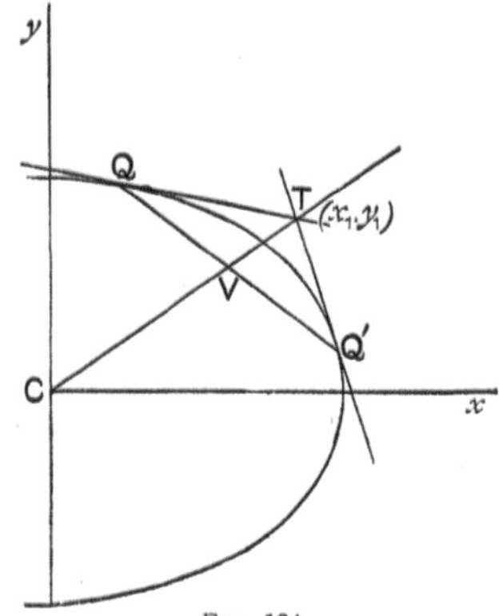

FIG. 134.

Let TQ, TQ' be the tangents, and $(x_1,\ y_1)$ the co-ordinates of T.

The equation of CT is

$$\frac{y}{y_1} = \frac{x}{x_1}. \quad\dots\dots\dots(1)$$

The equation of the chord of contact QQ' is

$$\frac{xx_1}{a^2} + \frac{yy_1}{b^2} = 1. \quad\dots\dots\dots(2)$$

The product of the slopes of (1) and (2) is

$$\frac{y_1}{x_1} \times \left(-\frac{b^2 x_1}{a^2 y_1} \right) = -\frac{b^2}{a^2}.$$

$\therefore$ CT bisects QQ' (Art. 208), which proves the proposition.

231. *The portion of the tangent intercepted between the point of contact and the directrix subtends a right angle at the corresponding focus.*

If $(a\cos\theta,\ b\sin\theta)$ are the co-ordinates of P, the equation of the tangent is

$$\frac{x\cos\theta}{a} + \frac{y\sin\theta}{b} = 1.$$

At the point R, where this meets the directrix, $x = \dfrac{a}{e}$,

$$\therefore \frac{\cos\theta}{e} + \frac{y\sin\theta}{b} = 1,$$

whence

$$y\,(=\mathsf{RX}) = \frac{b}{e\sin\theta}(e-\cos\theta).$$

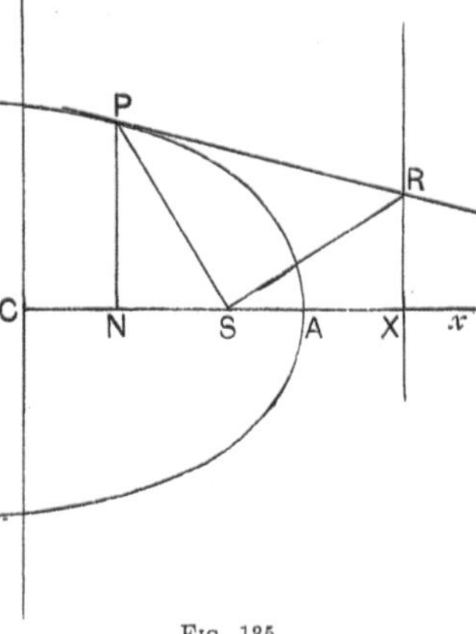

FIG. 135.

$$\therefore \ \tan RSX = \frac{RX}{SX} = \frac{\dfrac{b}{e \sin \theta}(e - \cos \theta)}{\dfrac{a}{e} - ae} = \frac{b(e - \cos \theta)}{a(1 - e^2) \sin \theta}$$

$$= \frac{ab(e - \cos \theta)}{b^2 \sin \theta}$$

$$= \frac{ae - a \cos \theta}{b \sin \theta} = \frac{CS - CN}{PN} = \frac{SN}{PN}$$

$$= \cot PSN.$$

$\therefore \ \angle^s\, RSX,\ PSN$ are complementary.

$\therefore \ \angle\, PSR = a\ rt.\ \angle.$ Q.E.D.

232. The above proposition may also be proved geometrically.

Let the chord PQ meet the directrix at R. Produce PS to P′ and draw PM, QK perpendicular to the directrix.

$$\frac{SP}{SQ} = \frac{e \cdot PM}{e \cdot QK} = \frac{PM}{QK} = \frac{PR}{QR}$$

from the similar $\triangle^s$ PRM, QRK.

$\therefore$ SR bisects the exterior angle QSP′.

Let the chord turn about the point P, so that the point Q moves up to P, and ultimately coincides with it.

Throughout the process

$$\angle\, RSQ = \angle\, RSP'.$$

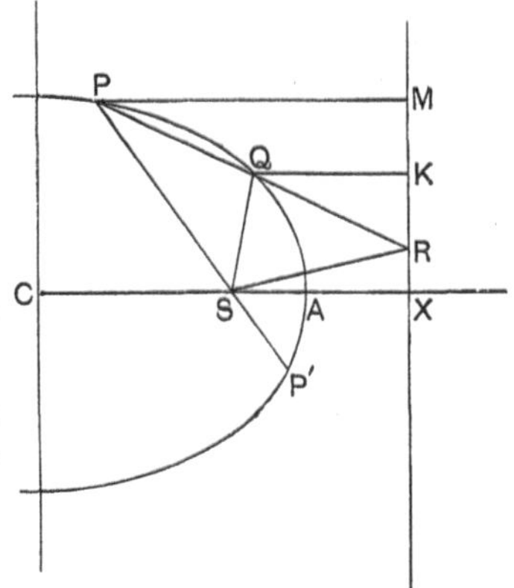

Fig. 186.

But when Q coincides with P, PR becomes a tangent, and $\angle$ PSR = $\angle$ P′SR.

$\therefore$ each of these angles is then a right angle.

$\therefore$ PR, which is then a tangent, subtends a right angle at the focus S.

233. *In an ellipse, tangents at the ends of a focal chord intersect on the corresponding directrix.*

This may be proved as in Art. 166 for the parabola, or as follows :

The co-ordinates of any point on the directrix, may be taken to be $\left(\dfrac{a}{e}, y_1\right)$.

The equation of the polar of this point is

$$\frac{xa}{a^2 e} + \frac{yy_1}{b^2} = 1,$$

$$\text{or } \frac{x}{ae} + \frac{yy_1}{b^2} = 1.$$

This passes through the point $(ae, 0)$ the corresponding focus, and this proves the proposition.

234. *If from* T, *any point on the tangent at* P, *perpendiculars* TR, TM, *are drawn to the focal distance* SP, *and the directrix,* SR $= e$. TM. (Adams' Proposition.)

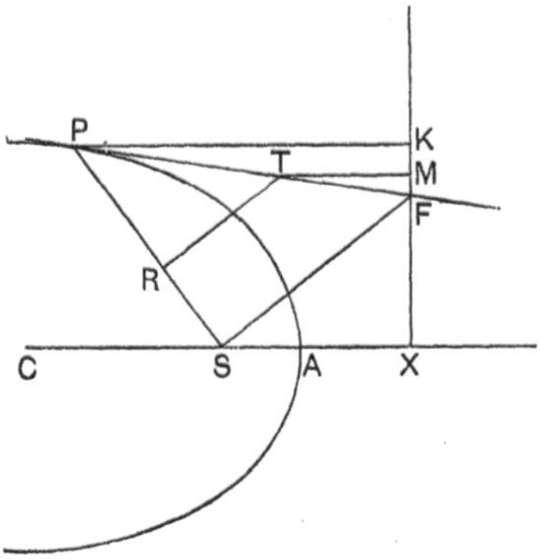

Let the tangent meet the directrix at F. Join SF.

Draw PK perpendicular to the directrix.

$$\angle FSP = \text{a rt. } \angle.$$

$\therefore$ RT is parallel to SF,

$$\therefore \frac{SR}{SP} = \frac{FT}{FP} = \frac{TM}{PK},$$

from the similar $\triangle^s$ FMT, FKP.

Fig. 187.

But $\qquad$ SP $= e$. PK by definition.

$$\therefore \text{ SR} = e \text{ . TM.} \qquad \text{Q.E.D.}$$

235. *Tangents to an ellipse subtend equal angles at the focus.*

If **TP, TQ** are the tangents, and **TR, TR′** are perpendicular to **SP**, and **SQ** respectively,

$$SR = e \cdot TM = SR'.$$

Also in the △ˢ **TSR, TSR′** the ∠ˢ at **R** and **R′** are rt. ∠ˢ,

$$\therefore \angle TSR = \angle TSR',$$

which proves the proposition.

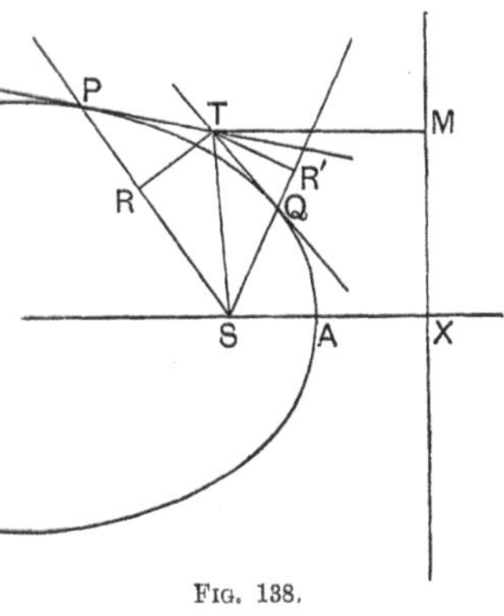

Fig. 138.

236. *To draw a tangent at a given point* **P** *on an ellipse.*

First method. Take either focus **S**, and join **SP**. Draw **ST** at rt. ∠ˢ to **SP** to meet the corresponding directrix at **T**. Join **PT**.

PT subtends a rt. ∠ at the focus **S**, ∴ **PT** is the tangent at **P**.

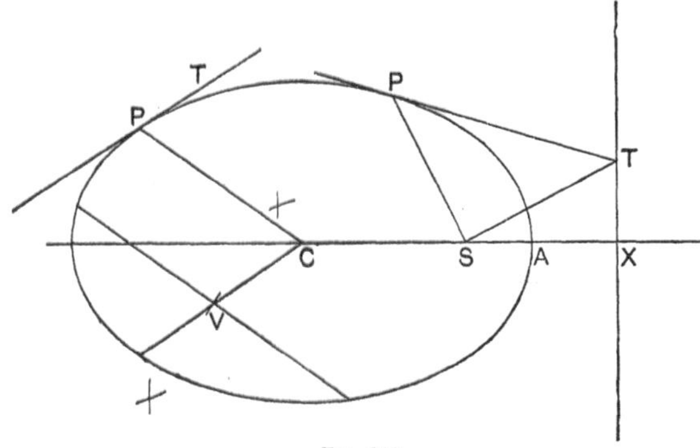

Fig. 189.

Second method. Take the centre **C**, and join **CP**.

Draw any chord parallel to **CP** and bisect it at **V**. Join **CV**, and draw **PT** parallel to **CV**. **PT** is the tangent at **P**, for it is parallel to **CV** which is the diameter conjugate to **CP**.

237. *To draw tangents to an ellipse from an external point* T.

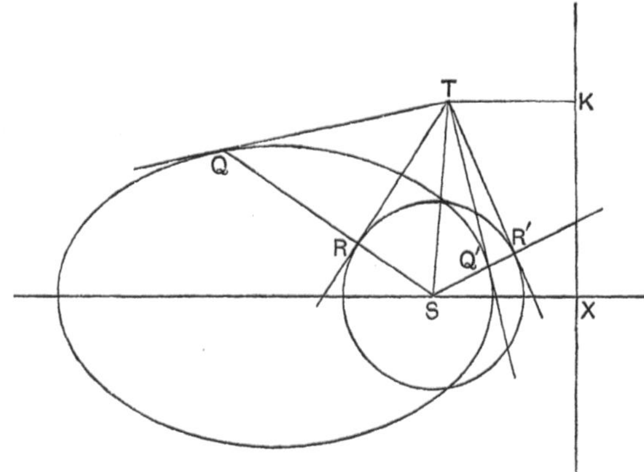

FIG. 140.

First method. Draw TK perpendicular to the directrix, and with centre S (the corresponding focus), and radius $e \cdot$ TK, describe a circle.

From T, draw tangents TR, TR′ to this circle.

Join SR, SR′ and let them meet the ellipse at Q, Q′ respectively. TQ, TQ′ are tangents at Q and Q′ respectively.

The proof is similar to that in Art. 171.

Second method. Draw the auxiliary circle. On ST as diameter describe a circle cutting the auxiliary circle at Y and Z. Join TY, TZ. TY, TZ are tangents to the ellipse.

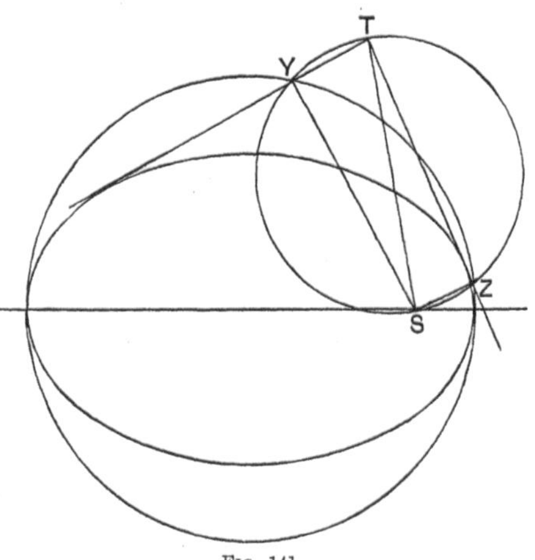

FIG. 141.

Y is a point on the auxiliary circle and $\angle$ SYT is the angle in a semi-circle and therefore a right angle.

$\therefore$ TY is a tangent. (Art. 227.)

Similarly TZ also is a tangent.

238. *Given a curve, which is known to be an ellipse, find its centre, the positions and lengths of its principle axes, and its foci.*

Draw any two parallel chords and bisect them at V and V′. Let VV′ produced meet the curve at P and P′.

PP′ is a diameter, for it bisects parallel chords.

$\therefore$ C the middle point of PP′ is the centre.

On PP′ as diameter describe a circle meeting the ellipse at Q. $\angle$ P′QP is a rt. $\angle$, and PQ, P′Q are supplemental chords.

$\therefore$ the diameters ACA′, BCB′ parallel to P′Q and PQ are conjugate diameters at right angles, *i.e.* they are the principal axes.

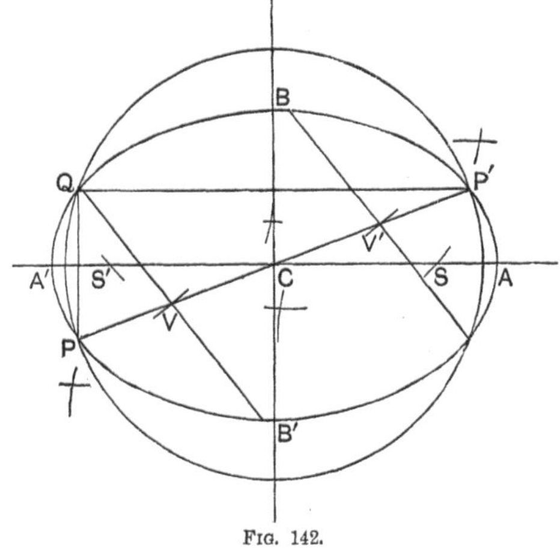

Fig. 142.

With centre B and radius equal to CA describe a circle cutting AA′ at S and S′. BS $=$ CA $=$ BS′.

$\therefore$ S and S′ are the foci.

239. *If* PCP′, DCD′ *are two conjugate diameters of an ellipse*

$$CP^2 + CD^2 = a^2 + b^2.$$

Let θ be the eccentric angle of P; then $\theta + \dfrac{\pi}{2}$ is the eccentric angle of D, and the co-ordinates of D are $(-a\sin\theta,\ b\cos\theta)$;

$\therefore$ $CP^2 = a^2\cos^2\theta + b^2\sin^2\theta$ and $CD^2 = a^2\sin^2\theta + b^2\cos^2\theta$;

$\therefore$ $CP^2 + CD^2 = a^2 + b^2.$

240. *If* CP, CD *are conjugate semi-diameters,* SP . S′P $=$ CD².

Let θ be the eccentric angle of P; then $\theta + \dfrac{\pi}{2}$ is the eccentric angle of D, and the co-ordinates of D are $(-a\sin\theta,\ b\cos\theta)$;

$$\therefore\ \ \mathrm{CD}^2 = a^2\sin^2\theta + b^2\cos^2\theta.$$

$$\mathrm{SP} = a - ae\cos\theta\ \ \text{and}\ \ \mathrm{S'P} = a + ae\cos\theta;\quad\text{(Art. 187)}$$

$$\therefore\ \ \mathrm{SP}\,.\,\mathrm{S'P} = a^2 - a^2e^2\cos^2\theta = a^2 - (a^2 - b^2)\cos^2\theta$$
$$= a^2\sin^2\theta + b^2\cos^2\theta = \mathrm{CD}^2. \qquad \text{Q.E.D.}$$

241. *If* CP, CD *are conjugate semi-diameters, and tangents* PT, DT *are drawn at* P *and* D, *the area of the parallelogram* PCDT $=$ AC . BC.

Let $(a\cos\theta,\ b\sin\theta)$ be the co-ordinates of P, then $(-a\sin\theta,\ b\cos\theta)$ are „ „ D.

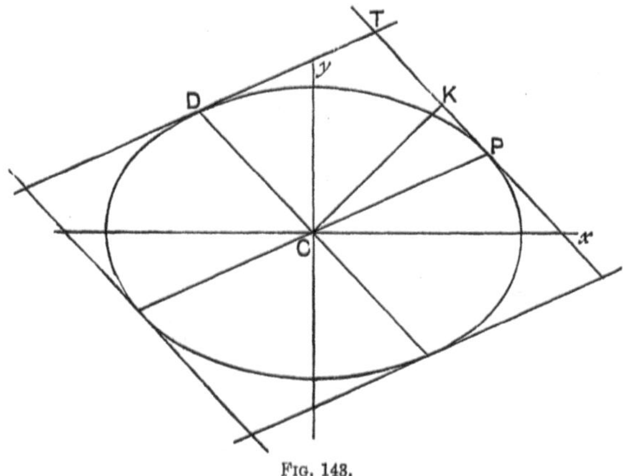

Fig. 148.

Draw CK perpendicular to the tangent PT, whose equation is

$$\frac{x\cos\theta}{a} + \frac{y\sin\theta}{b} = 1 \quad\text{or}\quad bx\cos\theta + ay\sin\theta = ab.$$

$$\mathrm{CK} = \frac{ab}{\sqrt{b^2\cos^2\theta + a^2\sin^2\theta}} \quad \left(p = \frac{\mathrm{A}x_1 + \mathrm{B}y_1 + \mathrm{C}}{\sqrt{\mathrm{A}^2 + \mathrm{B}^2}}\right)$$

$$\mathrm{CD}^2 = a^2\sin^2\theta + b^2\cos^2\theta;\quad \therefore\ \mathrm{CD} = \sqrt{b^2\cos^2\theta + a^2\sin^2\theta};$$

$$\therefore\ \text{the area of } \mathrm{PCDT} = \mathrm{CD}\,.\,\mathrm{CK} = ab = \mathrm{AC}\,.\,\mathrm{BC}. \qquad \text{Q.E.D.}$$

COROLLARY. *The area of the parallelogram formed by tangents at the ends of a pair of conjugate diameters $= 4ab$.*

The parallelogram $= 4 \cdot$ PCDT $= 4ab$.

242. *If the normal at* P *meets the major axis in* G, *and the diameter parallel to the tangent at* P *in* F, PF $\cdot$ PG $=$ BC^2.

Let $(a \cos \theta,\ b \sin \theta)$ be the co-ordinates of P.

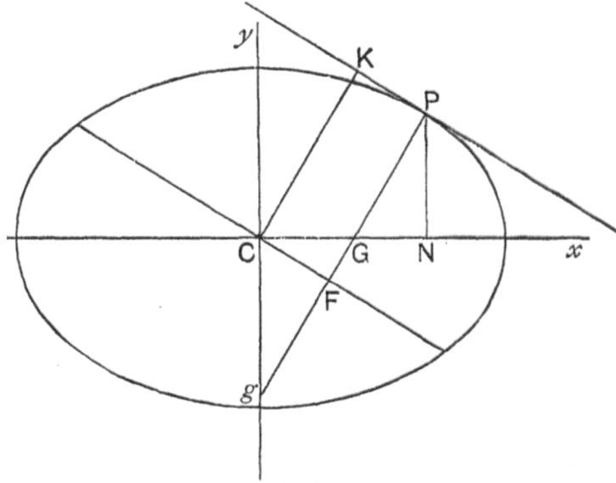

FIG. 144.

As in Art. 225,   CG $= ae^2 \cos \theta$;

$$\therefore\ \ \mathsf{PG}^2 = \mathsf{PN}^2 + \mathsf{GN}^2 = \mathsf{PN}^2 + (\mathsf{CN} - \mathsf{CG})^2$$

$$= b^2 \sin^2\theta + (a \cos\theta - ae^2\cos\theta)^2 = b^2\sin^2\theta + a^2\cos^2\theta(1 - e^2)^2$$

$$= b^2\sin^2\theta + a^2\cos^2\theta \cdot \frac{b^4}{a^4} = \frac{b^2}{a^2}(a^2\sin^2\theta + b^2\cos^2\theta). \ \ldots\ldots\ldots\ldots(1)$$

Draw CK perpendicular to the tangent at P.

The equation of the tangent is

$$\frac{x \cos \theta}{a} + \frac{y \sin \theta}{b} = 1, \quad \text{or} \quad bx \cos \theta + ay \sin \theta = ab.$$

PF $=$ CK $=$ the perpendicular from the origin on the tangent

$$= \frac{ab}{\sqrt{b^2\cos^2\theta + a^2\sin^2\theta}} ;$$

$\therefore$ from (1)   PF $\cdot$ PG $= b^2 =$ BC^2.   Q.E.D.

COROLLARY. *In the same way if the normal meets the minor axis in* g, *we can prove that* PF $\cdot$ Pg $=$ CA^2

***243.** *If from* T, *a point in the diameter* P'CP *produced, tangents* TQ . TQ' *are drawn to the ellipse, and* QQ' *meets* CT *in* V,

$$CV \cdot CT = CP^2.$$

Draw CD the semi-diameter conjugate to CP, and take CP, CD as axes of co-ordinates.

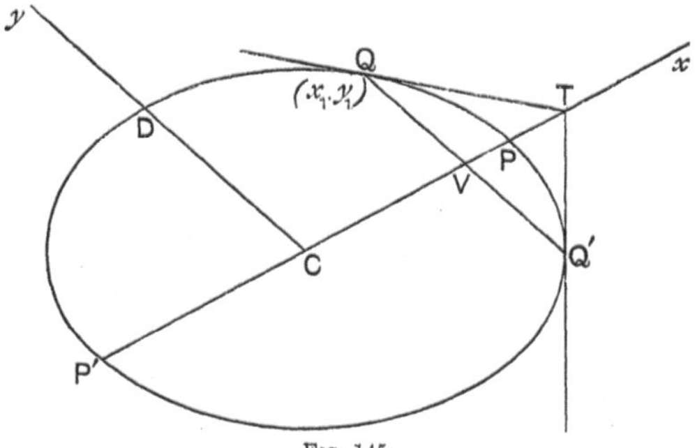

FIG. 145.

$\dfrac{x^2}{a'^2} + \dfrac{y^2}{b'^2} = 1$ is the equation of the ellipse where CP $= a'$ and CD $= b'$.

QQ' is bisected at V (Art. 230), and therefore is parallel to CD. Let (x_1, y_1) be the co-ordinates of Q, so that $x_1 =$ CV, $y_1 =$ QV. The equation of QT, the tangent at Q, is

$$\frac{xx_1}{a'^2} + \frac{yy_1}{b'^2} = 1.$$

At T, a point on this line, $y = 0$;

$$\therefore \quad \frac{xx_1}{a'^2} = 1 \quad \text{or} \quad CT \cdot CV = CP^2. \qquad \text{Q.E D.}$$

***244.** *Through any point* O (x_1, y_1) *a chord is drawn to the ellipse* $\dfrac{x^2}{a^2} + \dfrac{y^2}{b^2} = 1$ *to meet the curve in* Q *and* Q', *making an angle* θ *with the axis of* x : *to find the value of the rectangle* OQ . OQ'.

We may take as the equation of the chord

$$\frac{x - x_1}{\cos \theta} = \frac{y - y_1}{\sin \theta} = r.$$

$$x = x_1 + r \cos \theta, \quad y = y_1 + r \sin \theta,$$

$\therefore$ where the line meets the ellipse we have by substitution,

$$\frac{(x_1 + r \cos \theta)^2}{a^2} + \frac{(y_1 + r \sin \theta)^2}{b^2} = 1,$$

or $\quad r^2\left(\frac{\cos^2\theta}{a^2} + \frac{\sin^2\theta}{b^2}\right) + 2r\left(\frac{x_1 \cos \theta}{a^2} + \frac{y_1 \sin \theta}{b^2}\right) + \frac{x_1{}^2}{a^2} + \frac{y_1{}^2}{b^2} - 1 = 0.$

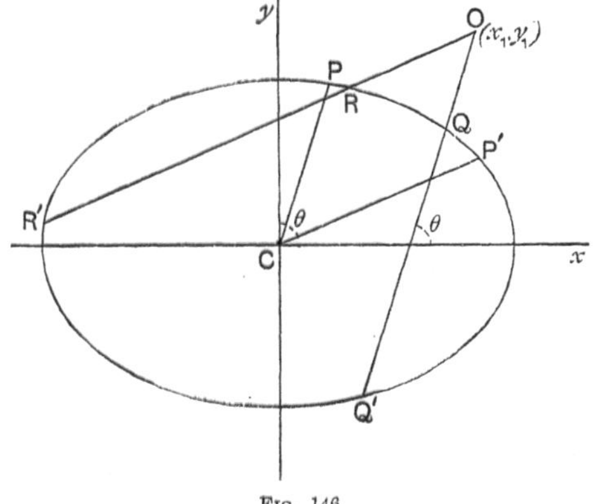

Fig. 146.

OQ, OQ′ are the roots of this quadratic,

$$\therefore \ \text{OQ . OQ}' = \text{the product of the roots} = \frac{\dfrac{x_1{}^2}{a^2} + \dfrac{y_1{}^2}{b^2} - 1}{\dfrac{\cos^2\theta}{a^2} + \dfrac{\sin^2\theta}{b^2}}.$$

N.B. The point O may be within or without the ellipse.

245. *If a chord through* O(x_1, y_1) *meets the ellipse* $\dfrac{x^2}{a^2} + \dfrac{y^2}{b^2} = 1$ *at* Q *and* Q′, *and* CP *is a semi-diameter parallel to the chord*

$$\frac{\text{OQ . OQ}'}{\text{CP}^2} = \frac{x_1{}^2}{a^2} + \frac{y_1{}^2}{b^2} - 1.$$

Or, the ratio OQ . OQ′ : CP^2 *is constant for all directions of* CP *and* QQ′. *(See fig. in the preceding article.)*

As in the previous article, $OQ \cdot OQ' = \dfrac{\dfrac{x_1^2}{a^2} + \dfrac{y_1^2}{b^2} - 1}{\dfrac{\cos^2\theta}{a^2} + \dfrac{\sin^2\theta}{b^2}}$. ....(1)

Now $\angle PCA = \theta$; $\therefore$ the co-ordinates of P are

$$(CP \cdot \cos\theta,\ CP \cdot \sin\theta),$$

and P is on the ellipse $\dfrac{x^2}{a^2} + \dfrac{y^2}{b^2} = 1$;

$$\therefore\ \frac{CP^2 \cos^2\theta}{a^2} + \frac{CP^2 \sin^2\theta}{b^2} = 1,\ \text{ or }\ CP^2 = \frac{1}{\dfrac{\cos^2\theta}{a^2} + \dfrac{\sin^2\theta}{b^2}} ;$$

$$\therefore\ \text{from (1)}\ \ \frac{OQ \cdot OQ'}{CP^2} = \frac{x_1^2}{a^2} + \frac{y_1^2}{b^2} - 1.$$

COROLLARY 1. If OQQ', ORR' meet an ellipse in Q, Q', R, R', and CP, CP' are semi-diameters parallel to QQ', and RR' respectively, $\quad \dfrac{OQ \cdot OQ'}{CP^2} = \dfrac{x_1^2}{a^2} + \dfrac{y_1^2}{b^2} - 1 = \dfrac{OR \cdot OR'}{CP'^2}.$

COROLLARY 2. If from $O(x_1,\ y_1)$ a tangent OQ is drawn to the ellipse and CP is the parallel semi-diameter,

$$\frac{OQ^2}{CP^2} = \frac{x_1^2}{a^2} + \frac{y_1^2}{b^2} - 1.$$

*246. *To find the equation of the pair of tangents drawn to the ellipse* $\dfrac{x^2}{a^2} + \dfrac{y^2}{b^2} = 1$ *from the point* $(x_1,\ y_1).$

Let OQ, OQ' be the tangents drawn from the point $O(x_1,\ y_1)$. Join QQ'. Let $P(x,\ y)$ be any point on either tangent, and draw PK, OL perpendicular to QQ'.

From the similar triangles QKP, QLO,

$$\frac{PK}{OL} = \frac{PQ}{OQ}. \quad ....(1)$$

Now $PK =$ the perpendicular from the point $(x,\ y)$ on QQ', whose equation is

$$\frac{xx_1}{a^2} + \frac{yy_1}{b^2} = 1 ;$$

$$\therefore \ \mathsf{PK} = \frac{\dfrac{xx_1}{a^2} + \dfrac{yy_1}{b^2} - 1}{\sqrt{\dfrac{x_1^{\,2}}{a^4} + \dfrac{y_1^{\,2}}{b^4}}}.$$

In the same way, $\mathsf{OL} = \dfrac{\dfrac{x_1^{\,2}}{a^2} + \dfrac{y_1^{\,2}}{b^2} - 1}{\sqrt{\dfrac{x_1^{\,2}}{a^4} + \dfrac{y_1^{\,2}}{b^4}}} \ ;$

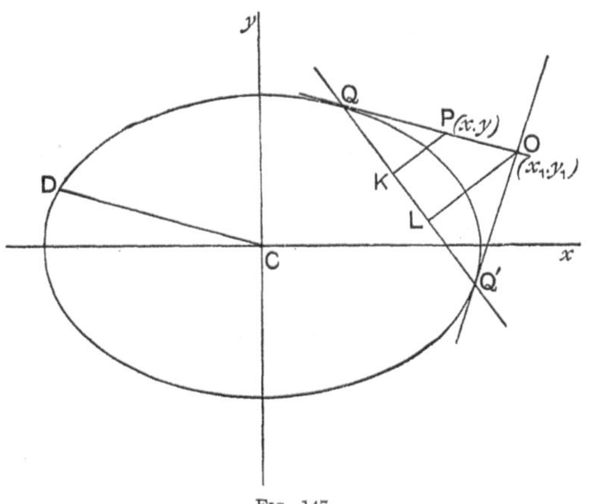

FIG. 147.

$$\therefore \ \frac{\mathsf{PK}}{\mathsf{OL}} = \frac{\dfrac{xx_1}{a^2} + \dfrac{yy_1}{b^2} - 1}{\dfrac{x_1^{\,2}}{a^2} + \dfrac{y_1^{\,2}}{b^2} - 1}. \quad \dotsb \dotsb (2)$$

Also if CD is the semi-diameter parallel to OPQ,

$$\frac{\mathsf{PQ}^2}{\mathsf{OQ}^2} = \frac{\dfrac{\mathsf{PQ}^2}{\mathsf{CD}^2}}{\dfrac{\mathsf{OQ}^2}{\mathsf{CD}^2}} = \frac{\dfrac{x^2}{a^2} + \dfrac{y^2}{b^2} - 1}{\dfrac{x_1^{\,2}}{a^2} + \dfrac{y_1^{\,2}}{b^2} - 1}. \quad (\text{Art. 245.})$$

$$\therefore \text{ from (1) and (2) } \quad \dfrac{\dfrac{xx_1}{a^2}+\dfrac{yy_1}{b^2}-1}{\dfrac{x_1^{\,2}}{a^2}+\dfrac{y_1^{\,2}}{b^2}-1}=\sqrt{\dfrac{\dfrac{x^2}{a^2}+\dfrac{y^2}{b^2}-1}{\dfrac{x_1^{\,2}}{a^2}+\dfrac{y_1^{\,2}}{b^2}-1}}.$$

Squaring and multiplying up, we have

$$\left(\frac{xx_1}{a^2}+\frac{yy_1}{b^2}-1\right)^2=\left(\frac{x^2}{a^2}+\frac{y^2}{b^2}-1\right)\left(\frac{x_1^{\,2}}{a^2}+\frac{y_1^{\,2}}{b^2}-1\right).$$

But (x, y) is any point on either tangent.

$\therefore$ this is the equation of the two tangents OQ, OQ'.

*247. *If* PCP', DCD' *are conjugate diameters, and* QQ' *a chord parallel to* DD' *cuts* PP' *at* V, $\dfrac{QV^2}{PV\,.\,VP'}=\dfrac{CD^2}{CP^2}.$

Take CP, CD as axes of co-ordinates, and let $CP=a'$, $CD=b'$.

Then $\dfrac{x^2}{a'^2}+\dfrac{y^2}{b'^2}=1$ is the equation of the ellipse.

QQ', being parallel to CD, is bisected by CP at V.

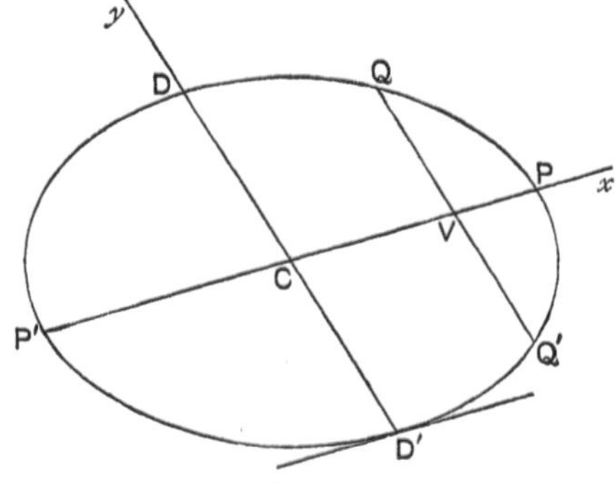

FIG. 148.

Let $(a'\cos\theta,\ b'\sin\theta)$ be the co-ordinates of Q.

$$\text{Then}\quad \frac{QV^2}{PV\,.\,VP'}=\frac{QV^2}{CP^2-CV^2}=\frac{b'^2\sin^2\theta}{a'^2-a'^2\cos^2\theta}=\frac{b'^2\sin^2\theta}{a'^2\sin^2\theta}$$

$$=\frac{CD^2}{CP^2}. \qquad\qquad \text{Q.E.D.}$$

Examples X. g.

1. Draw the ellipse whose axes are 6 inches and 4 inches long. Find its foci, and measure their distances from the centre. Verify your measurements by calculation.

2. If a focal chord of an ellipse meets the auxiliary circle at Q and Q', prove that $SQ \cdot SQ' = BC^2$.

3. Prove that the acute angle between two conjugate diameters of an ellipse is a minimum when they are equal.

4. If the diameter conjugate to the semi-diameter CP of an ellipse cut the focal distances of P in the points L, M, then $PL = PM$.

5. Find the co-ordinates of the intersection of tangents to the ellipse $\dfrac{x^2}{a^2} + \dfrac{y^2}{b^2} = 1$ at the points whose eccentric angles are $\theta + a$, $\theta - a$, and if a is constant deduce the fact that their tangents meet on the ellipse

$$\frac{x^2}{a^2 \sec^2 a} + \frac{y^2}{b^2 \sec^2 a} = 1.$$

6. Given the axes ACA', BCB' of an ellipse, and PP' which is known to be a tangent to the curve, find the foci and the point of contact of the tangent. State the steps of your construction.

7. In the ellipse $4x^2 + 2xy + y^2 = 12$, find the equation of the diameter conjugate to the axis of x; and if a and b be the semi-axes of the curve, prove that $a^2 + b^2 = 20$.

8. Find the foci and directrices of the ellipse
$$x^2 + y^2 = e^2(x \cos a + y \sin a - p)^2.$$

9. A chord of an ellipse moves parallel to itself; show that the sum of the eccentric angles at its extremities is constant.

10. Find the point in which the line joining the positive focus of the ellipse $\dfrac{x^2}{a^2} + \dfrac{y^2}{b^2} = 1$ to the positive end of the minor axis meets the curve again.

11. Show that if all chords of an ellipse parallel to the minor axis are lengthened in the same properly chosen ratio, their middle points being held fixed, their extremities lie on a circle.

12. A and B are two fixed points 4 inches apart, and a point P moves so that the sum of its distance from A and B is 5 inches. Take as axes of co-ordinates the line AB and the line bisecting AB perpendicularly, and find the equation of the locus of P. Find where the curve cuts each axis.

13. Find (i) the co-ordinates of the centre, (ii) the co-ordinates of the foci, (iii) the length of the major axis, (iv) the eccentricity of the ellipse $2x^2 + 5y^2 = 10x$. Draw the ellipse to scale, unit one inch.

14. If P be a point on an ellipse, and A, A' the extremities of the major axis, show that the tangent at P intersects the diameter parallel to AP on the tangent at A'.

B.A.G. P

15. In an ellipse, if a line be drawn through a focus making a constant angle with a tangent, prove that the locus of the point of intersection with the tangent is a circle.

16. From P, a point on an ellipse an ordinate PM is drawn parallel to the minor axis. On PM a point Q is taken such that QC=PM. Find the locus of Q.

17. If PM be the perpendicular upon the directrix from any point P on an ellipse, then MS, drawn through the adjacent focus S, meets the normal at P on the minor axis.

18. If the tangent and normal at a point P on an ellipse meet the minor axis at t and g respectively, prove that the points S, P, t, g, S′ are concyclic.

19. Show that $\dfrac{x^2}{a^2+k}+\dfrac{y^2}{b^2+k}=1$, where k may have different values, represents a system of confocal ellipses.

20. Find the locus of the pole of the straight line $lx+my=1$ with respect to the system of confocal ellipses represented by $\dfrac{x^2}{a^2+k}+\dfrac{y^2}{b^2+k}=1$.

21. The normal at a point $P(x_1, y_1)$ of the ellipse $\dfrac{x^2}{a^2}+\dfrac{y^2}{b^2}=1$ meets the diameter conjugate to CP at a point F. Show that

$$CF \cdot CD=\left(\frac{a}{b}-\frac{b}{a}\right) x_1 y_1.$$

22. If CP, CD be conjugate semi-diameters of an ellipse, and the normal at P cuts the major axis at G, prove that

$$\frac{PG}{CD}=\frac{a}{b}.$$

23. If m is the slope of a normal to the ellipse $\dfrac{x^2}{a^2}+\dfrac{y^2}{b^2}=1$, show that its equation may be written $(mx-y)\sqrt{a^2+b^2m^2}=m(a^2-b^2)$.

24. Find a common tangent to the ellipses $\dfrac{x^2}{a^2+b^2}+\dfrac{y^2}{b^2}=1$ and $\dfrac{x^2}{a^2}+\dfrac{y^2}{a^2+b^2}=1$, and show that it is parallel to a diagonal of the rectangle whose sides are their directrices.

25. Express the length of the normal (terminated by the major axis) to the ellipse $\dfrac{x^2}{a^2}+\dfrac{y^2}{b^2}=1$ in terms of the inclination of the normal to the major axis.

26. Show that the sum of the squares of the normals drawn at the extremities of conjugate semi-diameters and terminated by the major axis is

$$a^2(e^2-1)(e^2-2).$$

27. Find the intercepts on the axes made by the normal at any point (x_1, y_1) on the ellipse $\dfrac{x^2}{a^2}+\dfrac{y^2}{b^2}=1$.

Hence prove that if M and N be the feet of the perpendiculars upon the axes from any point on the ellipse, the line MN will be always normal to a concentric and similar ellipse.

28. In an ellipse of eccentricity $\tan\beta$, a focal chord is inclined at an angle a to the major axis; show that the tangents at the extremities of the chord include an angle $\tan^{-1}(\tan 2\beta \sin a)$.

29. C is the centre of the ellipse $\dfrac{x^2}{a^2}+\dfrac{y^2}{b^2}=1$, BCB′ its minor axis, S the positive focus; if B′S is produced to meet the curve at P, show that CP makes an angle ϕ with the major axis such that

$$2e \tan\phi = (1-e^2)^{\frac{3}{2}}.$$

30. If one focus of an ellipse inscribed in a triangle be at the circumcentre, prove that the other is at the ortho-centre of the triangle.

31. In an ellipse S, S′ are the foci, and P, Q any two points on the curve. If the tangents at P and Q intersect at T, show that the angles QTS, PTS′ are equal.

32. Show that if a polar to an ellipse touches the circle described on the semi-minor axis as diameter, the pole will lie on a parabola whose vertex is at the end of the minor axis.

33. Find the length of the polar chord of the point (x_1, y_1) with respect to the ellipse $\dfrac{x^2}{a^2}+\dfrac{y^2}{b^2}=1$.

Revision questions on the Ellipse.

[These may be taken orally, or the answers may be written down without any working.]

What are the lengths of the axes of each of the following ellipses :—

1. $\dfrac{x^2}{4}+\dfrac{y^2}{9}=1$.

2. $4x^2+9y^2=1$.

3. $5x^2+4y^2=25$.

4. $ax^2+by^2=1$.

5. Describe the position of the ellipse $\dfrac{(x-1)^2}{4}+\dfrac{(y-2)^2}{16}=1$.

How do you know that this equation represents an ellipse?

6. What are the co-ordinates of the centre of the ellipse $4x^2-8x+9y^2=0$?

7. $y=mx\pm\sqrt{a^2m^2+b^2}$ is a tangent to the ellipse $\dfrac{x^2}{a^2}+\dfrac{y^2}{b^2}=1$.

Deduce the 'm' equation of a tangent to $\dfrac{(x-h)^2}{a^2}+\dfrac{(y-k)^2}{b^2}=1$.

8. If $y-k=m(x-h)$ is a tangent to a curve at the point (h, k), what is the equation of the normal at the same point?

9. What is the condition that $ax^2+hxy+by^2+gx+fy+c=0$ may represent an ellipse?

10. What is the equation of the tangent at (x_1, y_1) to the curve
$$ax^2+by^2=1?$$

11. What is the slope of the tangent in the previous question?

12. What is the equation of the normal at (x_1, y_1) to $ax^2+by^2=1$?

13. If $(a\cos\theta, b\sin\theta)$ are the co-ordinates of one extremity of a diameter, what are the co-ordinates of the extremities of the conjugate diameter?

14. Give the equation of an ellipse whose centre is at the point $(2, 4)$, whose axes are parallel to the axes of co-ordinates, and of lengths 8 and 6 respectively.

15. $gx+fy=0$ is a tangent to the ellipse $ax^2+by^2+2gx+2fy=0$. Why? What are the co-ordinates of its point of contact? What is the equation of the normal at the same point?

Describe the ellipses represented by the following equations :—

16. $(x-x_1)^2+(y-y_1)^2=c^2x^2$.

17. $(x-x_1)^2+(y-y_1)^2=\dfrac{y^2}{9}$.

18. $(x-x_1)^2+(y-y_1)^2=\dfrac{(x\cos a+y\sin a-p)^2}{4}$.

19. Interpret the straight line $\dfrac{xx_1}{a^2}+\dfrac{yy_1}{b^2}=1$

    (i) when (x_1, y_1) is a point on the ellipse $\dfrac{x^2}{a^2}+\dfrac{y^2}{b^2}=1$,

    (ii)   ,,   ,,   ,,   outside   ,,   ,,
    (iii)   ,,   ,,   ,,   inside   ,,   ,,

20. As the eccentricity e of an ellipse diminishes, $ae(=\mathsf{CS})$ diminishes if a remains constant. What do you deduce when $e=0$, remembering that $b^2=a^2(1-e^2)$?

21. If θ is the eccentric angle of P on an ellipse, and $\mathsf{DCD'}$ is the diameter conjugate to CP, what are the co-ordinates of D (in the second quadrant) and $\mathsf{D'}$ (in the fourth quadrant)?

22. If the tangent $y=mx\pm\sqrt{a^2m^2+b^2}$ to the ellipse $\dfrac{x^2}{a^2}+\dfrac{y^2}{b^2}=1$ passes through the point $(c, 0)$ where c is positive, which sign would you take for the square root

    (i) when the line touches the curve in the first quadrant?
    (ii)   ,,   ,,   ,,   ,,   fourth   ,,

23. If $\dfrac{xx_1}{a^2}+\dfrac{yy_1}{b^2}=1$ is a tangent to the ellipse $\dfrac{x^2}{a^2}+\dfrac{y^2}{b^2}=1$ at the point (x_1, y_1), and $lx+my=n$ is a tangent at the same point, what do you deduce?

24. If the straight line $\dfrac{xx_1}{a^2}+\dfrac{yy_1}{b^2}=1$ passes through the point $(ae,\ 0)$, what can you deduce as to the position of the point $(x_1,\ y_1)$?

25. What is the condition that the ellipse

$$ax^2 + 2hxy + by^2 + 2gx + 2fy + c = 0$$

may touch (i) the axis of x, (ii) the axis of y ?

26. What is the general equation of an ellipse whose centre is at the origin ?

27. What is the distance of the point $(x_1,\ y_1)$ on the ellipse $\dfrac{x^2}{a^2}+\dfrac{y^2}{b^2}=1$ from the positive focus :

(i) when $(x_1,\ y_1)$ is in the first quadrant ?

(ii)   ,,   ,,   ,, second   ,,

28. What is the slope of the diameter of the ellipse $\dfrac{x^2}{7}+\dfrac{y^2}{3}=1$ which is conjugate to the diameter $y=2x$?

29. What is the value of c if the straight line $y=mx+c$ touches the ellipse $\dfrac{x^2}{a^2}+\dfrac{y^2}{b^2}=1$?

30. What are the co-ordinates of the centre of the ellipse

$$\frac{x^2 - ax}{a^2}+\frac{y^2 - by}{b^2}=0\ ?$$

CHAPTER XI.

REVISION PAPERS.

Revision Paper XI. a.

1. Prove that if $\frac{x}{c}+\frac{y}{d}=1$ be any line through the intersection of $\frac{x}{a}+\frac{y}{b}=1$ and $\frac{x}{b}+\frac{y}{a}=1$, then $\frac{1}{c}+\frac{1}{d}=\frac{1}{a}+\frac{1}{b}$.

2. Find the equations of the circles which pass through the point $(-1, 2)$ and touch both axes of co-ordinates.

3. A chord through the focus of the parabola $y^2=4ax$ cuts the curve in Q_1 and Q_2. If y_1, y_2 are the ordinates of these points prove that
$$y_1y_2 = -4a^2.$$

4. A focal chord of the parabola $y^2=4ax$ makes an angle θ with the axis of x; determine the equation of the circle described on this chord as diameter, and show that the circle touches the directrix whatever be the value of θ.

5. Prove that the sum of the squares of the perpendiculars on any tangent to the ellipse $\frac{x^2}{a^2}+\frac{y^2}{b^2}=1$ from the points $(0, ae)$, $(0, -ae)$ is $2a^2$.

6. Find the equations of the common tangents of the conics $y^2=4\sqrt{2}x$ and $3x^2+2y^2=6$.

7. If PSQ is a focal chord of an ellipse of eccentricity e, and θ, ϕ are the eccentric angles of P, Q respectively, prove that $\cos\frac{\phi-\theta}{2}=e\cos\frac{\phi+\theta}{2}$.

Revision Paper XI. b.

1. Form a single equation to represent the two straight lines $3y-2x=1$, $5y+4x=2$.

Also find the equation of the two straight lines joining the origin to the points where these lines are met by the line $y=\frac{2}{11}$.

2. Find the length of the portion of the line $x+2y=a$ intercepted by the circle $x^2+y^2=a^2$.

3. Find the co-ordinates of the intersection of the tangents at the points where the parabola $y^2=4ax$ is cut by the chord $8x-3y=9a$, and show that it lies on the diameter corresponding to this chord.

4. PNP′ is a chord of a parabola perpendicular to the axis AN, and the normal at P cuts the axis at G. QGQ′ is a chord parallel to PP′ of the circle whose diameter is PP′: show that the locus of Q for different positions of PP′ is another parabola.

5. Find at what points of the ellipse $3x^2 + 4y^2 = 12$, the normal makes an angle of $30°$ with the axis of x.

6. Find the equation of the ellipse whose focus is at the point $(3\sqrt{2}, 0)$, whose directrix is the line $y + x = 0$, and whose eccentricity is $\frac{1}{2}$. Find the length of its major axis and draw a rough sketch of the curve.

7. In an ellipse, the normals drawn at P, D, the extremities of conjugate diameters, intersect in Q. Show that the diameter through Q is perpendicular to PD.

Revision Paper XI. c.

1. Find the values of A and B if $5x - 7y + 12 = 0$ and $Ax + By + 4 = 0$ represent the same straight line.

2. Find the angle between the straight lines joining the points of intersection of $2x + 2y = a$ and $x^2 + y^2 = a^2$ to the origin.

3. Find the length of the chord intercepted by the parabola $y^2 = 4x$ on the line $y = 2x - 3$.

4. If t be a variable quantity, prove that the locus of the point (x, y) when $x = at + bt^2$, $y = bt$, is a parabola, whose axis is parallel to the axis of x.

Find the equation of the tangent at the origin.

5. Find the equations of the tangents drawn to the ellipse $\dfrac{x^2}{16} + \dfrac{y^2}{9} = 1$ from the point $(4, 9)$.

6. Find the equation of the ellipse whose foci are at the points $(0, 2)$, $(0, -2)$ and whose eccentricity is $\frac{1}{3}$.

7. Find a number of points on, and draw, the ellipse whose semi-axes are $2\frac{1}{2}$ and $1\frac{1}{2}$ inches long respectively.

Revision Paper XI. d.

1. One side (a) of a square is inclined to the axis of x at the angle a, and one of its extremities is at the origin; prove that the equations of its diagonals are

$$y(\cos a - \sin a) - x(\cos a + \sin a) = 0,$$
$$y(\cos a + \sin a) + x(\cos a - \sin a) = a.$$

2. Find the equation of the circle cutting orthogonally the circles $x^2 + y^2 = 1$, $x^2 + y^2 - 2x = 0$, $x^2 + y^2 - 6y + 8 = 0$.

3. Find the equations of the tangents to the parabola $y^2 = 4x$ at the points whose x co-ordinates are each equal to 1.

Show that they are at right angles, and intersect where the directrix cuts the axis.

4. Prove that if two normals to a parabola are at right angles, they intersect on a fixed parabola.

Find the dimensions and position of this parabola, and show it in a figure.

5. Find, from first principles, the equation of the tangent at the point (x_1, y_1) to the ellipse $ax^2 + by^2 + 2gx + 2fy + c = 0$.

6. A chord of the ellipse $7x^2 + 16y^2 = 112$ is bisected at the point $(3, 1)$: find its equation.

Find the co-ordinates of the pole of the chord.

7. Find the equation of the diameter of the ellipse $3x^2 + 5xy + 7y^2 = 120$, which is conjugate to the diameter $y = 2x$.

Revision Paper XI. e.

1. In any triangle ABC take AB, AC as axes of x and y respectively, and draw straight lines from the angles bisecting the sides. Find the co-ordinates of their point of intersection. (Oblique.)

2. Find the equations of the tangents drawn to the circle $x^2 + y^2 = a^2$ from the point $(h, 0)$.

3. If G be the foot of the normal at a point P of a parabola, R the middle point of SG (S being the focus), and X the foot of the directrix; show that the difference between the squares on RX and RP is equal to the square on the semi-latus rectum.

4. Draw the parabola $y^2 + 12 = 4x$ on squared paper. Draw the tangents at the points whose abscissae are 5, explaining your construction. Verify by calculation.

5. Tangents are drawn to the ellipse $\dfrac{x^2}{a^2} + \dfrac{y^2}{b^2} = 1$ from the point (h, k) to make angles θ_1, θ_2 with the axes of x. Prove that $\tan\theta_1 + \tan\theta_2 = \dfrac{2hk}{h^2 - a^2}$.

6. Find the co-ordinates of the middle point of the portion of the straight line $2x + 3y = 6$ intercepted by the ellipse $9x^2 + 18y^2 = 162$.

7. A circle is drawn with its centre at the centre of an ellipse, and touching any tangent of the ellipse. If a straight line, PQ, drawn through a focus S, parallel to the tangent, cut the circle in P and Q, PQ is equal to the minor axis.

Revision Paper XI. f.

1. Find the equations of the straight lines bisecting the angles between $y - b = (x - a)\tan\theta$ and $y - b = (x - a)\tan 2\theta$.

2. On squared paper draw the circle $x^2 + y^2 = 4$, and several chords through the point $(-1, 1)$. Draw tangents at the extremities of these chords, and so determine the locus of the intersections of these pairs of tangents. Write down the equation of the locus.

3. Find the equation of the locus of the centres of circles passing through the point $(2, 1)$ and touching the straight line $y = 2x$. Indicate, by drawing a rough diagram, the position and nature of the curve.

4. Find the angle at which the parabolas $x^2 = ay$, $y^2 = 8ax$ intersect.

5. Tangents are drawn from the origin to the ellipse

$$3x^2 + 6y^2 + 2x + 10y + 4 = 0:$$

find their equations.

6. An ellipse touches a given straight line at P, and S is one of its foci: find the locus of the other focus.

7. Find, from first principles, the equation of a tangent to the ellipse $7x^2 + 11y^2 = 231$, in terms of its slope.

Deduce the locus of the intersection of perpendicular tangents to the curve.

Revision Paper XI. g.

1. Show that the straight lines given by the equations $6x^2 - 5xy - 6y^2 = 0$, and $6x^2 - 5xy - 6y^2 + x + 5y - 1 = 0$ lie along the sides of a square.

2. Find the equation of the circle inscribed in the triangle formed by the axes of co-ordinates and the straight line $\dfrac{x}{a} + \dfrac{y}{b} = 1$.

3. Tangents are drawn to the parabola $y^2 = 4ax$ making angles of $30°$ and $60°$ respectively with the axis of x: show that the line joining the points of contact passes through the foot of the directrix.

4. A normal is drawn to the parabola $y^2 = 4ax$ at the point (x_1, y_1) : find the co-ordinates of the middle point of the normal chord.

5. Find the co-ordinates of the foci of the ellipse $4x^2 + y^2 + 16x = 0$.

6. Prove that the equation $a^2y^2 + kxy - b^2x^2 = 0$ represents a pair of conjugate diameters of the ellipse $b^2x^2 + a^2y^2 = a^2b^2$ for all values of k.

7. Find the equation of a normal to the ellipse $\dfrac{x^2}{a^2} + \dfrac{y^2}{b^2} = 1$ in terms of its slope or gradient, and deduce the fact that in general four normals can be drawn to an ellipse from a given point.

Revision Paper XI. h.

1. Find the value of a if the equation $axy - 9y + 20x - 15 = 0$ represents two straight lines.

2. Find the equation of the circle whose centre is on the line of centres of the circles $x^2 + y^2 = 2x$, $x^2 + y^2 - 8x - 10y + 30 = 0$, and which cuts them both orthogonally.

3. Prove that the straight line $4x + 7y - 8 = 0$ touches the parabola $(4x - 7y)^2 = 4x + 7y - 8$, and find the co-ordinates of its point of contact.

4. Prove that the straight line $y = 3x - 33$ is a normal to the parabola $y^2 = 4x$, and find the ordinates of the points where it meets the curve.

5. Find the equation of the ellipse whose focus is at the point $(0, b)$ whose directrix is the axis of x, and eccentricity e.

6. Find the pole of the straight line $y = m(x - ae)$ with respect to the ellipse $\dfrac{x^2}{a^2} + \dfrac{y^2}{b^2} = 1$, e being its eccentricity.

Deduce a geometrical property of the ellipse.

7. Find the locus of the intersection of tangents to the ellipse $\dfrac{x^2}{a^2} + \dfrac{y^2}{b^2} = 1$, when the eccentric angles of their points of contact differ by $120°$.

CHAPTER XII.

THE HYPERBOLA.

248. *Def.*   A **hyperbola** is a curve traced out by a point which moves so that its distance from a fixed point, called the focus, is in a constant ratio, greater than unity, to its perpendicular distance from a fixed straight line, which is called the directrix.

To find the equation of a hyperbola.

Let **S** be the focus, **KX** the directrix, e the eccentricity.

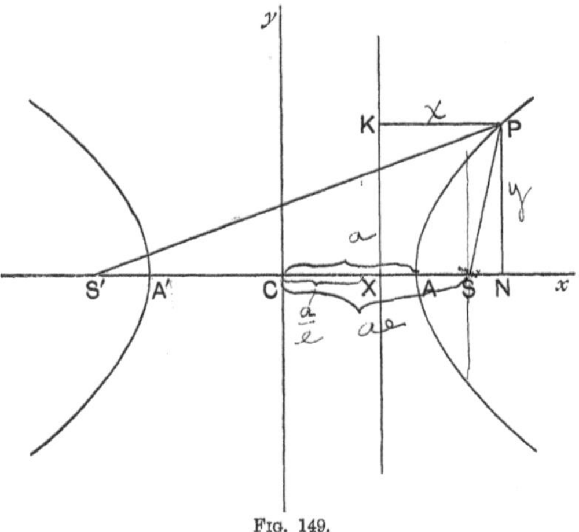

Fig. 149.

Draw **SX** perpendicular to the directrix, and in it take a point **A** such that

$$\frac{SA}{AX} = e.$$

By definition, **A** is a point on the curve.

Since a line can be divided internally and externally in the same ratio, we can find a point **A'**, in **SX** produced so that $\dfrac{SA'}{A'X} = e.$

The point A′ is also on the curve, by definition.

Bisect AA′ at C, and let $CA = a = CA'$.

$$SA' = e \ldots X, \quad \text{and} \quad SA = e \cdot AX;$$
$$\therefore \; SA' + SA = e(A'X + AX)$$
$$\text{or} \quad (CS + CA') + (CS - CA) = eAA' = e \cdot 2CA;$$
$$\therefore \; \mathbf{CS} = \mathbf{ae}. \quad \ldots\ldots\ldots\ldots\ldots\ldots\ldots\ldots(1)$$

Also
$$SA' - SA = e(A'X - AX)$$
$$\text{or} \quad AA' = e\big[(CX + CA') - (CA - CX)\big];$$
$$\therefore \; 2a = 2e \cdot CX;$$
$$\therefore \; \mathbf{CX} = \frac{\mathbf{a}}{\mathbf{e}}. \quad \ldots\ldots\ldots\ldots\ldots\ldots\ldots\ldots(2)$$

Now take CX as axis of x, and Cy, at right angles to it, as axis of y.

Let P(x, y), be any point on the curve, and draw PK perpendicular to the directrix, and PN perpendicular to CX.

By definition,
$$SP = e \cdot PK = e \cdot NX;$$
$$\therefore \; SP^2 = e^2 \cdot NX^2$$
$$SN^2 + PN^2 = e^2 \cdot NX^2.$$

But
$$SN = CN - CS = x - ae, \quad PN = y,$$
$$\text{and} \quad NX = CN - CX = x - \frac{a}{e},$$
$$\therefore \; (x - ae)^2 + y^2 = e^2\left(x - \frac{a}{e}\right)^2.$$

Re-arranging,
$$x^2(e^2 - 1) - y^2 = a^2(e^2 - 1)$$
$$\text{or} \quad \frac{x^2}{a^2} - \frac{y^2}{a^2(e^2 - 1)} = 1. \quad \ldots\ldots\ldots\ldots\ldots\ldots\ldots(3)$$

Let
$$a^2(e^2 - 1) = b^2.$$

[Note that $a^2(e^2 - 1)$ is positive.]

Equation (3) may then be written, $\dfrac{x^2}{a^2} - \dfrac{y^2}{b^2} = 1,$

and this is the equation of the curve.

We notice that the **terms of the second degree,** $\dfrac{x^2}{a^2} - \dfrac{y^2}{b^2}$, are in the form of the difference of two squares, and therefore **have real and different factors.**

This will be found to be the distinguishing characteristic of the equation of any hyperbola.

FORM OF THE HYPERBOLA.

249. The equation may be written $\dfrac{y^2}{b^2} = \dfrac{x^2}{a^2} - 1 = \dfrac{x^2 - a^2}{a^2}$,

$$\text{or} \quad \frac{y}{b} = \pm \sqrt{\frac{x^2 - a^2}{a^2}}.$$

$\therefore$ for any value of x, we have two values of y, equal but opposite in sign.

This proves that **the curve is symmetrical about the axis of x.**

Also if the value of x lies between a and $-a$, $\dfrac{x^2 - a^2}{a^2}$ is negative, and the values of y are imaginary.

This proves that no part of the curve lies between the two straight lines drawn at A and A' at right angles to AA'.

The equation may also be written, $\dfrac{x}{a} = \pm \sqrt{\dfrac{y^2 + b^2}{b^2}}$.

$\therefore$ for any value of y, we have two values of x, equal but opposite in sign.

This proves that the curve is **symmetrical about the axis of y.**

We also see that there are no limitations to the value of y. As y increases, x increases with it, and if y is infinitely large, x is also infinitely large.

The curve therefore consists of **two branches**, each infinite in extent, as shown in the figure.

AA' is called the *transverse axis*.

If BCB' is drawn at right angles to AA' so that

$$BC = B'C = b = a\sqrt{e^2 - 1},$$

BCB is called the *conjugate axis*.

We notice that the hyperbola does not meet its conjugate axis.

As in the ellipse, the symmetry of the curve about its axes shows that it has two foci, and two directrices.

As in the ellipse (Art. 183) we can show that every chord through C is bisected at C. C is therefore called the **centre**.

Any straight line drawn through the centre is called a **diameter**.

It will be seen later on that some diameters do not meet the curve.

250. If in the equation $\dfrac{x^2}{a^2} + \dfrac{y^2}{b^2} = 1$ of an ellipse, we write $-b^2$ for b^2, we obtain the hyperbola $\dfrac{x^2}{a^2} - \dfrac{y^2}{b^2} = 1$.

It follows therefore that many properties of the hyperbola may be proved by the methods employed for the ellipse.

The point (x_1, y_1) is outside or inside the hyperbola $\dfrac{x^2}{a^2} - \dfrac{y^2}{b^2} = 1$, *according as* $\dfrac{x_1^2}{a^2} - \dfrac{y_1^2}{b^2} - 1$ *is greater or less than zero.*   (See Art. 184.)

251. *To find the equation of the hyperbola whose focus is at the point (h, k), directrix the straight line* $Ax + By + C = 0$, *and eccentricity* e. (See Art. 185.)

$e > 1$, and hence it will be seen that the terms of the second degree have real and different factors.

If S *and* S′ *are the foci of the hyperbola* $\dfrac{x^2}{a^2} - \dfrac{y^2}{b^2} = 1$, *and* P (x, y) *is any point on the curve,* SP $= ex - a$, S′P $= ex + a$, *and* S′P $\sim$ SP $= 2a$, *the transverse axis.*

With the figure of Art. 248.

$$\text{SP} = e \cdot \text{PK} = e \cdot \text{NX} = e(\text{CN} - \text{CX}) = e\left(x - \frac{a}{e}\right) = ex - a.$$

$$\text{S′P} = e \cdot \text{PK′} = e \cdot \text{NX′} = e(\text{CN} + \text{CX′}) = e\left(x + \frac{a}{e}\right) = ex + a.$$

By subtraction,   S′P $\sim$ SP $= 2a$.

As in the ellipse (Art. 188) the length of the latus rectum $= \dfrac{2b^2}{a}$.

Examples XII. a.

Find the eccentricity, the distance between the foci, and the latus rectum in each of the following curves :

1. $16x^2 - 9y^2 = 144$.

2. $4x^2 - 3y^2 = 24$.

3. $x^2 - y^2 = 16$.

4. $\dfrac{y^2}{25} - \dfrac{x^2}{144} = 1$.   Draw the curve.

Find the co-ordinates of the foci in the following curves :

5. $\dfrac{x^2}{36} - \dfrac{4y^2}{81} = 1.$ **6.** $16y^2 - 9x^2 = 144.$

7. Find the equation of the hyperbola whose focus is at the point $(-c,\ 0)$, whose eccentricity is e, and whose directrix is the axis of y. Draw one branch of the curve freehand.

8. Draw the hyperbola $\dfrac{(x-2)^2}{4} - \dfrac{y^2}{16} = 1.$

9. Draw the hyperbola $\dfrac{(y-3)^2}{9} - \dfrac{x^2}{4} = 1.$

10. Find the equation of the hyperbola whose focus is at the point $(4,\ 0)$, whose eccentricity is 2, and whose directrix is the line $x - y = 0$. Draw the curve freehand.

11. P is any point on a hyperbola whose axes are equal : prove that $SP . S'P = CP^2.$

12. The eccentricity of the hyperbola whose transverse axis is $2a$, whose axes are the axes of co-ordinates, and which passes through the point $(x_1,\ y_1)$ is

$$\left\{ \frac{x_1^2 + y_1^2 - a^2}{x_1^2 - a^2} \right\}^{\frac{1}{2}}.$$

13. The eccentricity of a hyperbola being $\sqrt{2}$, and the distance of its focus from the directrix $\dfrac{a}{\sqrt{2}}$, obtain its equation in its simplest form with the centre as origin.

252. *To find the equation of the tangent to the hyperbola* $\dfrac{x^2}{a^2} - \dfrac{y^2}{b^2} = 1$ *at the point* $(x_1,\ y_1)$.

The proof in Art. 189 holds if $-b^2$ is written for b^2.

The required equation is $\dfrac{xx_1}{a^2} - \dfrac{yy_1}{b^2} = 1.$

To find the equation of the normal to the hyperbola $\dfrac{x^2}{a^2} - \dfrac{y^2}{b^2} = 1$ *at the point* $(x_1,\ y_1)$.

In Art. 191 write $-b^2$ for b^2, and the proof holds.

The equation is $y - y_1 = -\dfrac{a^2 y_1}{b^2 x_1}(x - x_1)$

or $\dfrac{y - y_1}{\dfrac{y_1}{b^2}} + \dfrac{x - x_1}{\dfrac{x_1}{a^2}} = 0.$

To find the equation of a tangent to the hyperbola $\dfrac{x^2}{a^2} - \dfrac{y^2}{b^2} = 1$ in terms of its slope m.

In Art. 192 write $-b^2$ for b^2.

The required equation is $y = mx \pm \sqrt{a^2 m^2 - b^2}$.

253. *Def.* The circle described on the transverse axis, AA′, of a hyperbola as diameter is called the **auxiliary circle**.

Take any point P(x, y), on the hyperbola and draw the ordinate PN.

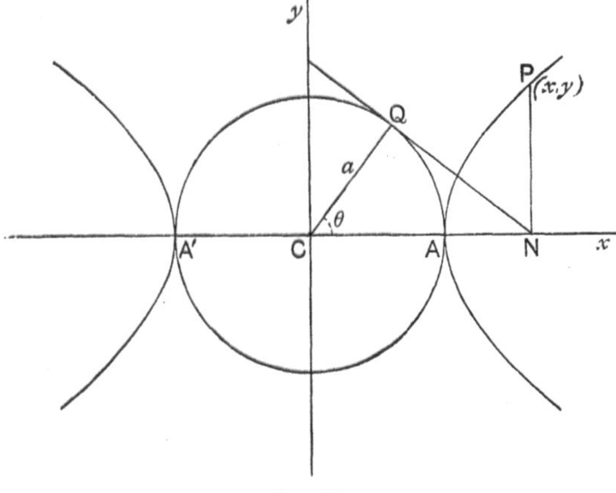

Fig. 150.

From N draw NQ to touch the auxiliary circle at Q.

Join CQ, and let $\angle$ QCN $= \theta$.

$$x = \text{CN} = \text{CQ} \sec \theta = a \sec \theta.$$

Also $\dfrac{x^2}{a^2} - \dfrac{y^2}{b^2} = 1,$ $\therefore$ $\dfrac{y^2}{b^2} = \dfrac{x^2}{a^2} - 1 = \sec^2\theta - 1 = \tan^2\theta$;

$$\therefore \ y = b \tan \theta.$$

$\therefore$ **the co-ordinates of P, any point on the hyperbola $\dfrac{x^2}{a^2} - \dfrac{y^2}{b^2} = 1$**

may be taken to be (a sec θ, b tan θ), when $\theta = \angle$ QCN.

The $\angle$ QCN in the hyperbola corresponds to the eccentric angle in the ellipse.

254. *To find the equation of the tangent to the hyperbola $\dfrac{x^2}{a^2} - \dfrac{y^2}{b^2} = 1$
at the point $(a \sec \theta,\ b \tan \theta)$.*

Take the point $[a \sec (\theta + a),\ b \tan (\theta + a)]$ near to the point
$(a \sec \theta,\ b \tan \theta)$.

The equation of the chord joining these points is

$$\frac{y - b \tan \theta}{b[\tan (\theta + a) - \tan \theta]} = \frac{x - a \sec \theta}{a[\sec (\theta + a) - \sec \theta]}.$$

i.e.
$$\frac{y - b \tan \theta}{b[\sin(\theta + a)\cos \theta - \cos(\theta + a)\sin \theta]} = \frac{x - a \sec \theta}{a[\cos \theta - \cos(\theta + a)]},$$

i.e.
$$\frac{y - b \tan \theta}{b \sin a} = \frac{x - a \sec \theta}{2a \sin \left(\theta + \dfrac{a}{2}\right)\sin \dfrac{a}{2}},$$

i.e.
$$\frac{y - b \tan \theta}{2b \sin \dfrac{a}{2} \cos \dfrac{a}{2}} = \frac{x - a \sec \theta}{2a \sin \left(\theta + \dfrac{a}{2}\right)\sin \dfrac{a}{2}}.$$

Now let the two points move up to one another, and coincide.
a becomes zero, the chord becomes a tangent, and its equation is

$$\frac{y - b \tan \theta}{b} = \frac{x - a \sec \theta}{a \sin \theta}.$$

This may be written, (multiplying both sides by $\tan \theta$)

$$\frac{\mathbf{x} \sec \theta}{\mathbf{a}} - \frac{\mathbf{y} \tan \theta}{\mathbf{b}} = \mathbf{1}.$$

Note. This equation might be deduced from $\dfrac{xx_1}{a^2} - \dfrac{yy_1}{b^2} = 1$, by writing
$a \sec \theta$ for x_1, and $b \tan \theta$ for y_1.

255. *To find the condition that the straight line $lx + my + n = 0$ may
touch the hyperbola $\dfrac{x^2}{a^2} - \dfrac{y^2}{b^2} = 1$.*

Suppose that it touches the hyperbola at the point

$$(a \sec \theta,\ b \tan \theta).$$

The equation of the tangent at this point is

$$\frac{x \sec \theta}{a} - \frac{y \tan \theta}{b} = 1 ;$$

$\therefore$ this equation is identical with

$$lx + my = -n,$$

for the two equations represent the same straight line.

$\therefore$ comparing coefficients,

$$\frac{\sec\theta}{al} = -\frac{\tan\theta}{bm} = -\frac{1}{n}.$$

$$\sec\theta = -\frac{al}{n}, \quad \tan\theta = \frac{bm}{n}.$$

But $\sec^2\theta - \tan^2\theta = 1$;

$$\therefore \frac{a^2l^2}{n^2} + \frac{b^2m^2}{n^2} = 1,$$

or $\quad a^2l^2 + b^2m^2 = n^2$ is the reqd. condition.

256. *Given that* $lx + my + n = 0$ *is a tangent to the hyperbola* $\dfrac{x^2}{a^2} - \dfrac{y^2}{b^2} = 1$, *to find the co-ordinates of its point of contact.*

Employ the method of Art. 197.

257. *To find the equation of the normal at the point* $(a\sec\theta,\, b\tan\theta)$ *to the hyperbola* $\dfrac{x^2}{a^2} - \dfrac{y^2}{b^2} = 1.$

The equation of the tangent at the point $(a\sec\theta,\, b\tan\theta)$ is

$$\frac{x\sec\theta}{a} - \frac{y\tan\theta}{b} = 1.$$

The normal passes through the point $(a\sec\theta,\, b\tan\theta)$ and is perpendicular to the tangent, therefore its equation is

$$(x - a\sec\theta)\frac{a}{\sec\theta} + (y - b\tan\theta)\frac{b}{\tan\theta} = 0. \quad [aa' + bb' = 0.]$$

This may be written $\dfrac{\mathbf{a x}}{\mathbf{\sec}\,\theta} + \dfrac{\mathbf{b y}}{\mathbf{\tan}\,\theta} = \mathbf{a}^2 + \mathbf{b}^2.$

258. *The equation of the chord joining the points* (x_1, y_1), (x_2, y_2) *on the hyperbola* $\dfrac{x^2}{a^2} - \dfrac{y^2}{b^2} = 1$ *may be written in the form*

$$\frac{(x - x_1)(x_1 + x_2)}{a^2} - \frac{(y - y_1)(y_1 + y_2)}{b^2} = 0.$$

In Art. 198 write $-b^2$ for b^2, and the proof holds.

B.A.G.   Q

259. *To find the equation of the chord joining the points*

$$(a \sec \theta, \ b \tan \theta), \ (a \sec \phi, \ b \tan \phi).$$

The equation of the chord is

$$\frac{y - b \tan \phi}{b(\tan \theta - \tan \phi)} = \frac{x - a \sec \phi}{a(\sec \theta - \sec \phi)},$$

i.e.
$$\frac{y - b \tan \phi}{b \sin(\theta - \phi)} = \frac{x - a \sec \phi}{a(\cos \phi - \cos \theta)},$$

i.e.
$$\frac{y - b \tan \phi}{2b \sin \frac{\theta - \phi}{2} \cos \frac{\theta - \phi}{2}} = \frac{x - a \sec \phi}{2a \sin \frac{\theta + \phi}{2} \sin \frac{\theta - \phi}{2}},$$

i.e.
$$\frac{y}{b} \sin \frac{\theta + \phi}{2} - \frac{x}{a} \cos \frac{\theta - \phi}{2} = \tan \phi \sin \frac{\theta + \phi}{2} - \sec \phi \cos \frac{\theta - \phi}{2}$$

$$= \frac{1}{2 \cos \phi}\left[2 \sin \phi \sin \frac{\theta + \phi}{2} - 2 \cos \frac{\theta - \phi}{2}\right]$$

$$= \frac{1}{2 \cos \phi}\left[\cos \frac{\phi - \theta}{2} - \cos \frac{\theta + 3\phi}{2} - 2 \cos \frac{\theta - \phi}{2}\right]$$

$$= -\frac{1}{2 \cos \phi}\left[\cos \frac{\theta + 3\phi}{2} + \cos \frac{\theta - \phi}{2}\right]$$

$$= -\frac{2 \cos \frac{\theta + \phi}{2} \cos \phi}{2 \cos \phi},$$

$\therefore$ the required equation is

$$\frac{x}{a} \cos \frac{\theta - \phi}{2} - \frac{y}{b} \sin \frac{\theta + \phi}{2} = \cos \frac{\theta + \phi}{2}.$$

Note. The equation of the tangent at the point $(a \sec \theta, \ b \tan \phi)$ might be deduced from this by letting ϕ become equal to θ.

Examples XII. b.

1. Find the equations of the tangents to the hyperbola $3x^2 - 4y^2 = 15$ which are parallel to the line $y = 2x + k$.

2. Find the length of the subnormal at the point $(4, 4)$ of the hyperbola $16x^2 - 9y^2 = 112$.

3. A tangent to the hyperbola $9x^2 - 4y^2 = 9$ makes an angle $\tan^{-1}\frac{15}{8}$ with the axis of x. Find the co-ordinates of its point of contact.

4. Prove that the straight line $21x + 5y = 116$ touches the hyperbola $7x^2 - 5y^2 = 232$, and find its point of contact.

5. Tangents are drawn to the hyperbola $3x^2 - 2y^2 = 25$ from the point $(0, 2\frac{1}{2})$.   Find their equations.

6. Tangents are drawn to the hyperbola $4x^2 - y^2 = 256$ from the point $(8, 12)$.   Find their equations.

7. Find the condition that the line $y = mx + \dfrac{p}{m}$ may touch both the parabola $y^2 = 4px$ and the hyperbola $\dfrac{x^2}{a^2} - \dfrac{y^2}{b^2} = 1$.

8. Tangents are drawn to the hyperbola from the point (h, k), and make angles θ and θ' with the axis of x : prove that

$$\tan \theta + \tan \theta' = \frac{2hk}{h^2 - a^2}.$$

9. Find the length of the straight line $x - 3y = 2$ intercepted by the hyperbola $x^2 - 4y^2 = 1$.

10. P is any point on the curve $\left(\dfrac{x}{4}\right)^2 - \left(\dfrac{y}{3}\right)^2 = 1$.   Q is the point $(5, 0)$, and PM the perpendicular on the line $x = 3 \cdot 2$.   Calculate the lengths PQ, PM, and show that their ratio is independent of the position of P on the curve.   What can you deduce?

11. If the tangent at (x', y') to the hyperbola $\dfrac{x^2}{a^2} - \dfrac{y^2}{b^2} = 1$ cuts the auxiliary circle at points whose ordinates are y_1, y_2, show that

$$\frac{1}{y_1} + \frac{1}{y_2} = \frac{2}{y'}.$$

12. Prove that $x \cos a + y \sin a = p$ is a tangent to the hyperbola

$$\frac{x^2}{a^2} + \frac{y^2}{b^2} = 1$$

if $p = \sqrt{a^2 \cos^2 a - b^2 \sin^2 a}$.

13. If a tangent to the hyperbola $\dfrac{x^2}{a^2} + \dfrac{y^2}{b^2} = 1$ meets the transverse axis at T, and the conjugate axis at t, prove that

$$\frac{a^2}{\mathsf{CT}^2} - \frac{b^2}{\mathsf{C}t^2} = 1.$$

14. Find, from first principles, the equation of the tangent to the hyperbola $x^2 - y^2 = a^2$ at the point (x_1, y_1).

15. If the normal at the point $(a \sec \theta, b \tan \theta)$ meets the transverse axis at G, prove that $\mathsf{AG} \cdot \mathsf{A'G} = a^2(e^4 \sec^2\theta - 1)$.

16. Tangents to the hyperbola $3x^2 - 2y^2 = 6$ are drawn from the point (h, k) and make angles θ_1, θ_2 with the axis of x.   If $\tan \theta_1 \tan \theta_2 = 2$, prove that $k^2 = 2h^2 - 7$.

260. *Def.* An **asymptote** to a hyperbola is a straight line which is itself not altogether at infinity, but which meets the conic at *two* points at infinity.

To find the equations of the asymptotes to the hyperbola

$$\frac{x^2}{a^2} - \frac{y^2}{b^2} = 1.$$

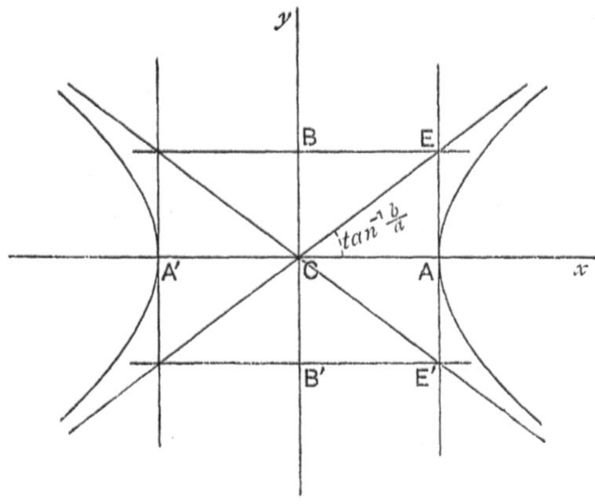

Fig. 151.

Where the straight line $y = mx + c$ meets the curve, we have, by substitution,

$$\frac{x^2}{a^2} - \frac{(mx+c)^2}{b^2} = 1,$$

or $\quad x^2(b^2 - a^2m^2) - 2ma^2cx - a^2c^2 - a^2b^2 = 0.$

Now if $y = mx + c$ is an asymptote, both roots of this equation must be infinite;

$$\therefore \ b^2 - a^2m^2 = 0 \ \text{ and } \ ma^2c = 0 \ \text{ (Art. 150)};$$

$$\therefore \ m = \pm\frac{b}{a} \ \text{ and } \ c = 0;$$

$$\therefore \ y = \pm\frac{bx}{a} \ \text{ are the equations of the asymptotes.}$$

They may be written

$$\frac{x}{a}+\frac{y}{b}=0 \quad \text{and} \quad \frac{x}{a}-\frac{y}{b}=0,$$

or, in one equation,

$$\frac{x^2}{a^2}-\frac{y^2}{b^2}=0.$$

The slopes of the asymptotes are $\pm\dfrac{b}{a}$;

$\therefore$ the angle between them $= 2\tan^{-1}\dfrac{b}{a}$.

We see that the asymptotes are the diagonals of the parallelogram formed by drawing straight lines through A, A', B, B' parallel to the axes.

261. *If $2a$ is the angle between the asymptotes of a hyperbola* $\sec a = e$, *the eccentricity.*

If CE, CE' are the asymptotes, when EAE' is the tangent at A,

$$CA = a, \quad AE = b ;$$

$$\therefore \quad \sec^2 a = \frac{CE^2}{CA^2} = \frac{a^2+b^2}{a^2} = e^2 ;$$

$$\therefore \quad \sec a = e.$$

262. We know that $y = mx + \sqrt{a^2m^2 - b^2}$ is a tangent to the hyperbola $\dfrac{x^2}{a^2} - \dfrac{y^2}{b^2} = 1$ for all values of m.

When $a^2m^2 - b^2 = 0$ or $m = \pm\dfrac{b}{a}$, this equation becomes $y = \pm\dfrac{bx}{a}$, which represents the asymptotes.

$\therefore$ we may look upon the asymptotes as tangents whose points of contact are at an infinite distance.

263. *A straight line drawn parallel to an asymptote meets the hyperbola at one finite point and at one point at infinity.*

$y = \dfrac{bx}{a} + c$ is parallel to an asymptote to the hyperbola

$$\frac{x^2}{a^2} - \frac{y^2}{b^2} = 1.$$

Where it meets the curve, we have by substitution,

$$\frac{x^2}{a^2} - \frac{\left(\frac{bx}{a} + c\right)^2}{b^2} = 1,$$

or $\quad 0 \cdot x^2 - \frac{2cx}{ab} - \frac{c^2}{b^2} - 1 = 0,$

a quadratic with one finite root, and one infinite root (Art. 150). This proves the proposition.

264. The student must be careful to distinguish between a straight line meeting a curve at *infinity* and at *imaginary points*.

Take the straight line $\dfrac{mx}{a} - \dfrac{y}{b} = 0$ where $m > 1$.

Where it meets the curve $\dfrac{x^2}{a^2} - \dfrac{y^2}{b^2} = 1$, we have, by substitution,

$$\frac{x^2}{a^2}[1 - m^2] = 1,$$

$$x = \frac{a}{\sqrt{1 - m^2}}.$$

Now $m^2 > 1$; $\therefore \sqrt{1 - m^2}$ is imaginary, and the line does not meet the curve at real points.

In the case of the asymptote $\dfrac{x}{a} - \dfrac{y}{b} = 0$, where it meets the curve, we have by substitution,

$$\frac{x^2}{a^2} - \frac{x^2}{a^2} = 1 \quad \text{or} \quad 1 = 0.$$

In this case the line meets the curve at *real* points, but those points are at an infinite distance.

265. *To prove that the asymptotes of a hyperbola continually approach the curve but only meet it at infinity.*

Draw an ordinate PN to the hyperbola $\dfrac{x^2}{a^2} - \dfrac{y^2}{b^2} = 1$, and let it meet the asymptote $\dfrac{x}{a} - \dfrac{y}{b} = 0$ at Q.

We shall show that the length PQ continually diminishes, but is never zero, as the abscissa CN is increased.

Since P is on the hyperbola, $\dfrac{CN^2}{a^2} - \dfrac{PN^2}{b^2} = 1$,

$$\text{whence } \frac{PN}{b} = \frac{\sqrt{CN^2 - a^2}}{a}.$$

Since Q is on the asymptote, $\dfrac{QN}{b} = \dfrac{CN}{a}$;

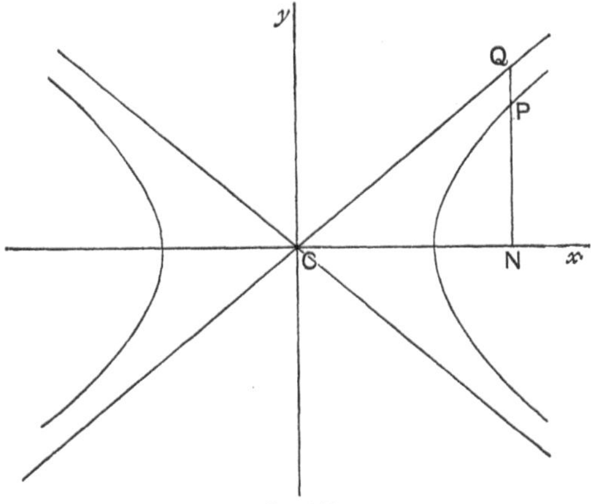

Fig. 152.

$$\therefore \ \frac{QN - PN}{b} = \frac{CN - \sqrt{CN^2 - a^2}}{a} = \frac{CN^2 - (CN^2 - a^2)}{a\left[CN + \sqrt{CN^2 - a^2}\right]}$$

$$= \frac{a}{CN + \sqrt{CN^2 - a^2}}.$$

$\therefore$ as CN increases $QN - PN$ decreases continually, but is never zero.   This proves the proposition.

266. Interpretation of the equation $O \cdot x + O \cdot y + C = 0$.

The intercepts of the straight line $Ax + By + C = 0$ on the axes are $-\dfrac{C}{A}$ and $-\dfrac{C}{B}$.

Let A and B be decreased indefinitely.

As this takes place, the intercepts $-\dfrac{C}{A}$, $-\dfrac{C}{B}$ increase indefinitely, and the straight line moves further and further from the origin until when $\mathsf{A} = \mathsf{B} = 0$, the intercepts are infinitely large, and the straight line is itself altogether at infinity. The equation is then $0 \cdot x + 0 \cdot y + \mathsf{C} = 0$.

Hence $\mathsf{C} = 0$ is said to represent a straight line at infinity.

The above holds for both rectangular and oblique axes.

EXAMPLES. Where the parallel straight lines

$$3x + 4y + 7 = 0,$$
$$3x + 4y - 2 = 0$$

meet, we have by subtraction, $9 = 0$.

$$\therefore \text{ parallel straight lines meet at infinity.}$$

Again, where the hyperbola $\dfrac{x^2}{a^2} - \dfrac{y^2}{b^2} = 1$ meets the straight lines

$$\frac{x^2}{a^2} - \frac{y^2}{b^2} = 0,$$

we have by subtraction, $\qquad 1 = 0.$

i.e. the hyperbola meets these straight lines at infinity only.

Hence $\dfrac{x^2}{a^2} - \dfrac{y^2}{b^2} = 0$ represents the asymptotes of the hyperbola $\dfrac{x^2}{a^2} - \dfrac{y^2}{b^2} = 1.$

267. *If $ax^2 + 2hxy + by^2 = 1$ represents a hyperbola, the equation of its asymptotes is $ax^2 + 2hxy + by^2 = 0$.*

Since $ax^2 + 2hxy + by^2 = 1$ represents a hyperbola, the expression $ax^2 + 2hxy + by^2$ has real linear factors (Art. 251).

$\therefore$ $ax^2 + 2hxy + by^2 = 0$ represents two straight lines not altogether at infinity.

Also where $ax^2 + 2hxy + by^2 = 1$, and $ax^2 + 2hxy + by^2 = 0$ meet, we have by subtraction, $1 = 0$, which proves that the straight lines meet the curve at infinity only.

$\therefore$ $ax^2 + 2hxy + by^2 = 0$ represents the asymptotes.

The above holds for both rectangular and oblique axes.

268. *The equation $(ax + by + c)(a'x + b'y + c') = k$ represents a hyperbola whose asymptotes are the straight lines $ax + by + c = 0$, and $a'x + b'y + c' = 0$.*

The equation $(ax + by + c)(a'x + b'y + c') = k$ represents a hyperbola, for the terms of the second degree $(ax + by)(a'x + b'y)$ have real and different factors.

Also where the curve meets the straight lines whose equation is

$$(ax + by + c)(a'x + b'y + c') = 0,$$

we have by subtraction, $\qquad k = 0.$

$\therefore$ the curve meets the straight lines at infinity only.

$\therefore$ $ax + by + c = 0$, and $a'x + b'y + c' = 0$ are the asymptotes of the hyperbola.

269. *To find the equation of the asymptotes of the hyperbola*

$$ax^2 + 2hxy + by^2 + 2gx + 2fy + c = 0. \quad\ldots\ldots\ldots\ldots(1)$$

Consider the equation

$$ax^2 + 2hxy + by^2 + 2gx + 2fy + \lambda = 0, \quad\ldots\ldots\ldots\ldots(2)$$

and as in Art. 49 find the condition that it may represent two straight lines.

The condition is

$$af^2 + bg^2 + ch^2 - 2fgh - ab\lambda = 0. \quad\ldots\ldots\ldots\ldots\ldots(3)$$

This is a simple equation for λ, and we thus see that *one* value of λ can be found for which equation (2) will represent two straight lines

Where (1) and (2) meet we have by subtraction,

$$\lambda - c = 0,$$

which shows that they meet at infinity only.

$\therefore$ equation (2) represents the asymptotes of the hyperbola (1), the value of λ being obtained from equation (3).

N.B. **The equations of a hyperbola and its asymptotes only differ in the constant terms.**

The above holds both for rectangular and oblique axes.

270. Example 1.　Find the asymptotes of the hyperbola,

$$10x^2 + 11xy - 6y^2 - 19x + 19y - 21 = 0. \quad\ldots\ldots\ldots\ldots\ldots(1)$$

Let λ have such a value that

$$10x^2 + 11xy - 6y^2 - 19x + 19y + \lambda = 0 \quad\ldots\ldots\ldots\ldots\ldots(2)$$

represent two straight lines.

We may proceed as in Art. 269, or as follows :

The equation (2) may be written

$$(2x + 3y)(5x - 2y) - 19x + 19y + \lambda = 0. \quad\ldots\ldots\ldots\ldots\ldots(3)$$

$\therefore$ the factors of the left-hand side must be of the form
$$2x+3y+a, \text{ and } 5x-2y+b.$$
$\therefore$ the equation (3) must be identical with
$$(2x+3y+a)(5x-2y+b)=0,$$
or $\qquad (2x+3y)(5x-2y)+x(2b+5a)+y(3b-2a)+ab=0.$
$\therefore$ equating co-efficients we have,
$$2b+5a=-19, \dots\dots\dots\dots\dots\dots\dots(4)$$
$$3b-2a=19, \dots\dots\dots\dots\dots\dots\dots(5)$$
$$ab=\lambda. \dots\dots\dots\dots\dots\dots\dots(6)$$
From (4) and (5), $\qquad a=-5, b=3$
$$\therefore \text{ from (6), } \lambda=-15.$$
$$\therefore 10x^2+11xy-6y^2-19x+19y-15=0$$
is the equation of the asymptotes.

The separate equations are
$$2x+3y-5=0, \text{ and } 5x-2y+3=0.$$

EXAMPLE 2.   Find the asymptotes of the hyperbola,
$$xy-3x+4y=0.$$
We have to find a value of λ such that the equation
$$xy-3x+4y+\lambda=0$$
represents two straight lines.

In other words, the expression $xy-3x+4y+\lambda$ has real factors.
Now $\qquad xy-3x+4y=x(y-3)+4(y-3)+12,$
$$\therefore xy-3x+4y-12=(y-3)(x+4),$$
$$\therefore -12 \text{ is the required value of } \lambda,$$
and $y-3=0, x+4=0$ are the equations of the asymptotes.

Examples XII. c.

Find the equations of the asymptotes of the following hyperbolas :

1. $2x^2+3xy-2y^2=6.$ $\qquad\qquad$ 2. $3x^2-4y^2-12x-8y-7=0.$
3. $2x^2-3y^2+xy-4x-y-10=0.$ $\qquad$ 4. $5x^2-3y^2-2xy-4x+4y-4=0.$

5. Find the equation of the hyperbola whose asymptotes are the straight lines $x-y=0, x+2y=0$, and which passes through the point (2, 1).

6. Find the equation of the hyperbola whose asymptotes are the straight lines $3y-4x-12=0, 4x+3y-12=0$, and which passes through the origin. Make a freehand drawing of the curve.

Find the equations of the asymptotes, and the eccentricity of the following hyperbolas :

7. $4x^2-y^2+8x=0.$ $\qquad\qquad$ 8. $4x^2-y^2+2y=0.$
9. $x^2-y^2+2x+4y=18.$ $\qquad$ 10. $6x^2-5xy-6y^2=11.$
11. $6x^2-13xy+6y^2-12=0.$ $\qquad$ 12. $6x^2-13xy+6y^2+12=0.$

DRAWING HYPERBOLAS.

271. *To draw a hyperbola having given its focus, directrix, and eccentricity.*

Using squared paper, let S be the focus, OXy the directrix, 2 the eccentricity.

Taking O, the origin, anywhere in the directrix, OXy as axis of y, and Ox at right angles to Oy as axis of x, draw Op, the graph of $y = 2x$.

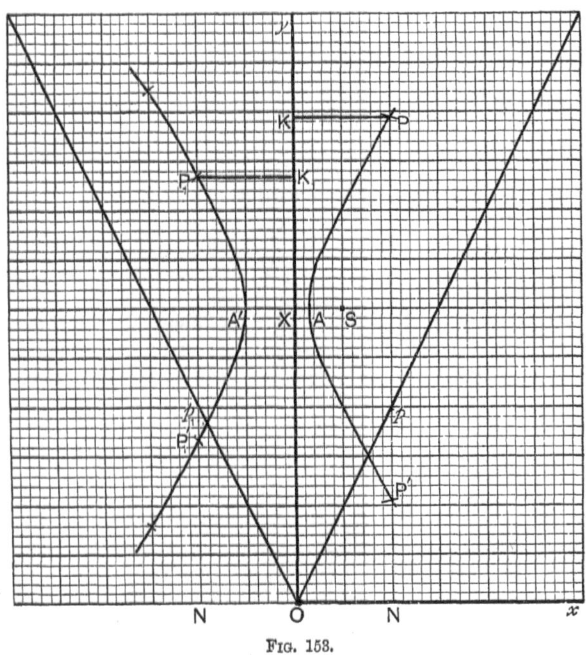

Fig. 153.

Take p any point on this line, and draw its ordinate pN.

With centre S and radius pN, describe a circle cutting Np produced in P, P′.

P, P′ are points on the hyperbola.

For
$$\frac{SP}{PK} = \frac{p\,N}{ON} = 2 \text{ by construction.}$$

In the same way we may obtain other points on the curve.

Joining them by an even curve, we have one branch of the hyperbola.

To draw the other branch, draw Op_1, the graph of $y = -2x$.

With centre S and radius p_1N_1, describe a circle cutting N_1p_1 produced in P_1 and P_1'.

P_1, P_1' are points on the hyperbola.

For
$$\frac{SP_1}{P_1K_1} = \frac{p_1N_1}{ON_1} = 2.$$

Finding other points on this branch in the same way, we obtain the curve by joining them.

272. *To trace a hyperbola by means of the property* $S'P - SP = 2a$.

Take a rod $S'D$ capable of turning in the plane of the paper about one end S'. Fasten one end of a string, which is shorter than $S'D$, to the end D, and the other end to the point S.

Keeping the string taut by means of a pencil point, and keeping the pencil point against the rod as at P, let the rod turn about the point S'.

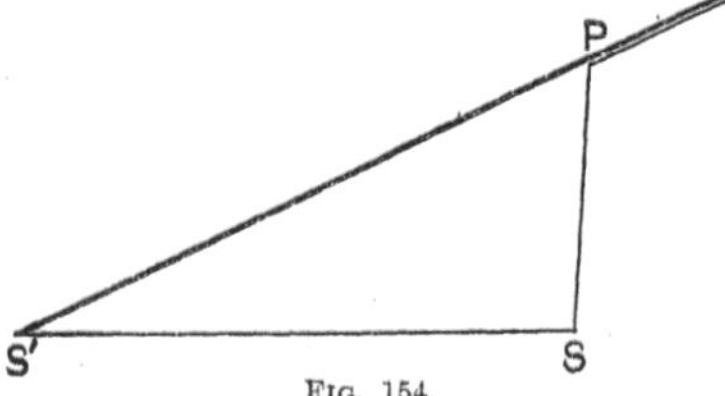

FIG. 154.

$$S'P - SP = S'D - PD - SP$$
$$= \text{length of rod} - \text{length of string};$$

∴ P traces a hyperbola whose foci are S and S', and whose transverse axis = length of rod − length of string.

The following method may also be used:

Draw the hyperbola having given that the distance between its foci is 6 cms. and the length of its transverse axis is 4 cms.

Take S, S' the foci 6 cms. apart.

With centre S and radius 4 cms., and centre S' and radius 8 cms., describe arcs cutting at P and Q.

P and Q are points on the hyperbola, for $SP \sim SP' = 4$ cms., and $SQ \sim S'Q = 4$ cms.

In the same way other points on the curve may be obtained; viz. by describing arcs of 7 and 3 cms. radii, 6 and 2 cms. radii, 5 and 1 cm. radii, and so on.

Joining the points by an even curve, we have the hyperbola.

273. *To draw the hyperbola* $9x^2 - 4y^2 + 54x + 8y + 41 = 0$.

The equation may be written $9(x^2 + 6x) - 4(y^2 - 2y) = -41$.

Completing the squares inside the brackets, the equation becomes

$$9(x+3)^2 - 4(y-1)^2 = 81 - 4 - 41 = 36,$$

$$\text{or} \quad \frac{(x+3)^2}{4} - \frac{(y-1)^2}{9} = 1.$$

If the origin is transferred to the point $(-3, 1)$ this equation becomes

$$\frac{x^2}{4} - \frac{y^2}{9} = 1.$$

$\therefore$ the given equation represents a hyperbola, whose centre is at the point $(-3, 1)$, and whose semi-axes are 2 and 3 in length. The transverse axis (length 4) is parallel to the axis of x.

The drawing of the curve is left to the student.

N.B. When the origin is transferred to $(-3, 1)$, the equations of the asymptotes are $\dfrac{x}{2} + \dfrac{y}{3} = 0$, and $\dfrac{x}{2} - \dfrac{y}{3} = 0$.

Examples XII. d.

1. Draw the hyperbola whose focus is half an inch from the directrix, and whose eccentricity is $\sqrt{2}$.

2. Given that $SS' = 2$ inches, and $SP \backsim S'P = 1$ inch, draw the hyperbola.

3. Draw the hyperbola whose focus is one inch from the directrix, and whose eccentricity is $\frac{5}{3}$.

4. Draw the hyperbola whose equation is $x^2 - 4y^2 + 8y = 0$.

5. Given that $SS' = 1$ inch, and $SP \backsim S'P = \frac{1}{2}$ inch, draw the hyperbola.

6. Draw the hyperbola whose equation is $9x^2 - 8y^2 + 18x + 16y - 71 = 0$.

7. Draw the hyperbola whose distance from focus to focus is 5 cms. and transverse axis 3 cms.

In each of the following cases draw the curve and its asymptotes :

8. $xy = 8$.   **9.** $16x^2 - 64x - 9y^2 = 0$.

10. $x^2 - 25y^2 - 50y = 0$.   **11.** $x^2 - y^2 = 4$.

12. Use the definition of a hyperbola to interpret the equation $(x-1)^2 + (y+2)^2 = 4y^2$. Draw a rough sketch of the curve.

13. Determine the equations of the asymptotes of the hyperbola $xy - 2x - 2y = 0$. Find several points on the curve and draw it.

14. A hyperbola passes through the origin, and $x - 2 = 0$, $y + 1 = 0$ are the equations of its asymptotes. Determine its equation, find several points on the curve and draw it.

15. The straight lines $2x - y = 0$, $y = 0$ are the asymptotes of a hyperbola, and it passes through the point $(2, 1)$. Find the equation of the curve, the co-ordinates of several points on it, and draw it.

274. *Tangents to the hyperbola* $\dfrac{x^2}{a^2} - \dfrac{y^2}{b^2} = 1$ *are drawn from the point* (x_1, y_1) *: to find the equation of their chord of contact.*

In Art. 204 write $-b^2$ for b^2 throughout.

To find the polar of the point (x_1, y_1) *with respect to the hyperbola* $\dfrac{x^2}{a^2} - \dfrac{y^2}{b^2} = 1.$

In Art. 205 write $-b^2$ for b^2 throughout.

If the polar of the point P *passes through the point* Q, *the polar of the point* Q *passes through the point* P.

This may be proved as in Art. 144.

Find the co-ordinates of the pole of the straight line $lx + my + n = 0$ *with respect to the hyperbola* $\dfrac{x^2}{a^2} - \dfrac{y^2}{b^2} = 1.$

Proceed as in Art. 206, writing $-b^2$ for b^2.

To find the equation of a chord of the hyperbola $\dfrac{x^2}{a^2} - \dfrac{y^2}{b^2} = 1$, *in terms of the co-ordinates of its middle point* (x_1, y_1).

Proceed as in Art. 207, writing $-b^2$ for b^2.

Conjugate diameters. As in the ellipse, two diameters are said to be conjugate to each other when each bisects chords parallel to the other.

If a series of parallel chords of the hyperbola $\dfrac{x^2}{a^2} - \dfrac{y^2}{b^2} = 1$ *make an angle* θ *with the transverse axis, the locus of their middle points is a straight line through the centre (a diameter) whose equation is*

$$\frac{x \cos \theta}{a^2} - \frac{y \sin \theta}{b^2} = 0.$$

To prove this, use the method of Art. 208.

If m, m' are the slopes of two conjugate diameters, $mm' = \dfrac{b^2}{a^2}$.

See Art. 208.

The tangents at the extremities of a diameter which meets the curve are parallel to the chords bisected by that diameter.

See Art. 209.

Supplemental chords of a hyperbola are parallel to a pair of conjugate diameters.

See Art. 212.

275. Rectangular hyperbola. When the two axes of a hyperbola are equal in length the hyperbola is said to be *rectangular*.

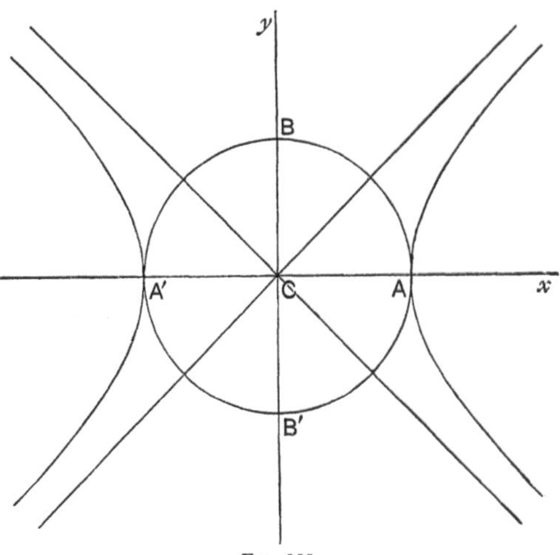

FIG. 155.

Referred to its axes its equation is $\dfrac{x^2}{a^2} - \dfrac{y^2}{a^2} = 1$,

$$\text{or} \quad \mathbf{x}^2 - \mathbf{y}^2 = \mathbf{a}^2.$$

$x^2 - y^2 = 0$ is the equation of its asymptotes.

These, the lines $x - y = 0$, $x + y = 0$, are at right angles; hence the name *rectangular* hyperbola.

276. *The Conjugate Hyperbola.*

At right angles to ACA' the transverse axis of the hyperbola $\dfrac{x^2}{a^2} - \dfrac{y^2}{b^2} = 1$, draw BCB', making $BC = B'C = b$.

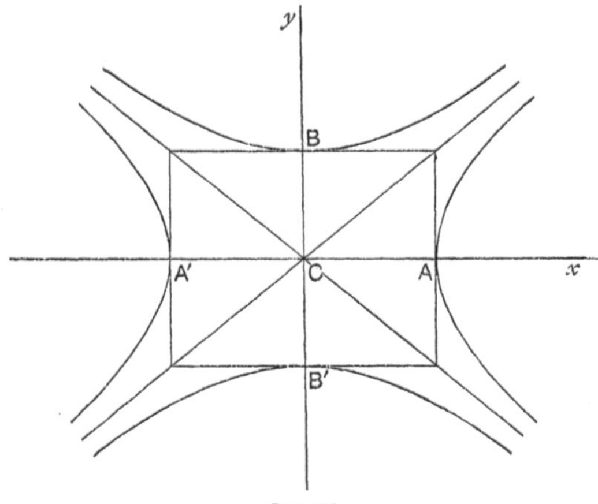

Fig 156.

The hyperbola whose transverse axis is BB', and conjugate axis AA', is said to be *conjugate* to the hyperbola $\dfrac{x^2}{a^2} - \dfrac{y^2}{b^2} = 1$.

The equation of this conjugate hyperbola is $\dfrac{y^2}{b^2} - \dfrac{x^2}{a^2} = 1$,

$$\text{or} \quad \frac{x^2}{a^2} - \frac{y^2}{b^2} = -1.$$

We notice that the point $(a \tan \theta,\ b \sec \theta)$ lies on this hyperbola for all values of θ, for $\tan^2 \theta - \sec^2 \theta = -1$.

We see that a hyperbola and its conjugate have the same asymptotes.

277. *If the diameter $y = mx$ meets the hyperbola $\dfrac{x^2}{a^2} - \dfrac{y^2}{b^2} = 1$ in real points, its conjugate diameter $y = m'x$ meets the curve in imaginary points.*

Let $y = mx$ meet the curve at the point (x_1, y_1), and $y = m'x$ at the point (x_2, y_2).

Where $y = mx$ meets the curve, we have by substitution,

$$x_1{}^2\left(\frac{1}{a^2} - \frac{m^2}{b^2}\right) = 1$$

$$\therefore \; x_1{}^2 = \frac{a^2b^2}{b^2 - a^2m^2}.$$

Similarly where $y = m'x$ meets the curve,

$$x_2{}^2 = \frac{a^2b^2}{b^2 - a^2m'^2}$$

$$= \frac{a^2b^2}{b^2 - \dfrac{b^4}{a^2m^2}}, \;\; \text{for} \;\; mm' = \frac{b^2}{a^2}$$

$$= \frac{a^4m^2}{a^2m^2 - b^2} = -\frac{a^4m^2}{b^2 - a^2m^2}.$$

$$\therefore \; \frac{x_1{}^2}{x_2{}^2} = -\frac{b^2}{a^2m^2}.$$

Hence, if x_1 is real, x_2 is imaginary. This proves the proposition.

If 2α is the length of a diameter of a hyperbola, and 2β the length of the conjugate diameter cut off by the conjugate hyperbola, 2α and 2β are often called the lengths of these two conjugate diameters of the original hyperbola.

278. *If a hyperbola has its centre at the origin, its equation contains no terms of the first degree.*

See Art. 213.

***279.** To find the equation of a hyperbola referred to two conjugate diameters as axes of co-ordinates.* (Oblique.)

The origin being at the centre of the curve, if (x, y) lies on the curve, $(-x, -y)$ also lies on the curve.

$\therefore$ its equation can contain no terms of the first degree.

We may therefore take

$$A x^2 + 2 H xy + B y^2 = 1 \quad \dots\dots\dots\dots\dots\dots(1)$$

to be the equation of the hyperbola.

Again if P is the point (x, y) and the chord PVP' is drawn parallel to CB' the axis of y, PV' = PV, or the co-ordinates of P', a point on the curve, are $(x, -y)$.

$\therefore$ $Ax^2 - 2Hxy + By^2 = 1$, by substitution.

$\therefore$ from (1) by subtraction, $H = 0$.

$\therefore$ the equation of the curve reduces to

$$Ax^2 + By^2 = 1.$$

Let the curve cut the axis of x at A' and let $CA' = a'$.   When

$$y = 0, \quad x = a'.$$

$$\therefore Aa'^2 = 1, \quad \text{and} \quad A = \frac{1}{a'^2}.$$

The axis of y does not meet the curve; $\therefore$ B must be negative.

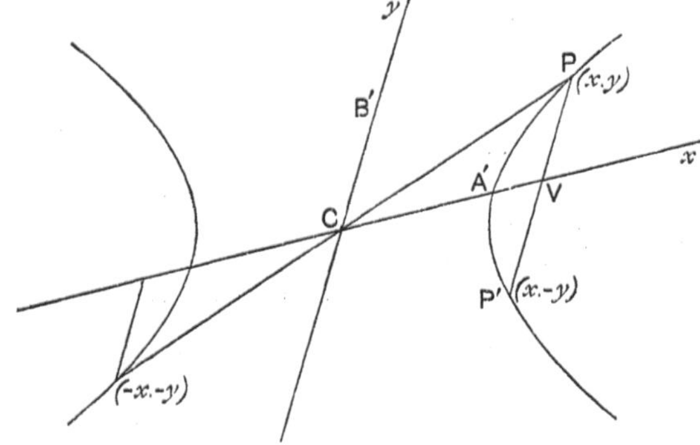

Fig. 157.

Taking it equal to $-\dfrac{1}{b'^2}$, the equation of the curve is

$$\frac{x^2}{a'^2} - \frac{y^2}{b'^2} = 1.$$

N.B. **The point ($a'\sec\theta$, $b'\tan\theta$) lies on the curve for all values of θ.**

Examples XII. e.

1. Find the pole of the chord $21x - 9y = 28$ with respect to the hyperbola $7x^2 - 12y^2 = 112$.

2. In the hyperbola $25x^2 - 16y^2 = 400$ find the equation of the diameter conjugate to $3y = x$.

3. Find the condition that $\dfrac{x}{l} + \dfrac{y}{m} = 1$ may be a tangent to the hyperbola $\dfrac{x^2}{a^2} - \dfrac{y^2}{b^2} = 1$.

4. In the hyperbola $\dfrac{x^2}{a^2} - \dfrac{y^2}{b^2} = 1$ find the condition that the chord joining the two points $(a \sec \theta,\ b \tan \theta)$, $(a \sec \phi,\ b \tan \phi)$ should subtend a right angle at the centre.

5. **P** is a point on a hyperbola, **Q** on the conjugate hyperbola; if **PCQ** is a right angle $\dfrac{1}{\mathbf{CP}^2} - \dfrac{1}{\mathbf{CQ}^2}$ is constant.

6. In any hyperbola the circle on **SP** as diameter touches the auxiliary circle.

7. Find the equation of the chord of the hyperbola $16x^2 - 9y^2 = 144$ which is bisected at the point $(12, 3)$.

8. Find the length of the semi-diameter conjugate to the diameter $y = 3x$ in the hyperbola $9x^2 - 4y^2 = 36$.

9. In the hyperbola $\dfrac{x^2}{a^2} - \dfrac{y^2}{b^2} = 1$ find the condition that the chord joining the points $(a \sec \theta,\ b \tan \theta)$, $(a \sec \phi,\ b \tan \phi)$ should subtend a right angle at a vertex.

10. Obtain the equations of the asymptotes of the hyperbola
$$x^2 - 4y^2 + x + 4y = 0.$$
Draw the curve to scale on squared paper, taking 2 inches as unit.

11. Through one of the vertices **A**, and the extremities **P**, **P′** of a double ordinate of a hyperbola, a circle is drawn cutting the axis again in **K**. If **G** be the foot of the normal at **P**, prove that **GK** is of constant length.

12. If the normal to the hyperbola $x^2 - y^2 = a^2$ at the point $(a \sec \theta,\ a \tan \theta)$ meets the curve again at the point (x_1, y_1), prove that
$$y_1 = \frac{a(1 + \cos^2 \theta)}{\sin \theta \cos \theta}.$$

13. Find the asymptotes of the hyperbola $xy + ax + by + c^2 = 0$.

14. In the rectangular hyperbola the straight line joining the centre with the pole makes the same angle with the transverse axis as the straight line drawn from the centre perpendicular to the polar, and the semi-axis is a mean proportional between the lengths of these straight lines.

15. If two hyperbolas are conjugate to each other, the polar of a point on either of them with respect to the other is a tangent to the first.

16. The perpendicular from the centre on any normal to an equilateral hyperbola meets the curve in imaginary points.

17. If h and k be the intercepts on the axis of any tangent to a hyperbola,
$$\frac{a^2}{h^2} - \frac{b^2}{k^2} = 1.$$

18. A perpendicular from the centre on any tangent to the hyperbola $x^2 - y^2 = a^2$ meets the tangent in **Z** and the curve in **Q**; prove that
$$\mathbf{CZ} \cdot \mathbf{CQ} = a^2.$$

19. Find the equation to the chord of the hyperbola $25x^2 - 16y^2 = 400$ which is bisected at the point (5, 3).

20. If the polar of any point with respect to the hyperbola

$$b^2x^2 - a^2y^2 = a^2b^2$$

passes through one end of the conjugate axis, prove that the pole will lie on the tangent to the conjugate hyperbola at the other end of that axis.

***280.** *The equation of a hyperbola referred to its asymptotes as axes of co-ordinates is* $\mathbf{4xy} = \mathbf{a}^2 + \mathbf{b}^2$.

The equation $(Ax + By + C)(A'x + B'y + C') = k^2$ represents a hyperbola whose asymptotes are the straight lines $Ax + By + C = 0$, and $A'x + B'y + C' = 0$.

$\therefore$ the equation $xy = k^2$ represents a hyperbola whose asymptotes are the lines $x = 0$ and $y = 0$, *i.e.* the axes of co-ordinates.

We now have to determine the value of the constant term k^2.

Take the vertex A, and draw its co-ordinates CN, AN.

$\angle\, \text{NAC} = \angle\, y\text{CA} = \angle\text{ACN}\,; \quad \therefore \quad \text{CN} = \text{AN}, \; i.e. \text{ at A}, \; x = y.$

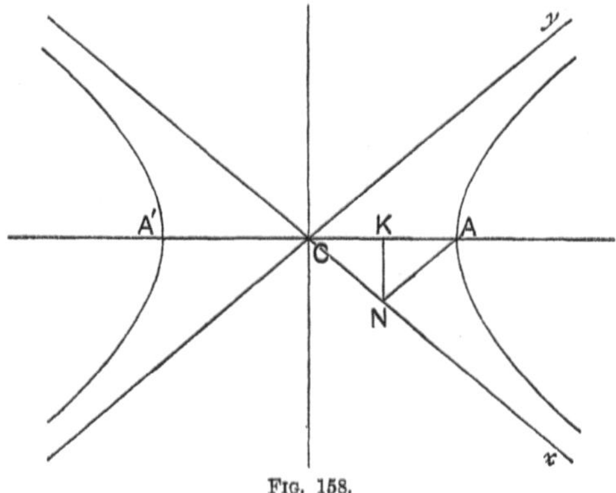

Fig. 158.

Draw NK perpendicular to the transverse axis, and let $2a$ be the angle between the asymptotes.

Then
$$x = \text{CN} = \text{CK} \sec a = \frac{\text{CA}}{2} \sec a = \frac{a \sec a}{2}\,;$$

$$\therefore \; k^2 = \text{CN}^2 = \frac{a^2 \sec^2 a}{4} = \frac{a^2 e^2}{4} = \frac{a^2 + b^2}{4}\,;$$

$$\therefore \; xy = \frac{a^2 + b^2}{4} \text{ is the equation required.}$$

***281.** *To find the equation of a tangent at the point (x_1, y_1) to the hyperbola $2xy = c^2$.*

Take a point $(x_1 + \Delta x_1, y_1 + \Delta y_1)$ on the curve, and near to the point (x_1, y_1).

(x_1, y_1) is on the curve, $\therefore 2x_1 y_1 = c^2$.

For the same reason $2(x_1 + \Delta x_1)(y_1 + \Delta y_1) = c^2$.

By subtraction $\quad x_1 \Delta y_1 + y_1 \Delta x_1 + \Delta x_1 \Delta y_1 = 0$;

and in the limit, when the points approach one another,

$$x_1 \Delta y_1 + y_1 \Delta x_1 = 0.$$

$$\therefore \frac{\Delta y_1}{\Delta x_1} = -\frac{y_1}{x_1} \text{ in the limit.}$$

$\therefore$ the equation of the tangent is

$$y - y_1 = -\frac{y_1}{x_1}(x - x_1),$$

$$\text{i.e. } x_1 y + y_1 x = 2x_1 y_1,$$

$$\mathbf{x_1 y + y_1 x = c^2}.$$

***282.** *The portion of a tangent to a hyperbola intercepted between the asymptotes is bisected at the point of contact.*

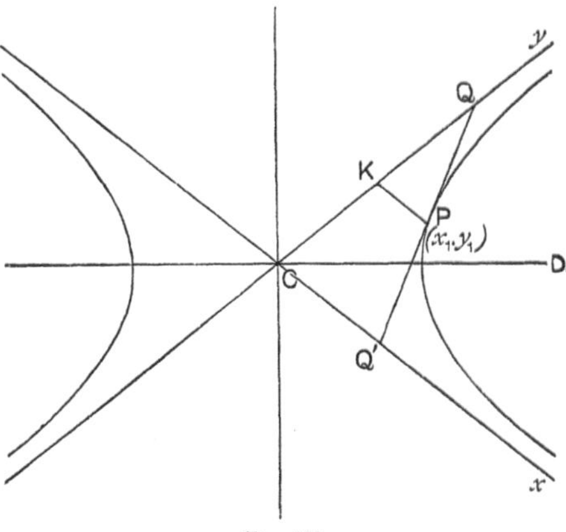

Fig. 159.

Let the equation of the hyperbola referred to its asymptotes as axes be $2xy = c^2$.

The equation of the tangent at the point P, (x_1, y_1), is

$$x_1 y + y_1 x = c^2. \quad \ldots\ldots\ldots\ldots\ldots\ldots\ldots\ldots(1)$$

Let the tangent meet the asymptotes at Q and Q′, and draw PK parallel to Cx to meet Cy at K.

At Q′ $y = 0$, $\therefore$ from (1) $y_1 \cdot$ CQ′ $= c^2$;

$$\therefore \ \text{CQ}' = \frac{c^2}{y_1} = \frac{2x_1 y_1}{y_1} = 2x_1 = 2\text{PK}.$$

Also PK is parallel to CQ′.

$$\therefore \ \text{PQ} = \text{PQ}'.$$

***283.** *The area of the triangle contained by a tangent to a hyperbola and its asymptotes is constant.*

Let the equation of the hyperbola referred to its asymptotes be $2xy = c^2$, and let $2a$ be the angle between the asymptotes.

At the point P(x_1, y_1) draw the tangent to meet the asymptotes in Q and Q′.

By Art. 282, CQ $= 2y_1$, and CQ′ $= 2x_1$.

$$\therefore \ \text{area of triangle QCQ}' = \tfrac{1}{2}\text{CQ} \cdot \text{CQ}' \sin 2a$$

$$= 2x_1 y_1 \sin 2a = c^2 \sin 2a,$$

which is constant for all positions of P.

<h2 align="center">*Examples XII. f.</h2>

In the rectangular hyperbola $xy = c^2$, show that :

1. The equation of the normal at $(x_1\ y_1)$ is $xx_1 - yy_1 = x_1{}^2 - y_1{}^2$.

2. The equation of a chord in terms of its middle point (x_1, y_1) is

$$\frac{x}{x_1} + \frac{y}{y_1} = 2.$$

3. If the tangent and normal to a rectangular hyperbola make intercepts a_1, a_2 on one asymptote, and b_1, b_2 on the other, prove that $a_1 a_2 + b_1 b_2 = 0$.

4. If four points on the rectangular hyperbola $xy = c^2$ lie on a circle, the product of their abscissae is equal to c^4.

5. Prove that the asymptotes of the hyperbola $xy = \beta x + ay$ are $x = a$, $y = \beta$. Draw the curve.

6. In the rectangular hyperbola $xy = c^2$, a tangent meets the asymptotes in T, T′, and the normal at the same point meets them in N, N′. Prove that TN . T′N′ varies as the fourth power of the central radius vector of the point.

7. Prove that the rectangle contained by the intercepts made by any tangent to a hyperbola on its asymptotes is constant.

8. Find the co-ordinates of the intersections of the rectangular hyperbola $xy = 1$ and $ny + n^4 = n^3 x + 1$, and show that at one of these points the hyperbola and the straight line cut at right angles. Draw the figure when $n = 2$, unit one inch.

9. Prove that the circle described on the line joining the origin to $\left(c^3, \dfrac{1}{c^3} \right)$ as diameter passes through the point $\left(\dfrac{1}{c}, c \right)$. Hence justify the following graphical construction for a cube root by means of a figure of the rectangular hyperbola $xy = 1$:

On Ox lay off OM, whose cube root is required. Draw the ordinate MP to meet the curve in P. On OP as diameter describe a circle meeting the curve again in Q. From Q draw QN perpendicular to Ox. Then QN is the required cube root.

10. If $(c \tan \theta,\ c \cot \theta)$ be the co-ordinates of a point on the hyperbola $xy = c^2$, show that the chord through the points θ and ϕ, where $\theta + \phi$ is constant, passes through a fixed point on the conjugate axis of the hyperbola.

11. Show that if a variable line form, with two fixed perpendicular lines, a triangle of constant area, the locus of a point, which divides the intercept made on the variable line in a given ratio, is a hyperbola.

LOCUS PROBLEMS ON THE HYPERBOLA.

284. *Tangents are drawn to the parabola $y^2 = 4ax$ making a constant angle α with one another: prove that the locus of their intersection is a hyperbola whose asymptotes are parallel to the lines $y = \pm x \tan \alpha$, and whose eccentricity is $\sec \alpha$.*

The straight line $y = mx + \dfrac{a}{m}$ is a tangent to the parabola $y^2 = 4ax$ for all values of m.

Multiplying up, we may write the equation

$$m^2 x - my + a = 0.$$

The roots of this quadratic for m are the slopes of the tangents which can be drawn from the point (x, y) to the parabola $y^2 = 4ax$.

Let $m_1,\ m_2$ be the roots.

$$m_1 + m_2 = \frac{y}{x} \quad \text{and} \quad m_1 m_2 = \frac{a}{x}.$$

$$\therefore\ (m_1 - m_2)^2 = \frac{y^2}{x^2} - \frac{4a}{x} = \frac{y^2 - 4ax}{x^2}\ ;$$

$$\therefore\ \tan^2 \alpha = \left(\frac{m_1 - m_2}{1 + m_1 m_2} \right)^2 = \frac{y^2 - 4ax}{x^2} \times \frac{x^2}{(a + x)^2}\ ;$$

whence we have

$$x^2 \tan^2 \alpha - y^2 + 2ax \tan^2 \alpha + 4ax + a^2 \tan^2 \alpha = 0.$$

This is the equation of the locus, and represents a hyperbola, for $x^2\tan^2a - y^2$, the terms of the second degree, has real and different factors. The asymptotes are parallel to the lines represented by

$$x^2\tan^2a - y^2 = 0 \quad\text{or}\quad y = \pm x\tan a. \quad (\text{Art. 268.})$$

From a diagram it is seen at once that $2a$ is the angle between the asymptotes, and therefore $\sec a = $ the eccentricity. (Art. 261.)

285. *Chords of a hyperbola subtend a right angle at one of the vertices : find the locus of the poles of the chords.*

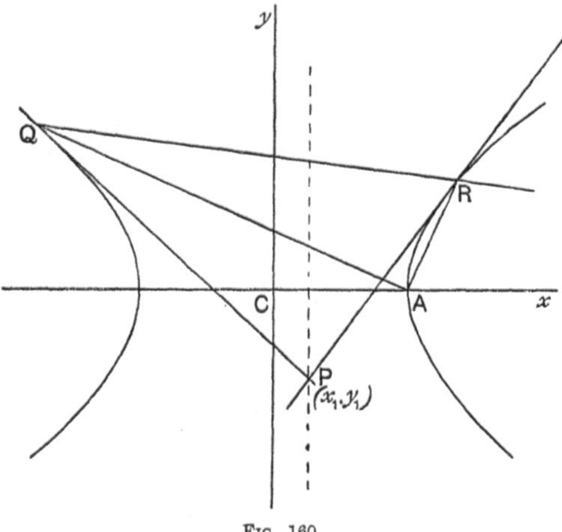

FIG. 160.

Let $\dfrac{x^2}{a^2} - \dfrac{y^2}{b^2} = 1$ be the hyperbola, $P(x_1, y_1)$ the pole of any one of the chords.

The equation of the polar QR is

$$\frac{xx_1}{a^2} - \frac{yy_1}{b^2} = 1.$$

Transferring the origin to the vertex $A(a, 0)$, the equation of the hyperbola becomes

$$\frac{(x+a)^2}{a^2} - \frac{y^2}{b^2} = 1 \quad\text{or}\quad \frac{x^2}{a^2} - \frac{y^2}{b^2} + \frac{2x}{a} = 0. \quad\ldots\ldots\ldots\ldots(1)$$

The equation of the polar becomes

$$\frac{(x+a)x_1}{a^2} - \frac{yy_1}{b^2} = 1 \text{ or } \frac{xx_1}{a^2} - \frac{yy_1}{b^2} = 1 - \frac{x_1}{a}. \quad \ldots\ldots\ldots\ldots(2)$$

Combining equations (1) and (2), so as to make the result homogeneous and of the second degree, we have

$$\left(\frac{x^2}{a^2} - \frac{y^2}{b^2}\right)\left(1 - \frac{x_1}{a}\right) + \frac{2x}{a}\left(\frac{xx_1}{a^2} - \frac{yy_1}{b^2}\right) = 0. \quad \ldots\ldots\ldots\ldots(3)$$

This equation being of the second degree throughout represents two straight lines through the origin. Also, it is formed by combining equations (1) and (2), and therefore its locus passes through their common points Q and R.

Therefore the equation represents AQ and AR.

But, by hypothesis, AQ and AR are at right angles.

$\therefore$ in equation (3) the sum of the coefficients of x^2 and y^2 is zero.

$$\therefore \quad \frac{1}{a^2}\left(1 - \frac{x_1}{a}\right) - \frac{1}{b^2}\left(1 - \frac{x_1}{a}\right) + \frac{2x_1}{a^3} = 0,$$

or

$$\frac{x_1}{a}\left(\frac{1}{a^2} + \frac{1}{b^2}\right) = \frac{1}{b^2} - \frac{1}{a^2},$$

or

$$x_1(a^2 + b^2) = a(a^2 - b^2).$$

But (x_1, y_1) is any point on the locus.

$\therefore$ suppressing the suffix, the equation of the locus is

$$x(a^2 + b^2) = a(a^2 - b^2),$$

a straight line at right angles to the transverse axis of the hyperbola.

Examples XII. g.

1. Find the equation of, and draw, the locus of the intersection of the straight lines $x+y=4t$, $tx - ty =4$, for all values of t.

2. The normal at P to a hyperbola meets the axes of the curve at G and g. Completing the rectangle gCGQ, find the equation of, and draw, the locus of Q.

3. Two tangents are drawn to the hyperbola $\dfrac{x^2}{a^2} - \dfrac{y^2}{b^2} = 1$ such that the product of their slopes is k^2: prove that they intersect on the curve $y^2 + b^2 = k^2(x^2 - a^2)$, and draw this curve.

4. Find the equation of the locus of the intersection of the tangent at any point of the hyperbola $x^2 - y^2 = a^2$ with the perpendicular upon it from the centre.

5. PNP' is a double ordinate of the hyperbola $\dfrac{x^2}{a^2} - \dfrac{y^2}{b^2} = 1$ whose vertices are at A and A'.   If AP, A'P' intersect in Q, prove that the locus Q is the ellipse $\dfrac{x^2}{a^2} + \dfrac{y^2}{b^2} = 1$.

6. Given a chord of a parabola and the direction of its axis, show that the locus of the focus is a hyperbola whose foci are at the extremities of the given chord.

7. Show that if $x = t + \dfrac{1}{t}$, and $y = t - \dfrac{1}{t}$, the locus of the point $(x,\ y)$ when t varies is a hyperbola.

8. Find the locus of the intersection of the straight lines
$$bx - ay - mab = 0, \quad mbx + may - ab = 0$$
where m is variable.

9. From a fixed point $(h,\ k)$ a line is drawn at right angles to the tangent to the parabola $y^2 = 4ax$ at P, and meeting the diameter through P in Q.   Prove that the locus of Q for different positions of P is the rectangular hyperbola $y(x + 2a - h) = 2ak$.   Draw the curve and its asymptotes.

10. A straight line PCM is drawn from a fixed point P to intersect a fixed straight line AB in C, and is produced to a point M such that a perpendicular MD to the fixed line intercepts a segment CD of given magnitude.   Show that M lies on an equilateral hyperbola, and find its centre and asymptotes.

11. Show that the locus of the vertex of a triangle constructed on a given base, one of whose base angles is double the other, is a hyperbola whose transverse axis is two-thirds of the base.

12. The co-ordinates of a point $(x,\ y)$ are given by the equations
$$x = a \tan(\theta + a), \quad y = b \tan(\theta + \beta)\ ;$$
θ being variable.   Show that the locus of the point is a hyperbola, and find the position of its asymptotes.

13. From a point P perpendiculars PM, PN are drawn to two fixed lines AM, AN.   If the area AMPN is constant, prove that the locus of P is a hyperbola.   (Oblique axes.)

14. A tangent to the parabola $x^2 = 4ay$ cuts the rectangular hyperbola $xy = k^2$ in two points, P and Q.   Show that the middle point of PQ lies on a fixed parabola.

15. Prove that the equation of the tangent to the parabola $(x + y)^2 = 4ax$ at the point $(x_1,\ y_1)$ is $(x + y)(x_1 + y_1) = 2a(x + x_1)$.

16. A tangent to the parabola in the previous example cuts the rectangular hyperbola $xy = k^2$ in two points P and Q.   Show that the locus of the middle point of PQ is the hyperbola $2y(x - y) = ax$.

17. A normal to the hyperbola $\dfrac{x^2}{a^2} - \dfrac{y^2}{b^2} = 1$ meets the axes in M and N : show that if perpendiculars MP, NP are drawn to the axes, the locus of P will be the hyperbola $a^2x^2 - b^2y^2 = (a^2 + b^2)^2$.

18. Prove that the locus of the point of intersection of tangents to the curve $x^2 - y^2 = a^2$, which are inclined to each other at an angle of 45°, is the curve $(x^2 + y^2)^2 = 4a^2(a^2 + y^2 - x^2)$.

19. A straight line touches the circle which has for its diameter the line joining the foci of the hyperbola $b^2x^2 - a^2y^2 = a^2b^2$. Show that the locus of its pole with respect to the hyperbola is an ellipse whose eccentricity is

$$\frac{\sqrt{a^4 - b^4}}{a^2}.$$

20. AB and AC are fixed straight lines at right angles to one another, and the straight line BOC passes through a fixed point O, the points B, C being otherwise undetermined. Prove that the locus of the middle point of BC is a rectangular hyperbola, and determine its asymptotes.

PROPERTIES OF THE HYPERBOLA.

286. *To prove that* $\dfrac{PN^2}{AN \cdot A'N} = \dfrac{BC^2}{AC^2}.$

Let $(a \sec \theta,\ b \tan \theta)$ be the co-ordinates of P.

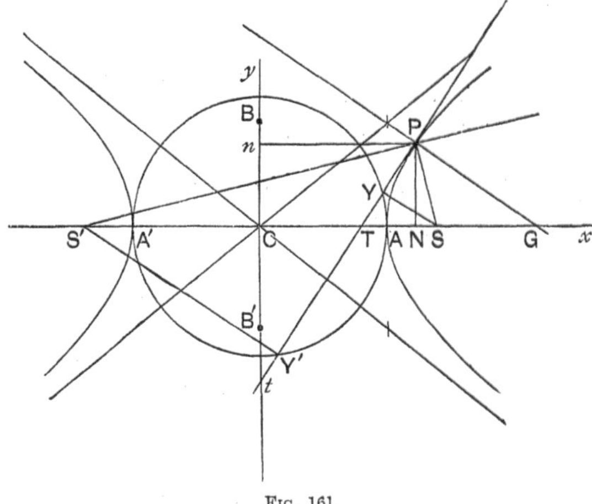

Fig. 161.

$$\frac{PN^2}{AN \cdot A'N} = \frac{PN^2}{CN^2 - CA^2} = \frac{b^2 \tan^2 \theta}{a^2 \sec^2 \theta - a^2} = \frac{b^2}{a^2}.$$

$$SP \sim S'P = the\ major\ axis.$$

This is proved in Art. 251.

To prove that $BC^2 = AS \cdot A'S$.

$$AS \cdot A'S = (CS + CA)(CS - CA) = CS^2 - CA^2 = a^2 e^2 - a^2 = b^2.$$

If PN *is the ordinate of* P, *and the tangent at* P *meets the transverse axis at* T, $\qquad CN \cdot CT = CA^2$.

Take $(a \sec \theta, \ b \tan \theta)$ for the co-ordinates of P, and use the method of proof for the ellipse in Art. 223.

If Pn *is drawn perpendicular to the conjugate axis, and the tangent at* P *meets the conjugate axis at* t, $Cn \cdot Ct = BC^2$.

Taking $(a \sec \theta, \ b \tan \theta)$ for the co-ordinates of P, use the method of proof for the ellipse in Art. 224.

It will be found that $Cn \cdot Ct = -b^2$, showing that Cn, and Ct are drawn in opposite directions.

If the normal at P *meets the major axis at* G, $CG = e^2 \cdot CN$.

Taking $\dfrac{ax}{\sec \theta} + \dfrac{by}{\tan \theta} = a^2 + b^2$ as the equation of the normal, use the method for the ellipse in Art. 225.

If the normal meets the transverse axis at G, $SG = eSP$. *Also the normal bisects the exterior angle, and the tangent the interior angle between the focal distances of* P.

Take $\dfrac{ax}{\sec \theta} + \dfrac{by}{\tan \theta} = a^2 + b^2$ as the equation of the normal, and use the method for the ellipse in Art. 226.

If perpendiculars SY, S'Y' *are drawn to any tangent,* Y *and* Y' *lie on the auxiliary circle.*

Taking $y = mx + \sqrt{a^2 m^2 - b^2}$ as the equation of the tangent, use the method for the ellipse in Art. 227.

If perpendiculars SY, S'Y' *are drawn to any tangent,* SY $\cdot$ S'Y' $= BC^2$.

Taking $y = mx + \sqrt{a^2 m^2 - b^2}$ as the equation of the tangent, use the method for the ellipse in Art. 228.

The portion of the tangent intercepted between the point of contact and the directrix subtends a right angle at the focus.

This can be proved as for the ellipse in Art. 231 writing $-b^2$ for b^2, or the geometrical method of Art. 232 may be used.

In a hyperbola, tangents at the ends of a focal chord intersect on the corresponding directrix.

Writing $-b^2$ for b^2 in Art. 233, the method holds.

If from T, *any point on the tangent at* P, *perpendiculars* TR, TM *are drawn to the focal distance* SP, *and the directrix,* SR $= e$ TM.

This can be proved as in Art. 234 with the necessary change in the figure.

Tangents to a hyperbola subtend equal angles at the focus.

As in the ellipse, this follows from the above.

To draw a tangent at a given point P *on a hyperbola.*

Both methods of Art. 236 hold.

To draw tangents to a hyperbola from an external point T.

Both methods of Art. 237 hold.

287. *If the diameter* PCP′ *meets the hyperbola* $\dfrac{x^2}{a^2} - \dfrac{y^2}{b^2} = 1$ *at* P *and* P′ *and* DCD′, *the diameter conjugate to* PCP′, *meets the conjugate hyperbola at* D *and* D′.

$$\mathbf{CP^2 - CD^2 = a^2 - b^2}.$$

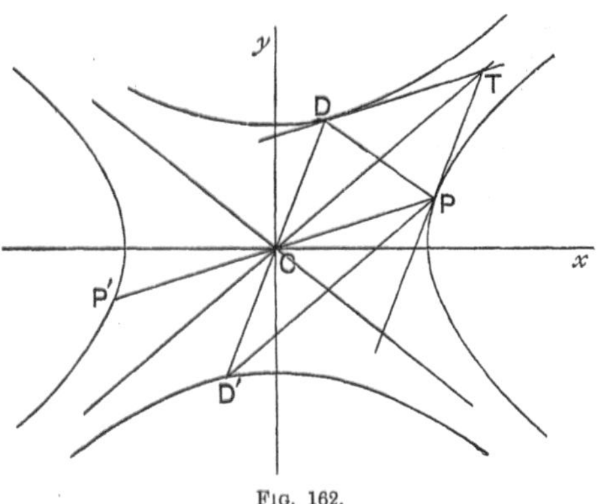

Fig. 162.

Let ($a \sec \theta$, $b \tan \theta$) be the co-ordinates of P, and ($a \tan \phi$, $b \sec \phi$) the co-ordinates of D.

The equations of CP and CD are

$$\frac{x}{a \sec \theta} = \frac{y}{b \tan \theta} \quad \text{and} \quad \frac{x}{a \tan \phi} = \frac{y}{b \sec \phi}.$$

These diameters are conjugate,

$$\therefore \quad \frac{b\tan\theta}{a\sec\theta} \cdot \frac{b\sec\phi}{a\tan\phi} = \frac{b^2}{a^2} \quad \left(mm' = \frac{b^2}{a^2}\right)$$

whence $\sin\phi = \sin\theta$;

$$\therefore \quad \sec\phi = \sec\theta, \text{ and } \tan\phi = \tan\theta;$$

$\therefore$ the co-ordinates of D are $(a\tan\theta, \ b\sec\theta)$.

$$\therefore \ \ \mathsf{CP}^2 - \mathsf{CD}^2 = (a^2\sec^2\theta + b^2\tan^2\theta) - (a^2\tan^2\theta + b^2\sec^2\theta)$$
$$= a^2 - b^2.$$

N.B. We have proved above that if $(a\sec\theta, \ b\tan\theta)$ are the co-ordinates of P, and the diameter conjugate to CP meets the conjugate hyperbola at D, then $(a\tan\theta, \ b\sec\theta)$ are the co-ordinates of D.

288. *If the diameter PCP′ meets a hyperbola at P and P′, and the conjugate diameter DCD′ meets the conjugate hyperbola at D and D′, the asymptotes bisect PD, and PD′.*

Let $(a\sec\theta, \ b\tan\theta)$ be the co-ordinates of P.
Then $(a\tan\theta, \ b\sec\theta)$ are the co-ordinates of D.   (Art. 287.)
Hence if (x, y) is the middle point of PD,

$$2x = a(\sec\theta + \tan\theta) \quad \left[\frac{x_1 + x_2}{2}, \ \frac{y_1 + y_2}{2}\right]$$
$$2y = b(\tan\theta + \sec\theta);$$
$$\therefore \ \frac{x}{a} = \frac{y}{b},$$

i.e. the middle point of PD lies on an asymptote.
Similarly, the middle point of PD′ lies on the other asymptote.

289. *If the diameter PCP′ meets the hyperbola at P and P′, and DCD′, the diameter conjugate to PCP′, meets the conjugate hyperbola at D and D′, the tangents to the conjugate curve at D and D′ are parallel to PCP′.*

The equation of the conjugate hyperbola is $\dfrac{x^2}{a^2} - \dfrac{y^2}{b^2} = -1$.

Let $(a\sec\theta, \ b\tan\theta)$ be the co-ordinates of P.
Then $(a\tan\theta, \ b\sec\theta)$ are the co-ordinates of D.

$\therefore$ the equation of the tangent at D is $\dfrac{x\tan\theta}{a} - \dfrac{y\sec\theta}{b} = -1$.

The slope of this line $= \dfrac{b \tan \theta}{a \sec \theta}$

$=$ the slope of CP.

$\therefore$ CP is parallel to the tangent at D.

290. *If* CD, *the semi-diameter conjugate to* CP, *meets the conjugate hyperbola at* D, *the tangents at* P *and* D *meet on an asymptote.*

Let $(a \sec \theta,\ b \tan \theta)$ be the co-ordinates of P.

Then $(a \tan \theta,\ b \sec \theta)$ are the co-ordinates of D.

The equation of the tangent at P is $\dfrac{x \sec \theta}{a} - \dfrac{y \tan \theta}{b} = 1.$

,,   ,,   D is $\dfrac{x \tan \theta}{a} - \dfrac{y \sec \theta}{b} = -1.$

Where these meet, we have by addition,

$$\frac{x}{a}(\sec \theta + \tan \theta) - \frac{y}{b}(\sec \theta + \tan \theta) = 0,$$

$$\text{or } \frac{x}{a} - \frac{y}{b} = 0.$$

But this is the equation of an asymptote, which proves the proposition.

If these tangents meet at T, CPTD is a parallelogram, by the preceding article.

***291.** With conjugate diameters as axes of co-ordinates, the equation of the asymptotes is $\dfrac{x^2}{a'^2} - \dfrac{y^2}{b'^2} = 0,$

or separately, their equations are $\dfrac{x}{a'} - \dfrac{y}{b'} = 0,$ and $\dfrac{x}{a'} + \dfrac{y}{b'} = 0.$

Let the axis of y meet the conjugate hyperbola at D, and draw tangents to the curves at P and D to meet at T.

T lies on the asymptote $\dfrac{x}{a'} - \dfrac{y}{b'} = 0.$

$$\therefore \frac{PT}{CP} = \frac{b'}{a'};\ \ \therefore PT = b',\ i.e.\ CD = b'.$$

Thus we see that the equation of the conjugate hyperbola referred to the same axes is

$$\frac{x^2}{a'^2} - \frac{y^2}{b'^2} = -1.$$

***292.** *If a chord* QQ′ *of a hyperbola meets the asymptotes at* R *and* R′, *the portions* QR, Q′R′ *intercepted between the curve and the asymptotes are equal.*

First method. Take the diameter CPV which bisects the chord, and its conjugate diameter as axes of co-ordinates, so that the equation of the curve is $\dfrac{x^2}{a'^2} - \dfrac{y^2}{b'^2} = 1$.

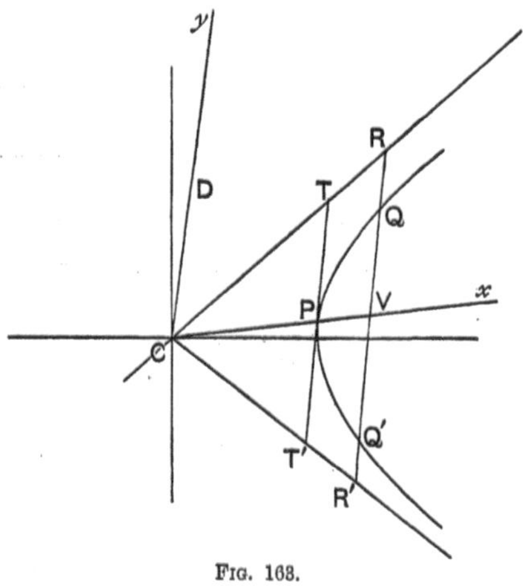

Fig. 168.

Let $(a'\sec\theta,\ b'\tan\theta)$ be the co-ordinates of Q.

The equation of the chord, which is parallel to the axis of y, is

$$x = a'\sec\theta.$$

At R, where it meets the asymptote $\dfrac{x}{a'} - \dfrac{y}{b'} = 0$,

$$\sec\theta - \frac{y}{b'} = 0\,;\ \ \therefore\ y = b'\sec\theta,\ \ i.e.\ \ \text{RV} = b'\sec\theta.$$

At R′ where it meets the other asymptote $\dfrac{x}{a'} + \dfrac{y}{b'} = 0$,

$$\sec\theta + \frac{y}{b'} = 0\,;\ \ \therefore\ y = -\,b'\sec\theta,\ \ i.e.\ \ \text{R′V} = -\,b'\sec\theta.$$

$$\therefore\ \text{V is the middle point of RR′.}$$

But $\qquad\qquad\qquad$ QV = Q′V ; $\ \therefore\ $ QR = Q′R′.

Second method.   Let CPV be the diameter which bisects QQ′, and let the tangent at P meet the asymptotes at T and T′.

TPT′ is parallel to the chord.

$\therefore$ from the similar $\triangle$s CVR, CPT, $\dfrac{VR}{PT} = \dfrac{CV}{CP}$

$$= \frac{VR'}{PT'} \text{ from the similar } \triangle s \text{ CVR}', \text{ CPT}'.$$

But $\qquad\qquad\qquad$ PT = PT′ ; $\therefore$ VR = VR′.

Also $\qquad\qquad\qquad$ QV = Q′V ; $\therefore$ QR = Q′R′.

293. *If the ordinate* PN *produced meets the asymptotes in* R *and* R′, PR . PR′ = BC².

Let (x_1, y_1) be the co-ordinates of P.

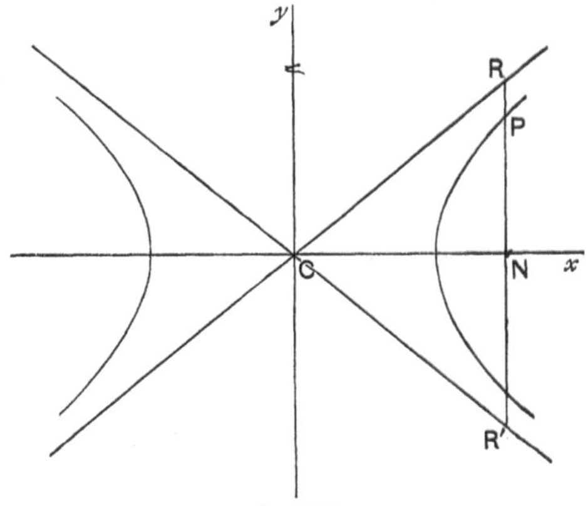

FIG. 164.

RR′ is bisected at N ;

$$\therefore \quad PR . PR' = RN^2 - PN^2 ;$$

$$\therefore \quad \frac{PR . PR'}{BC^2} = \frac{RN^2}{BC^2} - \frac{PN^2}{BC^2} = \frac{RN^2}{b^2} - \frac{y_1^2}{b^2}.$$

But the equation of CR is $\dfrac{x}{a} = \dfrac{y}{b}$; $\therefore$ $\dfrac{RN}{b} = \dfrac{x_1}{a}$;

$$\therefore \quad \frac{PR . PR'}{BC^2} = \frac{x_1^2}{a^2} - \frac{y_1^2}{b^2} = 1.$$

$$\therefore \quad PR . PR' = BC^2.$$
Q.E.D.

294. *If a chord be drawn through any point* Q *on a hyperbola to meet the asymptotes in* R *and* R′, *and* CD *the diameter parallel to the chord meets the conjugate hyperbola at* D, QR . QR′ = CD².

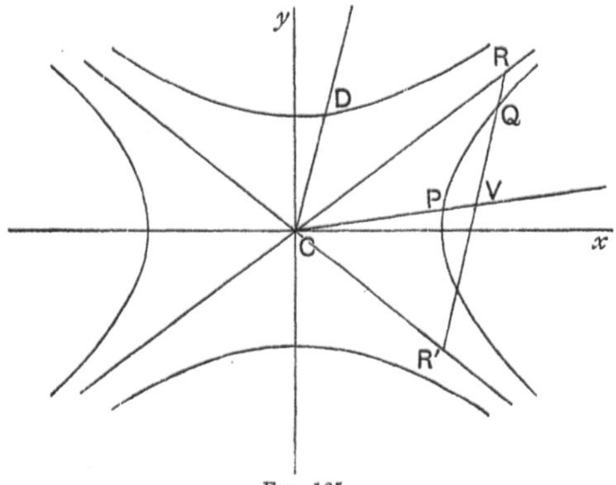

FIG. 165.

First Method.   If (x_1, y_1) are the co-ordinates of Q and the chord makes an angle θ with the transverse axis, the equation of the chord is

$$\frac{x - x_1}{\cos \theta} = \frac{y - y_1}{\sin \theta} = r.$$

Where this meets the asymptotes, $\dfrac{x^2}{a^2} - \dfrac{y^2}{b^2} = 0$, we have, by substitution,

$$\frac{(x_1 + r \cos \theta)^2}{a^2} - \frac{(y_1 + r \sin \theta)^2}{b^2} = 0,$$

or $$r^2 \left(\frac{\cos^2\theta}{a^2} - \frac{\sin^2\theta}{b^2} \right) + 2r \left(\frac{x_1 \cos \theta}{a^2} - \frac{y_1 \sin \theta}{b^2} \right) + \frac{x_1^2}{a^2} - \frac{y_1^2}{b^2} = 0.$$

Now QR, QR′ are the roots of this quadratic,

$$\therefore \quad \text{QR, QR′} = \text{the product of the roots}$$

$$= \frac{\dfrac{x_1^2}{a^2} - \dfrac{y_1^2}{b^2}}{\dfrac{\cos^2\theta}{a^2} - \dfrac{\sin^2\theta}{b^2}} = \frac{1}{\dfrac{\cos^2\theta}{a^2} - \dfrac{\sin^2\theta}{b^2}},$$

for (x_1, y_1) is on the hyperbola.

Thus we see that the rect. QR . QR′ is constant if θ is constant.

Let the chord move parallel to itself until it becomes a tangent,

If TPT′ the tangent meets the asymptotes in T and T′, the rectangle QR . QR′ becomes

$$PT . PT' = PT^2 = CD^2 \ (\text{Art. 290}).$$ Q.E.D.

Second Method. Take the diameter CPV which bisects QQ′ and the conjugate diameter CD as axes of co-ordinates.

Let $(a' \sec \theta, \ b' \tan \theta)$ be the co-ordinates of Q.

$$x = a' \sec \theta \text{ is the equation of } QQ'.$$

Where it meets the asymptote

$$\frac{x}{a'} - \frac{y}{b'} = 0, \quad \sec \theta - \frac{y}{b'} = 0, \quad \therefore \ RV = b' \sec \theta.$$

Now V is the middle point of RR′,

$$\therefore \ QR . QR' = RV^2 - QV^2 = b'^2 \sec^2\theta - b'^2 \tan^2\theta = b'^2 = CD^2.$$

295. *Tangents at the extremities of a diameter are parallel to one another.*

This may be proved as in the ellipse, Art. 229.

296. *If the normal at* P *meets the transverse axis in* G, *and the diameter parallel to the tangent at* P *in* F, PF . PG = BC².

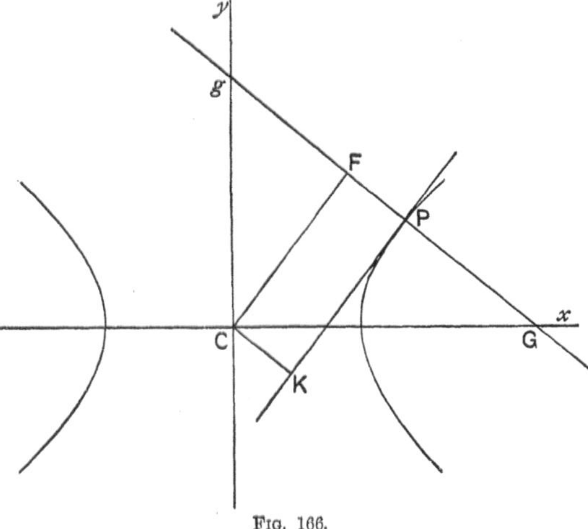

Fɪɢ. 166.

Let $(a \sec \theta, \ b \tan \theta)$ be the co-ordinates of P, and draw CK perpendicular to the tangent at P.

As in the ellipse (Art. 242),

$$PG^2 = \frac{b^2}{a^2} (a^2 \tan^2\theta + b^2 \sec^2\theta).$$

Also $PF = CK = \dfrac{ab}{\sqrt{a^2 \tan^2\theta + b^2 \sec^2\theta}}$, whence $PF \cdot PG = b^2$.

COROLLARY. In the same way, as for the ellipse, we can prove that
$$PF \cdot Pg = CA^2.$$

297. *If the diameter* PCP′ *meets the hyperbola at* P *and* P′, *and the conjugate diameter* DCD′ *meets the conjugate hyperbola at* D *and* D′,
$$CP^2 - CD^2 = a^2 - b^2.$$
See Art. 287.

With the assumption of the above proposition, SP $\cdot$ S′P $=$ CD^2.

$$SP \cdot S'P = (ae \sec\theta - a)(ae \sec\theta + a) = a^2e^2 \sec^2\theta - a^2$$
$$= (a^2 + b^2) \sec^2\theta - a^2 = a^2 \tan^2\theta + b^2 \sec^2\theta$$
$$= CD^2.$$

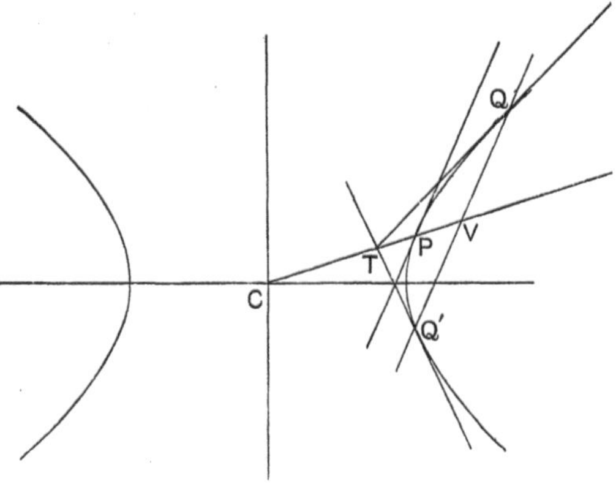

FIG. 167.

If CP, CD, *conjugate semi-diameters, meet the hyperbola and the conjugate hyperbola at* P *and* D *respectively, and tangents at* P *and* D *are drawn to meet at* T, *the area of the parallelogram* PCDT $=$ AC $\cdot$ BC.

Use the method of Art. 241, remembering that if $(a \sec\theta, b \tan\theta)$ are the co-ordinates of P, the co-ordinates of D are $(a \tan\theta, b \sec\theta)$.

Tangents at the ends of a chord intersect on the diameter which bisects that chord.

Use the method of Art. 230.

If from T, *a point in the diameter* PCP′, *tangents* TQ, TQ′ *are drawn to the hyperbola, and* QQ′ *meets* CT *in* V, CV . CT = CP^2. (See Fig. 167, p. 276.)

Use the method of Art. 243 with the necessary changes in the equations.

Writing $-b^2$ for b^2, Arts. 244-247 hold for the hyperbola.

Given a curve, which is known to be a hyperbola, find its centre, the positions and lengths of its principal axes, and its foci.

If both branches of the curve are given, the method of Art. 238 holds.

If only one branch is given, two pairs of parallel chords must be drawn. The point where the bisectors of these pairs meet is the centre.

If DCD′, *the diameter conjugate to* PRP′, *meets the conjugate hyperbola at* D *and* D′, *and* QQ′ *a chord of the hyperbola parallel to* DD′ *cuts* PP′ *at* V,

$$\frac{QV^2}{PV . VP'} = \frac{CD^2}{CP^2}.$$

Use the method of Art. 247, taking $(a' \sec \theta,\ b' \tan \theta)$ for the co-ordinates of P.

Examples XII. h.

1. Given the two branches of a hyperbola show how to draw the axes and the director circle. State your construction.

2. Prove that if two chords AB, CD of an equilateral hyperbola intersect at right angles at E, then EA . EB = EC . ED.

3. If the tangent at a point P of a hyperbola meet an asymptote in Q, and SP, drawn from the focus S which lies within the branch of the curve on which P lies, meet the same asymptote in R, show that the triangle RQS is isosceles.

4. Give a geometrical interpretation of the circumstance that the equation of a hyperbola results from the elimination of λ between the equations $ay = \lambda b(x+a),\ ay = \dfrac{b}{\lambda}(x-a)$.

5. Find the locus of the middle points of all chords of the curve $x^2 + 2xy - y^2 + 2ax + 2by = c^2$, which are parallel to the line $y = mx$, and apply the result to show that the sum of the inclinations to the axis of x of any two conjugate diameters is 135°.

6. Find the equations of the two straight lines conjugate to the co-ordinate axes of x and y respectively in the curve $Ax^2 + 2Bxy + Cy^2 = 1$. Find the condition that these straight lines should coincide, and interpret the result.

7. With a unit of one inch, draw the curve given by $x = \sec\theta$, $y = \tan\theta$, when θ varies.

8. Find the equations of the common tangents of the hyperbolas

$$\frac{x^2}{a^2} - \frac{y^2}{b^2} = 1, \quad \frac{y^2}{a^2} - \frac{x^2}{b^2} = 1.$$

9. A chord of an equilateral hyperbola $x^2 - y^2 = a^2$ touches the circle $x^2 + y^2 = a^2$ at Q. If P be the middle point of the chord, and α, β its co-ordinates, show that $(\alpha^2 - \beta^2)\sin QPC = a^2$.

10. Prove that the locus of the centres of the circles which can be drawn touching two equal circles consists of a straight line and a hyperbola, the transverse axis of which is equal to the diameter of either circle, and its eccentricity the ratio of the distance between the centres to the diameter.

11. Find the equations of the asymptotes of the hyperbola

$$2y^2 - xy - 3x^2 = 8.$$

Draw them, and hence draw a freehand sketch of the curve itself.

12. On squared paper draw two straight lines AB, AC, containing an angle of $35°$. Draw in several positions the base of a triangle of which two sides lie along AB and AC, and whose area is 4 sq. in., and so draw the curve which touches the base in all positions.

Draw also the curve which touches the base in all positions, when the two sides lie along AB and AC, and the length of the perpendicular from A on the base is $2\cdot3$ inches.

By laying your ruler against the two curves draw the base of the triangle which has an area of 4 sq. in., and has also the perpendicular from A on the base $2\cdot3$ inches long.

Also state how you could make a triangle having a vertical angle of $35°$, area 4 sq. in., and perpendicular from vertex on base $2\cdot3$ inches, without drawing any other loci than straight lines and circles.

13. If a portion of one branch of a hyperbola be given, show how to verify its hyperbolic nature.

14. If two sides of a triangle be given in position, and its perimeter given in magnitude, show that the locus of the point which divides the base in a given ratio is a hyperbola.

15. The extremities of one diagonal of a parallelogram, whose sides are parallel to the asymptotes of a hyperbola, lie on the curve: show that the other diagonal passes through the centre.

16. From any point P on a hyperbola a parallel to each asymptote is drawn to cut the other. These points of intersection being M and N, prove that the rectangle PM, PN is constant. Draw a hyperbola with perpendicular asymptotes, for which the value of the rectangle PM, PN is one square inch.

17. Find the equation of, and draw, the locus of the poles of tangents to the circle $x^2 + y^2 = a^2$ with respect to the parabola $y^2 = 4ax$.

18. Prove that any chord of an equilateral hyperbola subtends equal or supplementary angles at the extremity of any diameter.

19. If any straight line is drawn cutting a hyperbola and passing through the foot of a directrix, show that the focal radii vectores to the points of section make supplemental angles with the transverse axis.

20. Prove that the equation of the tangents from the point (x_1, y_1) to the rectangular hyperbola $xy = c^2$ may be written

$$4(x_1 y_1 - c^2)(xy - c^2) = (xy_1 - yx_1 - 2c^2)^2.$$

21. Find the equation of the locus of the foci of a rectangular hyperbola which passes through the point $(0, k)$ and has the line $y = 0$ for an asymptote.

22. If a right-angled triangle be inscribed in a rectangular hyperbola, prove that the perpendicular from the right angle on the hypotenuse touches the hyperbola.

23. If two tangents to the hyperbola $4xy = a^2 + b^2$ intersect at right angles, find the equation of the locus of their point of intersection.

24. Show that if either asymptote of the conic

$$ax^2 + 2hxy + by^2 + 2gx + 2fy + c = 0$$

passes through the origin, then $af^2 - 2fgh + bg^2 = 0$.

Examples XII. k.

MISCELLANEOUS EXAMPLES ON THE DRAWING OF CONIC SECTIONS.

1. On squared paper find the eccentricity of the conic with focus $(3, 0)$, directrix the axis of y, and passing through the point $(5, -3)$. Find out whether the conic would pass through the point $(6, 4)$.

Draw a tangent at the point $(5, -3)$. State your methods.

2. Take as directrix of a conic a straight line XY, and as focus a point Z, $1\frac{1}{2}$ inches from XY. One point on the curve is $1\frac{1}{2}$ inches from XY and $\frac{1}{2}$ inch from Z. Find a number of points on the curve and draw it.

3. Two conics L and M have the same focus and directrix, the length of the perpendicular SX from the focus on the directrix being $1\frac{1}{2}$ inches. The eccentricity of L is $\frac{1}{2}$ and that of M is 2. Draw a diagram showing the general shapes of the curve.

4. AB is a fixed straight line. A point S is taken $\frac{3}{4}$ in. from AB. A point P moves so that $SP = PM$, where M is a moving point on AB, and PM makes an angle of $60°$ with AB. Draw a part of the curve, and find its eccentricity to two decimal places.

5. XY is a straight line, and S and T two points on opposite sides of it, each $1\frac{1}{2}$ in. from XY and $3\frac{1}{2}$ in. from one another. From T draw tangents to the conic whose focus is S, directrix XY, and eccentricity $0\cdot5$. Roughly indicate the position of the curve.

6. Given a point on a conic, the tangent at the point, a focus, and the length of the semi-latus rectum, construct the direction of the transverse axis.

7. The problem is proposed to make a triangle with a base 12 cm. long, the sum of the other two sides 25 cm., and the vertical angle 50°. Make the base of the given length, and draw the locus of the vertex when the sum of the two sides has the given value, ignoring the condition as to the vertical angle. Also draw the locus of the vertex when the vertical angle is 50°, ignoring the other condition. From these two loci obtain the required triangle, and write down the lengths of its sides.

Write down the equations of the two loci referred to the base and its perpendicular bisector as axes. You may obtain any lengths you need from your drawing.

8. The length of SX, the focal perpendicular on the directrix of a conic is 5 cm. A tangent to the conic bisects SX and is inclined to it at an angle of 60°. Draw a figure from these data, and on it construct the point of contact of the tangent.

9. Draw two straight lines AOB, COD cutting at an angle of 45°; draw OF perpendicular to AB at O and one inch long. Find the points where CD cuts a conic whose focus is F, directrix AB, and eccentricity $\frac{6}{5}$. Also find one vertex of the curve and sketch the branch of the curve through that vertex.

10. The focus and directrix of a conic are given, and also one point, P, on the conic. How would you construct the tangent at P? Prove the correctness of your method.

11. With your instruments draw a triangle SPQ having $SP=5$ cm., $SQ=10$ cm., $PQ=12$ cm. Consider S the focus of a conic of eccentricity 2, and P and Q two points on the curve, and construct the directrix corresponding to S.

Give one solution, stating your construction. State how many solutions there are.

12. With mathematical instruments construct a triangle ABC in which $AB=7$ cm., $BC=10$ cm., and $CA=12$ cm. Draw the two conics which pass through A and have B and C as foci. Determine their eccentricities.

13. Carry out the following instructions :—Draw OB, BC at right angles to one another. Divide OB into any number of equal parts, and let A_1, A_2, A_3 ... be the points of division beginning from O. Along BC take points B_1, B_2 ... at equal distances apart, so that $BB_1=B_1B_2=B_2B_3....$ At A_1, A_2 ... draw perpendiculars to OB and let these intersect OB_1, OB_2 ... respectively at P_1, P_2 .... Join the points P_1, P_2 ... by an even curve, and prove that the curve is a parabola, whose vertex is O and axis parallel to BC.

14. Carry out the following instructions :—Draw two straight lines ACA', BCB' bisecting one another at right angles, and complete the rectangle $BCAD$. Divide CA into any number (say 9) of equal parts, and DA into the same number of equal parts. Let A_1, A_2, A_3 ... be the points of division of CA, beginning from C, and B_1, B_2, B_3 ... the points of division of DA, beginning from D. Let $B'A_1$ and BB_1 meet at P_1, $B'A_2$ and BB_2 at P_2, and so on.   Join P_1P_2 ... by an even curve.

Prove that the curve is an ellipse whose axes are AA' and BB'.

15. Carry out the following instructions :—Take any straight line OA, and produce it to any point N. Draw NQ at right angles to OAN, and complete the rectangle ANQM. Divide NQ into any number of equal parts, and let a_1, a_2, ... be the points of division, starting from N. Divide QM into the same number of equal parts, and let b_1, b_2, ... be the points of division, starting from M.

Take the points of intersection of Oa_1 and Ab_1, of Oa_2 and Ab_2 and so on. Join these points by an even curve and prove it is a hyperbola. Take OA $=1$ in., AN $=2$ in., NQ $=5$ in.

CHAPTER XIII.

POLAR EQUATION OF A CONIC SECTION.

298. *If the focus of a conic is taken as origin, and* **SX** *the perpendicular on the corresponding directrix as initial line, the equation of the conic is*

$$\frac{l}{r} = 1 + e \cos \theta,$$

where e is the eccentricity and l the semi-latus rectum.

If $(r,\ \theta)$ be the co-ordinates of any point P on the conic, SL the semi-latus rectum, and PM, PN be drawn perpendicular to the directrix and axis respectively,

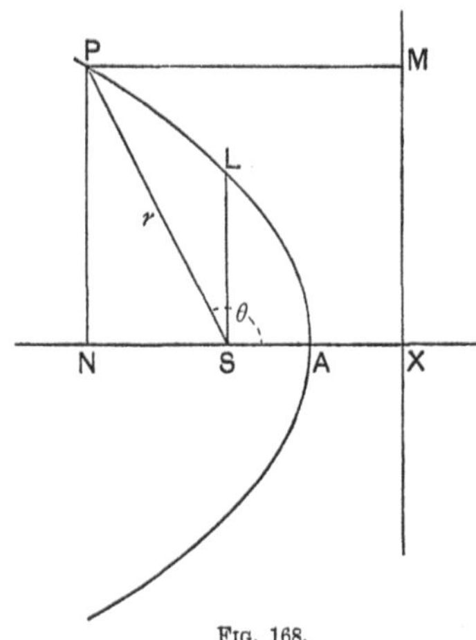

Fig. 168.

$$r = \text{SP} = e\,.\,\text{PM} = e\,.\,\text{NX}$$

$$= e(\text{SX} + \text{SN})$$

$$= \text{SL} + e\,.\,\text{SP} \cos \text{PSN}$$

(for $\text{SL} = e\,.\,\text{SX}$ by def.)

$$= l - er \cos \theta\,;$$

$$\therefore \quad l = r(1 + e \cos \theta),$$

$$\text{or} \quad \frac{l}{r} = 1 + e \cos \theta.$$

Q.E.D.

COROLLARY 1. If **XS** produced is taken as the initial line, the equation of the conic will be

$$\frac{l}{r} = 1 - e \cos \theta, \quad \text{for } \theta = \angle \text{PSN in this case.}$$

COROLLARY 2.   If the initial line passes through S, and SX makes an angle a with it, the equation of the conic will be

$$\frac{l}{r} = 1 + e \cos(\theta - a),$$

for in this case $\angle \text{PSA} = \theta - a.$

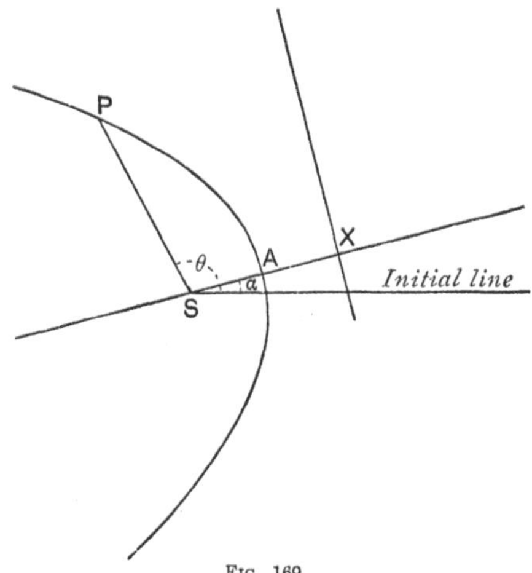

FIG. 169.

COROLLARY 3.   If the conic is a parabola, $e = 1$ and its equation is

$$\frac{l}{r} = 1 + \cos\theta,$$

or with the usual notation, $\dfrac{2a}{r} = 2\cos^2\dfrac{\theta}{2},$

$$i.e.\quad \frac{a}{r} = \cos^2\frac{\theta}{2}.$$

299. *With S as origin and SX as initial line, the equation of the directrix is*

$$\frac{l}{r} = e \cos\theta.$$

Let P(r, θ) be any point on the directrix.

$$\text{SX} = \text{SP}\cos\theta = r\cos\theta,$$

and $\text{SL} = e \,.\, \text{SX}$ by def. ;

$$\therefore \quad \frac{l}{e} = r \cos \theta,$$

$$i.e. \quad \frac{l}{r} = e \cos \theta \text{ is the equation of the directrix.}$$

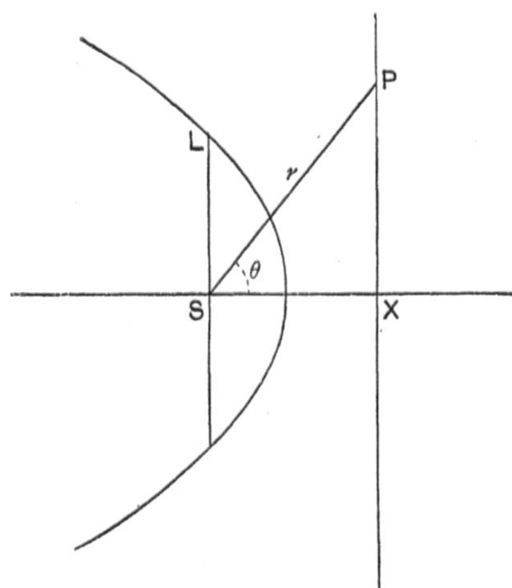

FIG. 170.

COROLLARY. $\theta = \dfrac{\pi}{2}$ is the equation of SL the upper portion of the latus rectum.

$\theta = -\dfrac{\pi}{2}$ or $\theta = \dfrac{3\pi}{2}$ is the equation of the other portion.

300. *To find the polar equation of the tangent to the conic* $\dfrac{l}{r} = 1 + e \cos \theta$ *at the point whose vectorial angle is* α.

Let $\dfrac{l}{r} = A \cos \theta + B \sin \theta$ be the equation of the chord through the points $(r_1, \; \alpha + \beta)$, $(r_2, \; \alpha - \beta)$ on the conic and near to one another.

The point $(r_1, \; \alpha + \beta)$ is on the conic, and on the chord;

$$\therefore \quad \frac{l}{r_1} = 1 + e \cos(\alpha + \beta), \quad \text{and} \quad \frac{l}{r_1} = A \cos(\alpha + \beta) + B \sin(\alpha + \beta);$$

$$\therefore \quad A\cos(\alpha+\beta)+B\sin(\alpha+\beta)=1+e\cos(\alpha+\beta)$$
$$\text{or} \quad (A-e)\cos(\alpha+\beta)+B\sin(\alpha+\beta)=1. \quad \ldots\ldots\ldots\ldots(1)$$

In the same way,

$$(A-e)\cos(\alpha-\beta)+B\sin(\alpha-\beta)=1. \quad \ldots\ldots\ldots\ldots(2)$$

Subtracting, $\quad -2(A-e)\sin\alpha\,\sin\beta+2B\cos\alpha\,\sin\beta=0$

$$\text{and} \quad (A-e)\sin\alpha-B\cos\alpha=0. \quad \ldots\ldots\ldots\ldots\ldots(3)$$

Now let the points move up to one another and coincide.

β becomes zero.

$\therefore$ from (1) or (2),

$$(A-e)\cos\alpha+B\sin\alpha-1=0. \quad \ldots\ldots\ldots\ldots\ldots(4)$$

Solving (3) and (4) [see Art. 23],

$$\frac{A-e}{\cos\alpha}=\frac{B}{\sin\alpha}=\frac{1}{\sin^2\alpha+\cos^2\alpha},$$
$$A=e+\cos\alpha, \quad B=\sin\alpha.$$

The chord $\dfrac{l}{r}=A\cos\theta+B\sin\theta$ has now become a tangent.

$\therefore$ the equation of the tangent is $\dfrac{l}{r}=(e+\cos\alpha)\cos\theta+\sin\alpha\sin\theta.$

It may be written $\quad \dfrac{l}{r}=\cos(\theta-\alpha)+e\cos\theta.$

301. *To find the polar equation of the normal to the conic* $\dfrac{l}{r}=1+e\cos\theta$ *at the point* $(r_1,\,\alpha)$.

The equation of the tangent at the point $(r_1,\,\alpha)$ is

$$\frac{l}{r}=\cos(\theta-\alpha)+e\cos\theta.$$

The equation of any straight line at right angles to this is

$$\frac{p}{r}=\cos\left(\theta+\frac{\pi}{2}-\alpha\right)+e\cos\left(\theta+\frac{\pi}{2}\right) \quad \text{(Art. 82.)}$$

$$\text{or} \quad \frac{p}{r}=\sin(\alpha-\theta)-e\sin\theta. \quad \ldots\ldots\ldots\ldots\ldots(1)$$

But if this straight line is a normal, it passes through the point (r_1, α).

$$\therefore \ \frac{p}{r_1} = -e \sin \alpha \ \dots\dots\dots\dots\dots\dots\dots\dots(2)$$

Also since (r_1, α) is on the conic,

$$\frac{l}{r_1} = 1 + e \cos \alpha \ \dots\dots\dots\dots\dots\dots\dots(3)$$

$\therefore$ from (2) and (3), $\quad \dfrac{p}{l} = \dfrac{-e \sin \alpha}{1 + e \cos \alpha}.$

$\therefore$ substituting in (1),

$$-\frac{l}{r} \cdot \frac{e \sin \alpha}{1 + e \cos \alpha} = \sin(\alpha - \theta) - e \sin \theta$$

is the equation of the normal.

This may be written

$$\frac{l}{r} \cdot \frac{e \sin \alpha}{1 + e \cos \alpha} = \sin(\theta - \alpha) + e \sin \theta$$

or $\quad \dfrac{e r_1 \sin \alpha}{r} = \sin(\theta - \alpha) + e \sin \theta, \ \ \text{for} \ \ \dfrac{l}{r_1} = 1 + e \cos \alpha.$

302. *To find the equation of the chord joining the two points on the conic $\dfrac{l}{r} = 1 + e \cos \theta$, whose vectorial angles are $\alpha + \beta$, $\alpha - \beta$.*

Let $\dfrac{l}{r} = A \cos \theta + B \cos \theta$ be the equation of the chord, and let r_1 be the radius vector of the point $\alpha + \beta$.

The point $(r_1, \alpha + \beta)$ is on the chord and on the conic.

$$\therefore \ \frac{l}{r_1} = A \cos(\alpha + \beta) + B \sin(\alpha + \beta) \ \ \text{and} \ \ \frac{l}{r_1} = 1 + e \cos(\alpha + \beta) ;$$

$$\therefore \ A \cos(\alpha + \beta) + B \sin(\alpha + \beta) = 1 + e \cos(\alpha + \beta)$$

$$\text{or} \ \ (A - e)\cos(\alpha + \beta) + B \sin(\alpha + \beta) - 1 = 0.$$

In the same way,

$$(A - e)\cos(\alpha - \beta) + B \sin(\alpha - \beta) - 1 = 0.$$

$$\therefore \; \frac{A - e}{-\sin(a+\beta)+\sin(a-\beta)} = \frac{-B}{-\cos(a+\beta)+\cos(a-\beta)}$$

$$= \frac{1}{\cos(a+\beta)\sin(a-\beta)-\sin(a+\beta)\cos(a-\beta)} \quad [\text{Art. 23.}]$$

$$-\frac{A-e}{2\cos a \sin \beta} = -\frac{B}{2\sin a \sin \beta} = -\frac{1}{\sin 2\beta} ;$$

$$A - e = \frac{\cos a}{\cos \beta} \quad \text{and} \quad B = \frac{\sin a}{\cos \beta} ;$$

$$\therefore \; \frac{l}{r} = \left(e + \frac{\cos a}{\cos \beta} \right) \cos \theta + \frac{\sin a}{\cos \beta} \sin \theta$$

is the equation of the chord.

It may be written, $\quad \dfrac{l}{r} = \sec \beta \cos (\theta - a) + e \cos \theta.$

303. *Tangents to a conic subtend equal angles at the focus.*

Let $a + \beta$, $a - \beta$ be the angular co-ordinates of the points P and Q on the conic $\dfrac{l}{r} = 1 + e \cos \theta.$

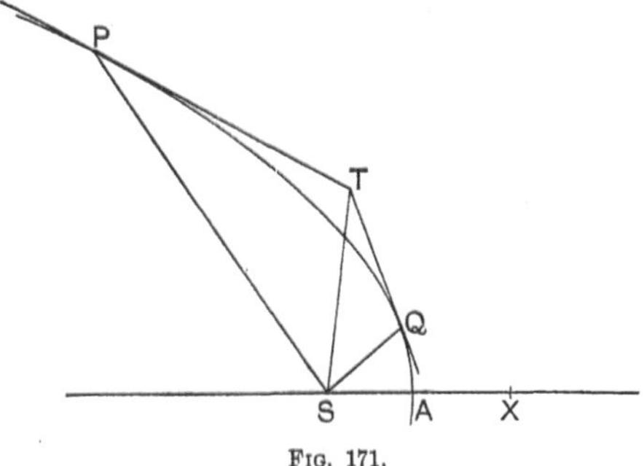

Fig. 171.

The equations of the tangents at P and Q are

$$\frac{l}{r} = \cos(\theta - a - \beta) + e \cos \theta,$$

$$\frac{l}{r} = \cos(\theta - a + \beta) + e \cos \theta,$$

At the point T where these meet, we have by subtraction,
$$\cos(\theta - a + \beta) = \cos(\theta - a - \beta).$$
The general solution of this equation is
$$\theta - a + \beta = 2n\pi \pm (\theta - a - \beta).$$
The only admissible solution in this case is
$$\theta - a + \beta = -(\theta - a - \beta);$$
$$\therefore \ \theta = a:$$
$$i.e. \ \angle TSX = a.$$
$$\therefore \ \angle TSQ = a - (a - \beta) = \beta,$$
$$\angle TSP = a + \beta - a = \beta;$$
$$\therefore \ \angle TSQ = \angle TSP. \qquad\qquad \text{Q.E.D.}$$

304. *To find the equation of the chord of contact of tangents drawn from the point* $(r_1, \ \theta_1)$ *to the conic*

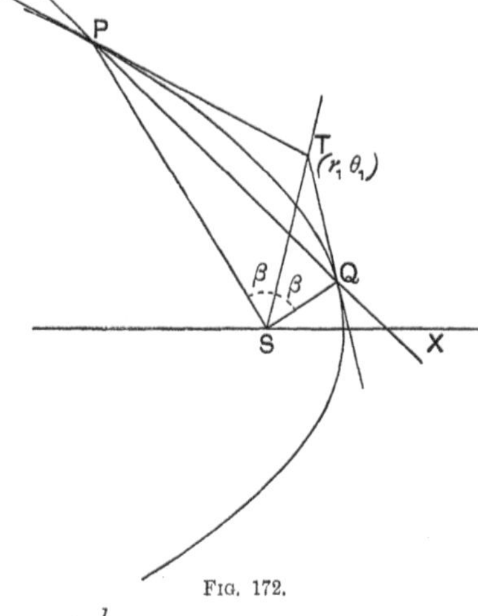

Fig. 172.

$$\frac{l}{r} = 1 + e \cos \theta.$$

Let T be the point (r_1, θ_1), and TP, TQ, tangents to the conic. It is required to find the equation of PQ.

Tangents subtend equal angles at the focus,
$$\therefore \ \angle TSP = \angle TSQ$$
$$= \beta \text{ suppose.}$$
Also $\angle TSX = \theta_1$,
$$\therefore \ \theta_1 - \beta \text{ is the angular}$$
co ordinate of Q,
and $\theta_1 + \beta$ is the angular co-ordinate of P.

$\therefore$ the equation of PQ may be written

$$\frac{l}{r} = \sec \beta \cos(\theta - \theta_1) + e \cos \theta \ldots\ldots\ldots(1) \quad \text{(Art. 302.)}$$

The equation of the tangent TQ is

$$\frac{l}{r} = \cos(\theta - \theta_1 + \beta) + e \cos \theta.$$

But this line passes through the point (r_1, θ_1);

$$\therefore \ \frac{l}{r_1} = \cos \beta + e \cos \theta_1.$$

Whence

$$\sec \beta = \frac{1}{\dfrac{l}{r_1} - e \cos \theta_1}.$$

$\therefore$ substituting this value of $\sec \beta$ in (1), the required equation

is

$$\frac{l}{r} = e \cos \theta + \frac{\cos(\theta - \theta_1)}{\dfrac{l}{r_1} - e \cos \theta_1}, \ \text{which may be written}$$

$$\left(\frac{l}{r} - e \cos \theta\right)\left(\frac{l}{r_1} - e \cos \theta_1\right) = \cos(\theta - \theta_1).$$

305. *Tangents to a conic at the ends of a focal chord intersect on the directrix.*

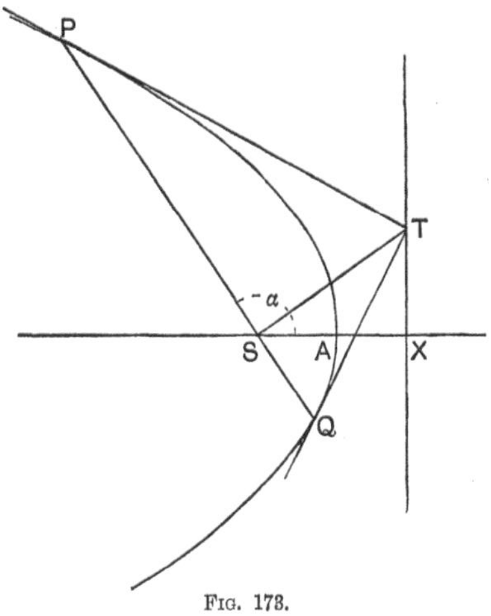

If **PSQ** is the focal chord, and α the vectorial angle of **P**, the vectorial angle of **Q** is $\pi + \alpha$.

$\therefore$ the equations of the tangents **TP** and **TQ** are

$$\frac{l}{r} = \cos(\theta - \alpha) + e \cos \theta, \quad (1)$$

$$\frac{l}{r} = \cos(\theta - \pi - \alpha) + e \cos \theta.$$

The second equation may be written

$$\frac{l}{r} = -\cos(\theta - \alpha) + e \cos \theta. \ (2)$$

At the point **T** where the tangents meet, we have from (1) and (2) by addition,

$$\frac{l}{r} = e \cos \theta.$$

But this is the equation of the directrix.

$\therefore$ the point **T** lies on the directrix. Q.E.D.

FIG. 173.

Corollary.   *In a parabola, tangents at the ends of a focal chord intersect* **at right angles** *on the directrix.*

$e = 1$ in the parabola.

$\therefore$ the equations of the tangents may be written

$$\frac{l}{r} = \cos(\theta - \alpha) + \cos\theta \quad \text{and} \quad \frac{l}{r} = -\cos(\theta - \alpha) + \cos\theta$$

i.e. $$\frac{l}{r} = 2\cos\left(\theta - \frac{\alpha}{2}\right)\cos\frac{\alpha}{2} \quad \text{and} \quad \frac{l}{r} = -2\sin\left(\theta - \frac{\alpha}{2}\right)\sin\frac{\alpha}{2}.$$

Now $$\sin\left(\theta - \frac{\alpha}{2}\right) = \cos\left(\frac{\pi}{2} + \theta - \frac{\alpha}{2}\right);$$

$\therefore$ the tangents are at right angles.   (Art. 82.)

306. *In a conic, the semi-latus rectum is a harmonic mean between the segments of any focal chord.*

Let $\dfrac{l}{r} = 1 + e\cos\theta$ be the equation of the conic, and let PSQ be any focal chord.

If α is the vectorial angle of the point P, $\pi + \alpha$ is the vectorial angle of Q.

The points P and Q are on the conic,

$$\therefore \quad \frac{l}{\mathsf{SP}} = 1 + e\cos\alpha,$$

$$\text{and} \quad \frac{l}{\mathsf{SQ}} = 1 + e\cos(\pi + \alpha) = 1 - e\cos\alpha\;;$$

$$\therefore \text{ by addition, } \frac{l}{\mathsf{SP}} + \frac{l}{\mathsf{SQ}} = 2,$$

$$\text{i.e.} \quad \frac{1}{\mathsf{SP}} + \frac{1}{\mathsf{SQ}} = \frac{2}{l}.$$

$\therefore l$ the semi-latus rectum is a harmonic mean between SP and SQ.

307. *The portion of any tangent to a conic intercepted between the point of contact and the directrix subtends a right angle at the corresponding focus.*

Let the tangent at P meet the directrix at K.

Let $\dfrac{l}{r} = 1 + e\cos\theta$ be the equation of the conic and α the vectorial angle of P.

The equation of the tangent at P is

$$\frac{l}{r} = \cos(\theta - a) + e\cos\theta.$$

The equation of the directrix is

$$\frac{l}{r} = e\cos\theta.$$

At the point K where these meet, we have by subtraction,

$$\cos(\theta - a) = 0.$$

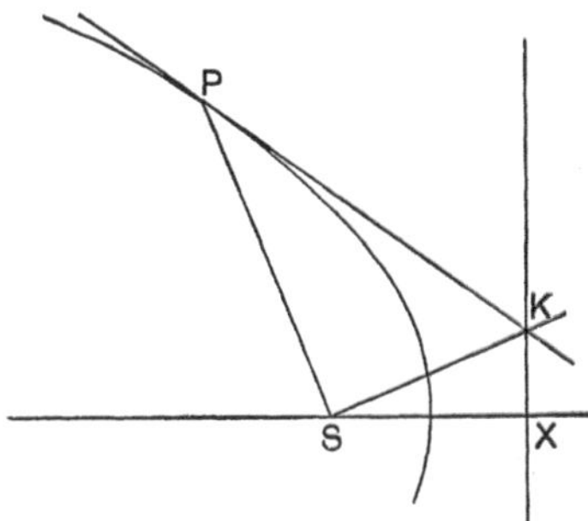

Fig. 174.

The only admissible solution of this equation is

$$\theta - a = -\frac{\pi}{2}.$$

$$\therefore\ a - \theta = \frac{\pi}{2},$$

$$i.e.\quad \angle\, \text{PSX} - \angle\, \text{KSX} = \frac{\pi}{2}\,;$$

$$\therefore\ \angle\, \text{PSK} = \text{a right angle.} \qquad \text{Q.E.D.}$$

Examples XIII.

1. If SP a focal radius vector of a conic is produced to Q, so that SQ$=k$. SP, when k is constant, prove that the locus of Q is a conic of equal eccentricity with the given conic, and latus rectum k times that of the given conic.

2. Find the equation of the directrix of the conic $\dfrac{l}{r} = 1 + e\cos(\theta - a)$ corresponding to the origin.

3. If PSQ is a focal chord of a conic, prove that the rectangle SP . SQ varies as PQ.

4. PP′ and QQ′ are two focal chords at right angles: show that $\dfrac{1}{PP'}+\dfrac{1}{QQ'}$ is constant.

5. A focal chord of the parabola $\dfrac{l}{r}=1+\cos\theta$ is drawn in direction α; find the co-ordinates of its middle point, and hence show that the locus of the middle points of the focal chords of a parabola is another parabola.

6. Find the polar equations of the circles passing through the points whose polar co-ordinates are $\left(a,\ \dfrac{\pi}{2}\right)$, $\left(b,\ \dfrac{\pi}{2}\right)$, and touching the line $\theta=0$.

7. Show that the polar equation $r=a\cos\theta+b\sin\theta$ represents a circle through the pole; and find the polar co-ordinates of its centre.
Find the equation of the circle which passes through the pole and through the two real intersections of the conics

$$\frac{l}{r}=1+\cos\theta,\quad \frac{2l}{r}=1+\sin\theta.$$

8. If a chord subtend a constant angle 2α at a focus, the locus of the point where it meets the internal bisector of that angle is

$$\frac{l}{r}=\sec\alpha+e\cos\theta.$$

Interpret this equation.

9. Find the angle between the radius vector and the tangent at any point of the conic $\dfrac{l}{r}=1+e\cos\theta$.

10. The tangents at $\alpha+\beta$, $\alpha-\beta$ to the parabola $\dfrac{2a}{r}=1+\cos\theta$ being drawn, show that the polar equation of the circle whose diameter is the line joining their intersection to the focus is

$$r(\cos\alpha+\cos\beta)=2a\cos(\theta-\alpha).$$

11. Tangents are drawn to the parabola $\dfrac{2a}{r}=1+\cos\theta$ from a point on the latus rectum, touching it at P and P′: show that $\dfrac{1}{SP}+\dfrac{1}{SP'}=\dfrac{1}{a}$.

12. Prove that perpendicular focal chords of a rectangular hyperbola are equal.

13. If the tangent and normal at a point P of a conic meet the transverse axis in T and G, show that $\dfrac{1}{SG}-\dfrac{1}{ST}$ is constant wherever P may be.

14. Find, and discuss fully, the locus of a point from which, if tangents be drawn to a conic, their points of contact subtend a constant angle at the focus.

15. Prove that tangents at the points $a+\dfrac{\pi}{3}$, $a-\dfrac{\pi}{3}$, to the parabola $\dfrac{2a}{r}=1+\cos\theta$ intersect on the confocal co-axial parabola whose latus rectum is $\dfrac{8a(1+\cos a)}{1+2\cos a}$.

16. A focus of an ellipse being taken as origin and the major axis as initial line, from the end of a radius vector SP of the ellipse a perpendicular is drawn upon a radius vector making a constant angle a with SP and in advance of SP. Prove that the locus of the foot of this perpendicular is a similar ellipse with linear dimensions in the ratio $1 : \cos a$.

17. Prove that $r^2=a^2\cos^2\theta+b^2\sin^2\theta$ is the polar equation of the locus of the foot of the perpendicular let fall from the centre upon the tangent to the ellipse $\dfrac{1}{r^2}=\dfrac{\cos^2\theta}{a^2}+\dfrac{\sin^2\theta}{b^2}$.

Prove also that if from any point P on this ellipse perpendiculars PM, PN be let fall upon the axes, the equation of the locus of the foot of the perpendicular let fall upon MN from the centre is $\dfrac{1}{r^2}=\dfrac{\sec^2\theta}{a^2}+\dfrac{\csc^2\theta}{b^2}$.

18. Draw the curves represented by the equations $r=a\cos\theta$, $r=a\sin\theta$, and find the polar equations of their common chord and one of their common tangents.

19. With the conic $\dfrac{l}{r}=1+e\cos\theta$, show that if $\tan a\tan\beta=\dfrac{e+1}{e-1}$, the points on the conic for which θ is equal to $2a$ and 2β respectively, are at the extremities of a diameter.

20. Determine what loci are represented by the following equations in polar co-ordinates :

    (i) $6\theta^2-5\pi\theta+\pi^2=0$.

    (ii) $4\sin^2\theta-2\sqrt{5+2\sqrt{6}}\sin\theta+\sqrt{6}=0$.

    (iii) $(r-a\cos\theta)^2+(r-a\sin\theta)^2=b^2$.

Find the form that the locus (iii) takes when $b=0$.

21. Find the polar equation of a parabola referred to the vertex as pole and the axis as initial line.

A circle passes through the vertex A of a parabola, and cuts it again at the points P, Q, R. Show that the sum of the cotangents of the angles made by AP, AQ, AR with the axis is zero.

What loci are represented by

    22. $r\cos\theta=a$.   **23.** $r^2\cos2\theta=a^2$.   **24.** $r^{\frac{1}{2}}\cos\dfrac{\theta}{2}=a^{\frac{1}{2}}$.

25. Find the polar equation of a rectangular hyperbola referred to its centre O as origin and one asymptote as initial line.

If P is a point on the hyperbola, and a circle is described with centre at P and passing through O ; prove that the locus of the ends of that diameter of this circle which is perpendicular to OP is a rectangular hyperbola whose transverse axis bears to the transverse axis of the given hyperbola the ratio $\sqrt{2}:1$.

CHAPTER XIV.

CURVE TRACING.   PARABOLA.

308. *The straight line $2gx + 2fy + c = 0$ is a tangent to the parabola*
$(ax + by + k)^2 + 2gx + 2fy + c = 0.$

[The curve represents a parabola because the terms of the second degree $(ax + by)^2$ form a perfect square.]

Where the straight line meets the curve, we have by substitution
$$(ax + by + k)^2 = 0,$$

$$\text{or} \quad \left[ax - b\left(\frac{2gx + c}{2f}\right) + k \right]^2 = 0, \quad \left(\text{for } y = -\frac{2gx + c}{2f}\right)$$

a perfect square, or a quadratic with equal roots.

$\therefore$ the line *touches* the curve.

The straight line $ax + by + k = 0$ is a diameter of the parabola
$(ax + by + k)^2 + 2gx + 2fy + c = 0.$

Where the straight line
$$ax + by + k = 0 \quad \dots\dots\dots\dots\dots\dots\dots(1)$$
meets the curve we have by substitution,
$$2gx + 2fy + c = 0. \quad \dots\dots\dots\dots\dots\dots(2)$$

Equations (1) and (2) have but one solution;

$\therefore$ the straight line $ax + by + k = 0$ meets the curve at but one finite point.

$$\therefore \text{ it is a diameter.} \quad \text{(Art. 151.)}$$

309. *To find the axis of the parabola $(3x + 4y)^2 + 8x - 156y - 96 = 0$.*
The axis is parallel to the line $3x + 4y = 0$.   (Art. 308.)
The given equation may be written,
$$(3x + 4y + \lambda)^2 = x(6\lambda - 8) + y(8\lambda + 156) + \lambda^2 + 96. \quad \dots\dots(1)$$
Let us find a value of λ which will make the lines
$$3x + 4y + \lambda = 0, \quad x(6\lambda - 8) + y(8\lambda + 156) + \lambda^2 + 96 \text{ at right angles.}$$

Such a value is given by the equation

$$3(6\lambda - 8) + 4(8\lambda + 156) = 0 ; \quad (aa' + bb' = 0)$$

$$\therefore\ 50\lambda = -600, \quad \lambda = -12.$$

Equation (1), which is the same as the given equation, may now be written

$$(3x + 4y - 12)^2 = -80x + 60y + 240,$$

$$\text{or}\ \ (3x + 4y - 12)^2 = \underline{20}\,(-4x + 3y + 12).$$

Now $-4x + 3y + 12 = 0$ is a tangent to the curve,

and $\qquad 3x + 4y - 12 = 0$ is a diameter,   (Art. 308)

and these lines are at right angles,

$\therefore\ 3x + 4y - 12 = 0$ is the equation of the axis of the parabola.

310. *To draw the parabola* $(3x + 4y)^2 + 8x - 156y - 96 = 0.$

Proceeding as in the previous article, we reduce the equation to the form

$$(3x + 4y - 12)^2 = 20\,(-4x + 3y + 12). \quad\ldots\ldots\ldots\ldots(2)$$

$$3x + 4y - 12 = 0 \text{ is the axis of the curve,}$$

$$-4x + 3y + 12 = 0 \text{ is the tangent at the vertex.}$$

Drawing these str. lines CD, EF, their pt. of intersection, A, is the vertex of the parabola.

Equation (2) may be written

$$\left(\frac{3x + 4y - 12}{5}\right)^2 = 4\left(\frac{-4x + 3y + 12}{5}\right). \quad\ldots\ldots\ldots\ldots(3)$$

Now $\dfrac{3x + 4y - 12}{5} =$ the length of the perpendicular from (x, y) on the line $3x + 4y - 12 = 0.$

And $\dfrac{-4x + 3y + 12}{5} =$ the length of the perpendicular from (x, y) on the line $-4x + 3y + 12 = 0.$

$\therefore$ if we make these lines the axes of x and y respectively,

$$\text{the new } y = \frac{3x + 4y - 12}{5},$$

$$\text{and the new } x = \frac{-4x + 3y + 12}{5}.$$

$\therefore$ equation (3) becomes $y^2 = 4x$.

To determine on which side of the line AEF the curve lies, examine the given equation.

When $y = 0$, $9x^2 + 8x - 96 = 0$.

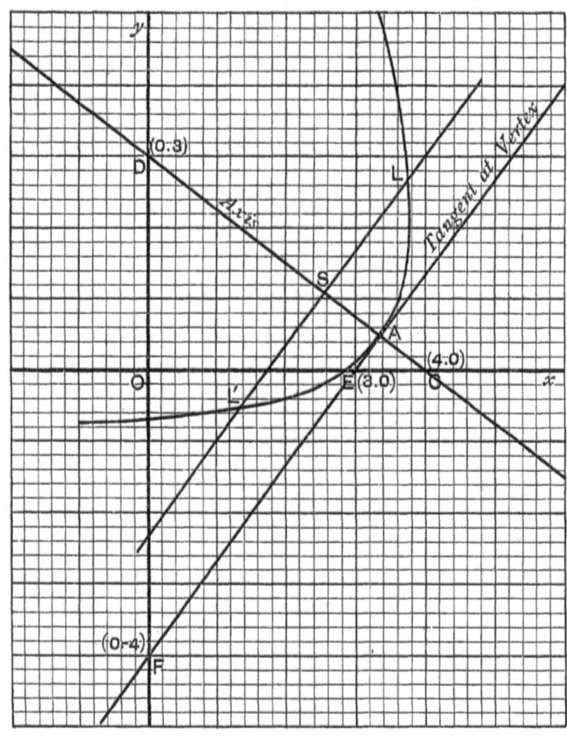

Fig. 175.

We see that the roots are real, but of opposite sign, the numerically greater being negative.

When $x = 0$, $16y^2 - 156y - 96 = 0$.

Here again the roots are real, but of opposite sign, the numerically greater being positive.

Therefore the curve must lie on the same side of AEF as the point D in the figure.

The latus-rectum $= 4$.

$\therefore$ measuring AS $= 1$, along AD, we have the focus S.

Drawing LSL$'$ at rt. $\angle$s. to the axis, such that SL $=$ SL$' = 2$, LSL$'$ is the latus rectum.

We can now draw the curve as shown in the figure.

311. *Trace the curve represented by the equation* $\sqrt{\dfrac{x}{a}} + \sqrt{\dfrac{y}{b}} = 1$, *and find the equation of its directrix and the co-ordinates of its focus.* (*Rectangular axes.*)

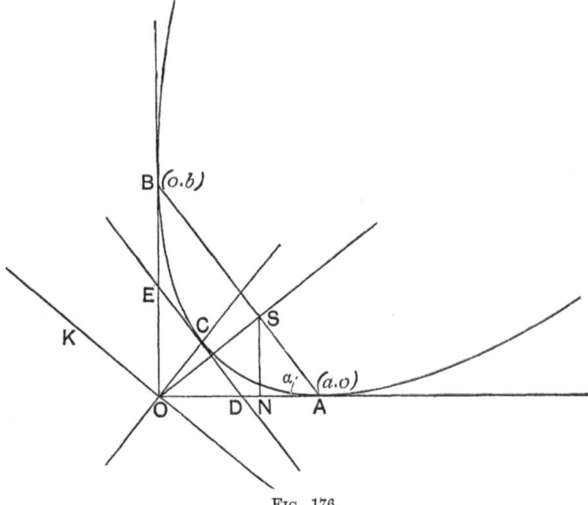

Fig. 176.

We first rationalise the equation.

$$\sqrt{\frac{x}{a}} = 1 - \sqrt{\frac{y}{b}}.$$

Squaring both sides

$$\frac{x}{a} = 1 + \frac{y}{b} - 2\sqrt{\frac{y}{b}},$$

$$\frac{x}{a} - \frac{y}{b} - 1 = -2\sqrt{\frac{y}{b}}.$$

Squaring both sides again and transposing,

$$\frac{x^2}{a^2} - \frac{2xy}{ab} + \frac{y^2}{b^2} - \frac{2x}{a} - \frac{2y}{b} + 1 = 0,$$

$$i.e. \ \left(\frac{x}{a} - \frac{y}{b}\right)^2 = \frac{2x}{a} + \frac{2y}{b} - 1 = 0. \ \dots\dots\dots\dots\dots(1)$$

The terms of the second degree form a perfect square,
∴ the equation represents a parabola.

$$\frac{2x}{a} + \frac{2y}{b} - 1 = 0 \ \text{is a tangent to the curve}$$

where
$$\frac{x}{a} - \frac{y}{b} = 0 \ \text{meets it.}$$

Also
$$\frac{x}{a} - \frac{y}{b} = 0 \ \text{is parallel to the axis.}$$

When
$$x = 0,$$
from equation (1) we have

$$\frac{y^2}{b^2} - \frac{2y}{b} + 1 = 0, \ \text{or} \ \left(\frac{y}{b} - 1\right)^2 = 0.$$

∴ the curve *touches* the axis of y at the point $(0, b)$.
Similarly it touches the axis of x at the point $(a, 0)$.
Plot the points $(a, 0)$, $(0, b)$, A and B in the figure.

Draw the straight line $\frac{x}{a} - \frac{y}{b} = 0$, OC in the figure.

Also draw the straight line $\frac{2x}{a} + \frac{2y}{b} - 1$, DE in the figure.

The tangents OA, OB are at right angles ;
  ∴ AB is a focal chord and O is a point on the directrix.
  ∴ drawing OS perpendicular to AB, S is the focus.
Also drawing OK perpendicular to OC, OK is the directrix, for
OC is parallel to the axis.

The equation of the directrix OK is $ax + by = 0$, for it is perpendicular to $\frac{x}{a} - \frac{y}{b} = 0$.

Let ∠ BAO $= a$, and draw SN perpendicular to Ox,

$$\tan a = \frac{b}{a}.$$

$$\text{SN} = \text{OS} \cos \text{OSN} = \text{OS} \cos \alpha = \text{OA} \sin \alpha \cos \alpha = a \sin \alpha \cos \alpha$$

$$= a \cdot \frac{b}{\sqrt{a^2+b^2}} \cdot \frac{a}{\sqrt{a^2+b^2}} = \frac{a^2 b}{a^2+b^2}.$$

$$\text{ON} = \text{OS} \sin \alpha = \text{OA} \sin^2 \alpha$$

$$= \frac{ab^2}{a^2+b^2} ;$$

$$\therefore \left(\frac{ab^2}{a^2+b^2}, \ \frac{a^2 b}{a^2+b^2} \right) \text{ are the co-ordinates of the focus S.}$$

312. Note the following carefully.

In the parabola $(ax + by + k)^2 + (bx - ay + c) = 0$,

the str. lines $ax + by + k = 0$, and $bx - ay + c = 0$ are at rt. $\angle^s$;

$$\therefore \ ax + by + k = 0 \text{ is the axis,}$$

and $\qquad bx - ay + c = 0$ is the tangent at the vertex.

$(x + y)^2 = k(x - y)$ represents a parabola passing through the origin.

$$x + y = 0, \ x - y = 0 \text{ are at rt. } \angle^s ;$$

$$\therefore \ x + y = 0 \text{ is its axis,}$$

$$x - y = 0 \text{ is the tangent at the vertex.}$$

If the left-hand side of the equation $(ax + by)^2 + 2gx + 2fy + c = 0$ has linear factors, they will be of the form

$$ax + by + k, \ ax + by + k'.$$

In this case, the equation represents two parallel straight lines. We thus see that two parallel straight lines are really a particular case of the parabola.

Examples XIV. a.

Find the equation of the axis and the co-ordinates of the vertex in each of the following parabolas, taking the axes of co-ordinates as rectangular in each case :

1. $x^2 + 2xy + y^2 - 3x - y = 0.$ $\qquad$ 2. $x^2 - 2xy + y^2 + 4x - 8y + 9 = 0.$

3. $25y^2 - 40xy + 16x^2 - 74y + 10x + 295 = 0.$

4. $4x^2 + 4xy + y^2 + 7x + 16y - 11 = 0.$

5. Prove that $y^2 = 2Ax + 2By + C$ represents a parabola whose axis is parallel to the axis of x. Find the equation of its axis, the co-ordinates of its vertex, and the length of its latus rectum.

6. Prove that $(Ax + By)^2 = Cy$ represents a parabola which has the axis of x for a tangent at the origin, and hence prove that all chords of a parabola which subtend a right angle at a given point of the curve intersect on the normal at that point.

7. Prove that the locus of a point, which moves so that its distance from a fixed line is equal to the length of the tangent drawn from it to a given circle, is a parabola.   Find the position of its vertex.

8. Prove that the locus of the centre of a circle, which intercepts a chord of given length $2a$ on the axis of x, and passes through a given point on the axis of y distant b from the origin is the curve $x^2 - 2by + b^2 = a^2$. Trace the curve.

9. Draw the curve given by the equations $4y = t^2$, $x = t + 1$, where t may have any value.

10. $x = 3(t^2 - 1)$, and $y = 6t$, where t may have any value.   Draw the curve given by these equations.

Draw the following parabolas to scale, taking rectangular axes in each case :

11. $x^2 + 4y - 8 = 0$.

12. $x^2 - 8x - 4y = 0$.   Find the equation of the directrix.

13. $y^2 + 2y - 3x + 7 = 0$.   Find the co-ordinates of the focus.

14. $x^2 + 6x - 8y + 25 = 0$.   Find the equation of the directrix.

15. $(x + y)^2 = 8\sqrt{2}(x - y)$.   **16.** $(x - y)^2 = -8\sqrt{2}(x + y)$.

17. $(x - y + 4)^2 = 4\sqrt{2}(x + y - 1)$.   **18.** $(2x + y)^2 = x - 2y - 1$.

19. $16x^2 - 24xy + 9y^2 - 60x - 80y = 0$.

20. $144x^2 - 120xy + 25y^2 + 1420x - 648y = 0$.

21. $\sqrt{\dfrac{x}{3}} + \dfrac{\sqrt{y}}{2} = 1$.   Find the co-ordinates of the focus, and the length of the latus rectum.

313. *Given the equation of an ellipse or hyperbola, to find the co-ordinates of its centre.*

Let $ax^2 + 2hxy + by^2 + 2gx + 2fy + c = 0$ be the equation of the curve, and let (x_1, y_1) be the co-ordinates of its centre.

Transferring the origin to this point, the equation of the curve becomes

$$a(x + x_1)^2 + 2h(x + x_1)(y + y_1) + b(y + y_1)^2 + 2g(x + x_1)$$
$$+ 2f(y + y_1) + c = 0. \quad \ldots\ldots\ldots\ldots(1)$$

The origin is now at the centre of the curve, therefore the equation of the curve contains no terms of the first degree, *i.e.* in equation (1) the coefficient of $x = 0$, and the coefficient of $y = 0$;

$$\therefore \quad 2ax_1 + 2hy_1 + 2g = 0, \quad ax_1 + hy_1 + g = 0,$$

and $\qquad\qquad 2hx_1 + 2by_1 + 2f = 0, \quad hx_1 + by_1 + f = 0$.

Solving these equations, $\dfrac{x_1}{hf - bg} = \dfrac{y_1}{hg - af} = \dfrac{1}{ab - h^2}$,

or $\qquad x_1 = \dfrac{hf - bg}{ab - h^2}, \quad y_1 = \dfrac{hg - af}{ab - h^2}.$

COROLLARY. If $ab = h^2$, the equation represents a parabola, and

$$x_1 = y_1 = \infty.$$

Thus we see that the centre of a parabola is at infinity.

N.B. In equation (1) the terms of the second degree are the same as in the given equation. Thus we see that if the origin is changed, the directions of the axes being unchanged, the terms of the second degree remain the same.

314. *The equation of the tangent at the point (x_1, y_1) to the conic*

$$ax^2 + 2hxy + by^2 + 2gx + 2fy + c = 0$$

is $\qquad axx_1 + h(x_1 y + y_1 x) + by y_1 + g(x + x_1) + f(y + y_1) + c = 0.$

First Method. Let $(x_1 + \Delta x_1,\ y_1 + \Delta y_1)$ be a point on the conic near to the point (x_1, y_1).

These two points are on the conic;

$$\therefore\ ax_1{}^2 + 2hx_1 y_1 + by_1{}^2 + 2gx_1 + 2fy_1 + c = 0, \quad\ldots\ldots\ldots\ldots(1)$$

and $\quad a(x_1 + \Delta x_1)^2 + 2h(x_1 + \Delta x_1)(y_1 + \Delta y_1) + b(y_1 + \Delta y_1)^2$

$$+\ 2g(x_1 + \Delta x_1) + 2f(y_1 + \Delta y_1) + c = 0.$$

Subtracting, and neglecting terms involving $(\Delta x_1)^2$, $(\Delta y_1)^2$, $(\Delta x_1)(\Delta y_1)$, we have

$$2ax_1 \Delta x_1 + 2h(y_1 \Delta x_1 + x_1 \Delta y_1) + 2by_1 \Delta y_1 + 2g \Delta x_1 + 2f \Delta y_1 = 0.$$

Whence, in the limit, $\dfrac{\Delta y_1}{\Delta x_1} = -\dfrac{ax_1 + hy_1 + g}{hx_1 + by_1 + f}$;

$$\therefore\ y - y_1 = -\frac{ax_1 + hy_1 + g}{hx_1 + by_1 + f}(x - x_1)$$

is the equation of the tangent.

This may be written

$$axx_1 + h(x_1 y + y_1 x) + by y_1 + gx + fy$$
$$= ax_1{}^2 + 2hx_1 y_1 + by_1{}^2 + gx_1 + fy_1$$
$$= -gx_1 - fy_1 - c \text{ from equation (1)};$$

$$\therefore\ axx_1 + h(x_1 y + y_1 x) + by y_1 + g(x + x_1) + f(y + y_1) + c = 0$$

is the equation of the tangent at the point (x_1, y_1).

Second Method, by means of the Differential Calculus.
Differentiating with respect to x, we have

$$(2hx + 2by + 2f)\frac{dy}{dx} + 2ax + 2hy + 2g = 0.$$

Whence $$\frac{dy}{dx} = -\frac{ax + hy + g}{hx + by + f};$$

$\therefore$ the slope of the tangent at the point (x_1, y_1) is

$$-\frac{ax_1 + hy_1 + g}{hx_1 + by_1 + f}.$$

We now proceed as in the first method.

**N.B.   From the equation of any conic we obtain the equation
of the tangent at the point $(\mathbf{x_1}, \mathbf{y_1})$ by writing**

$$\mathbf{xx_1} \text{ for } \mathbf{x^2},\ \mathbf{yy_1} \text{ for } \mathbf{y^2},\ \mathbf{x_1y + y_1x} \text{ for } \mathbf{2xy},$$
$$\mathbf{x + x_1} \text{ for } \mathbf{2x},\ \mathbf{y + y_1} \text{ for } \mathbf{2y};$$

the constant term remaining unchanged.

315. *If the axes of co-ordinates be turned through an angle θ, without
changing the origin, we must write*

$$\mathbf{x \cos \theta - y \sin \theta} \text{ in the place of } \mathbf{x},$$
and $$\mathbf{x \sin \theta + y \cos \theta} \qquad ,, \qquad ,, \qquad \mathbf{y}.$$

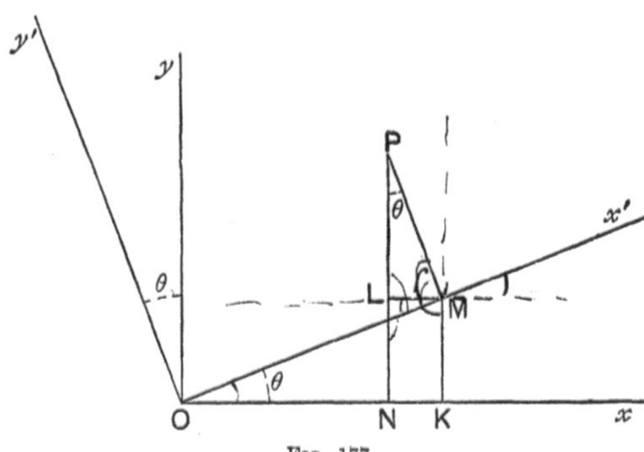

Fig. 177.

Let Ox', Oy' be the new axes.
Let (x, y) be the co-ordinates of P referred to the old axes, and
(x', y') its co-ordinates referred to the new, so that, in the figure,

$$ON = x,\ PN = y,\ OM = x',\ PM = y',$$

$$\angle\,\text{MPN} = 90° - \angle\,\text{PML} = \angle\,\text{LMO} = \angle\,\text{MOK} = \theta,$$

$x = \text{ON} =$ the algebraic sum of the projections of OM and MP on Ox

$$= \text{OM} \cos \theta - \text{PM} \sin \theta$$

$$= x' \cos \theta - y' \sin \theta.$$

$y = \text{PN} =$ the algebraic sum of the projections of OM and MP on PN

$$= \text{OM} \sin \theta + \text{PM} \cos \theta$$

$$= x' \sin \theta + y' \cos \theta.$$

$\therefore$ if the axes of co-ordinates are turned through an angle θ, we must write

$$x \cos \theta - y \sin \theta \text{ for } x, \text{ and } x \sin \theta + y \cos \theta \text{ for } y.$$

The same results may be obtained thus :

$$x = \text{ON} = \text{OK} - \text{KN} = \text{OK} - \text{LM} = \text{OM} \cos \theta - \text{PM} \sin \theta$$

$$= x' \cos \theta - y' \sin \theta,$$

$$y = \text{PN} = \text{LN} + \text{PL} = \text{MK} + \text{PL} = \text{OM} \sin \theta + \text{PM} \cos \theta$$

$$= x' \sin \theta + y' \cos \theta.$$

316. *If the equation* $ax^2 + 2hxy + by^2 + 2gx + 2fy + c = 0$ *represents a central conic, the* **directions** *of its axes are given by the equation*

$$\tan 2\theta = \frac{2h}{a - b}.$$

Let the origin be transferred to the centre, the directions of the axes of co-ordinates being unchanged. The terms of the first degree disappear and we may write the new equation thus :

$$ax^2 + 2hxy + by^2 = k. \quad \text{[See note to Art. 313.]}$$

Now turn the axes of co-ordinates through an angle θ. For x we write $x \cos \theta - y \sin \theta$, and for y, $x \sin \theta + y \cos \theta$.

The equation becomes

$$a(x \cos \theta - y \sin \theta)^2 + 2h(x \cos \theta - y \sin \theta)(x \sin \theta + y \cos \theta)$$
$$+ b(x \sin \theta + y \cos \theta)^2 = k,$$

which may be written

$$x^2(a \cos^2 \theta + 2h \sin \theta \cos \theta + b \sin^2 \theta)$$
$$+ xy[-2a \sin \theta \cos \theta + 2h(\cos^2 \theta - \sin^2 \theta) + 2b \sin \theta \cos \theta]$$
$$+ y^2(a \sin^2 \theta - 2h \sin \theta \cos \theta + b \cos^2 \theta) = k.$$

Equating the coefficient of xy to zero, we have

$$\sin 2\theta (b - a) + 2h \cos 2\theta = 0,$$

$$\text{or} \quad \tan 2\theta = \frac{2h}{a - b}.$$

Therefore if we turn the axes through an angle θ given by this equation, the coefficient of xy disappears, and the equation reduces to the form

$$a'x^2 + b'y^2 = k'.$$

Thus we see that the new axes of co-ordinates are the axes of the curve. This proves the proposition.

317. *To find the* **equations** *of the axes of the central conic*

$$ax^2 + 2hxy + by^2 + 2gx + 2fy + c = 0.$$

First find the centre (x_1, y_1) as in Art. 313.

The *directions* of the axes of the curve are given by the equation

$$\tan 2\theta = \frac{2h}{a - b}, \quad \text{or} \quad \frac{2 \tan \theta}{1 - \tan^2 \theta} = \frac{2h}{a - b}, \quad \text{(Art. 316)}$$

$$\text{i.e.} \quad h \tan^2 \theta + (a - b) \tan \theta - h = 0. \dots\dots\dots\dots\dots\dots(1)$$

Hence, if $\tan \theta_1$, $\tan \theta_2$ are the roots of this equation, they are the slopes of the axes of the conic.

$$\therefore \quad y - y_1 = (x - x_1) \tan \theta_1, \quad y - y_1 = (x - x_1) \tan \theta_2$$

are the equations of the axes, for they pass through the centre (x_1, y_1).

From equation (1) $\tan \theta_1 \tan \theta_2 = -1$.

$\therefore$ if $\tan \theta$ is either root of the equation,

$$y - y_1 = (x - x_1) \tan \theta, \quad y - y_1 = -(x - x_1) \cot \theta,$$

are the reqd. equations.

N.B. If (x, y) be any point on either axis, *referred to the centre as origin,* $\frac{y}{x} = \tan \theta$ is the equation of an axis, where $\tan \theta$ is a root of the equation $\frac{2 \tan \theta}{1 - \tan^2 \theta} = \frac{2h}{a - b}$.

$$\therefore \quad \frac{2\dfrac{y}{x}}{1 - \dfrac{y^2}{x^2}} = \frac{2h}{a - b} \quad \text{or} \quad \frac{xy}{x^2 - y^2} = \frac{h}{a - b}$$

is the equation of the two axes.

318. *To find the* **lengths** *of the axes of the conic* $ax^2 + 2hxy + by^2 = k$.

There are no terms of the first degree in the equation, therefore the centre is at the origin.

Taking the circle $x^2 + y^2 = r^2$, and combining its equation with that of the conic so as to make the result homogeneous and of the second degree, we have

$$r^2(ax^2 + 2hxy + by^2) = k(x^2 + y^2),$$
$$\text{or } x^2(ar^2 - k) + 2hr^2xy + y^2(br^2 - k) = 0.$$

This represents the two straight lines passing through the origin (the centre) and through the points of intersection of the conic and circle. (See figure.)

If these two straight lines coincide, the radius of the circle is one of the semi-axes.

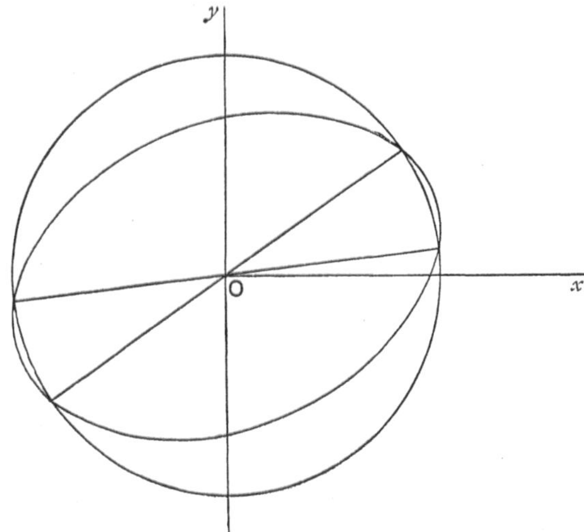

Fig. 178.

They coincide if $4h^2r^4 = 4(ar^2 - k)(br^2 - k)$. $(b^2 = 4ac)$

This may be written $(ab - h^2)r^4 - k(a + b)r^2 + k^2 = 0$.

This equation gives us the lengths of the semi-axes of the conic.

B.A.G.   U

319. If the left-hand side of the equation

$$(ax + by)(a'x + b'y) + 2gx + 2fy + c = 0$$

has linear factors they will be of the form

$$ax + by + k, \text{ and } a'x + b'y + k'.$$

In this case the equation represents two intersecting straight lines. Thus we see that two intersecting straight lines are really a particular case of the hyperbola.

320. *Draw the curve represented by*

$$36x^2 - 24xy + 29y^2 - 168x + 106y + 41 = 0. \qquad \ldots\ldots\ldots(1)$$

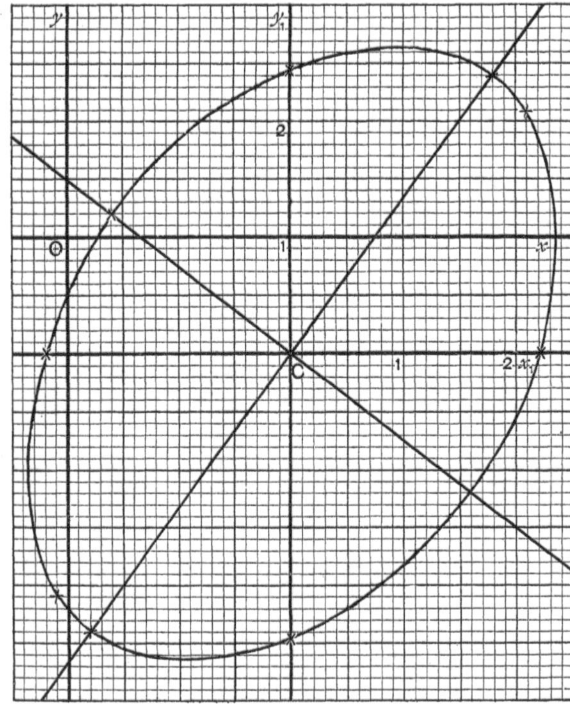

Fig. 179.

$24^2 < 4 \times 29 \times 36$, $\therefore$ the equation represents an ellipse.

Let $(x_1,\, y_1)$ be the co-ordinates of the centre, and transfer the origin to that point. The equation becomes,

$$36\,(x+x_1)^2 - 24\,(x+x_1)(y+y_1) + 29\,(y+y_1)^2$$
$$- 168\,(x+x_1) + 106\,(y+y_1) + 41 = 0. \quad\dots\dots\dots(2)$$

Since the origin is now at the centre, the coefficients of x and y are each equal to zero.

$$\therefore\ 72x_1 - 24y_1 - 168 = 0,\ \text{ or }\ 3x_1 - y_1 - 7 = 0, \quad\dots\dots(3)$$

and $\ -24x_1 + 58y_1 + 106 = 0,\ \text{ or }\ -12x_1 + 29y_1 + 53 = 0. \quad\dots\dots(4)$

Solving equations (3) and (4) $x_1 = 2,\ y_1 = -1$.

Substituting these values in equation (2) it becomes

$$36x^2 - 24xy + 29y^2 = 180.$$

When $y = 0$ here, $x = \pm\sqrt{5}$.

When $x = 0$ here, $y = \pm\sqrt{\dfrac{180}{29}} = \pm\,2{\cdot}5$ nearly, from a Slide Rule.

When $x = y$, $41x^2 = 180$, $x = y = \pm\sqrt{\dfrac{180}{41}} = \pm\,2{\cdot}1$ approx. from a Slide Rule.

Now turn the axes through an angle θ, where

$$\tan 2\theta = -\frac{24}{36-29} = -\frac{24}{7}\ \left(\tan 2\theta = \frac{2h}{a-b}.\quad \text{Art. 316.}\right)$$

$$\cos 2\theta = -\tfrac{7}{25},\ \ 2\cos^2\theta = 1 - \tfrac{7}{25},$$

$$\cos\theta = \tfrac{3}{5},\ \sin\theta = \tfrac{4}{5}.$$

We therefore write $\quad\dfrac{3x - 4y}{5}$ for x,

and $\dfrac{4x + 3y}{5}$ for y.

The equation becomes $20x^2 + 45y^2 = 180$, which may be written

$$\frac{x^2}{9} + \frac{y^2}{4} = 1.$$

The curve is shown in the figure.

The particular points determined above are marked with a cross (X) in the figure.

321. *Draw the conic $7x^2 + 8xy + y^2 + 6x + 6y - 9 = 0$.*

We see that the terms of the second degree have real and different factors, $7x + y$, and $x + y$.

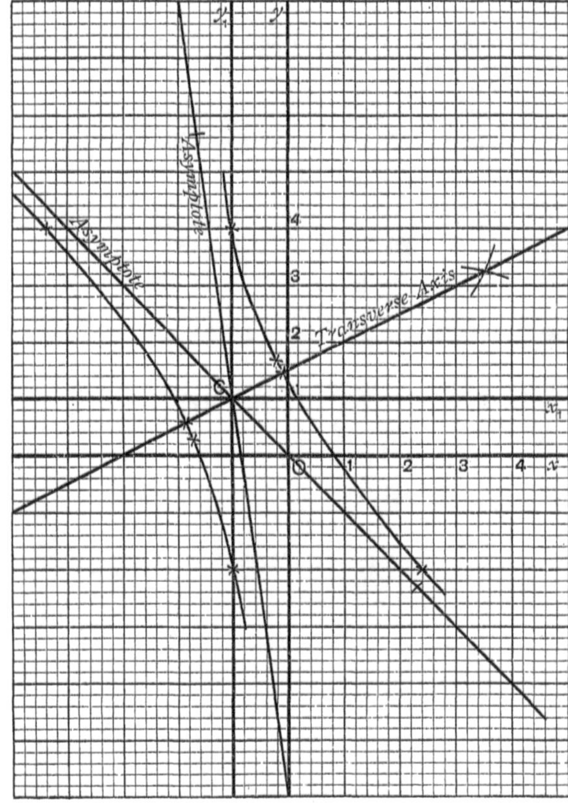

FIG. 180.

$\therefore$ the equation represents a hyperbola.

Let $(x_1,\ y_1)$ be the centre, and transfer the origin to that point.

The equation becomes

$$7(x+x_1)^2 + 8(x+x_1)(y+y_1) + (y+y_1)^2$$
$$+ 6(x+x_1) + 6(y+y_1) - 9 = 0. \quad \ldots\ldots(1)$$

Since the origin is now at the centre, the coefficients of x and y are each equal to zero.

$$\therefore \ \ 14x_1 + 8y_1 + 6 = 0, \ \text{ or } \ 7x_1 + 4y_1 + 3 = 0,$$
$$\text{and} \ \ \ 8x_1 + 2y_1 + 6 = 0, \ \text{ or } \ 16x_1 + 4y_1 + 12 = 0.$$

Whence $x_1 = -1$, and $y_1 = 1$.

Substituting these values in equation (1), it becomes

$$7x^2 + 8xy + y^2 = 9. \quad \ldots\ldots\ldots\ldots\ldots\ldots\ldots(2)$$

$7x^2 + 8xy + y^2 = 0$, or $(x+y)(7x+y) = 0$ represents the asymptotes and we draw these lines.

In equation (2), when $x = 0$, $y = \pm 3$.

> , $x = y$, $x = y = \pm\frac{3}{4} = \pm\cdot75$.
> , $y = 3$, $x = 0$, or $-\frac{24}{7} = -3\cdot4$.
> , $y = -3$, $x = 0$ or $\frac{24}{7} = 3\cdot4$.

Turn the axes through an angle θ, where

$$\tan 2\theta\left(=\frac{2h}{a-b}\right) = \tfrac{8}{6} = \tfrac{4}{3},$$

$$\cos 2\theta = \tfrac{3}{5}, \ \ 2\cos^2\theta = 1 + \tfrac{3}{5},$$

$$\cos\theta = \frac{2}{\sqrt{5}}, \ \ \sin\theta = \frac{1}{\sqrt{5}}.$$

We therefore write $\dfrac{2x-y}{\sqrt{5}}$ for x, and $\dfrac{x+2y}{\sqrt{5}}$ for y.

The equation becomes $\qquad 9x^2 - y^2 = 9$

$$\text{or } \ x^2 - \frac{y^2}{9} = 1.$$

This shows us that the semi-transverse axis = 1 unit in length.

Bisecting the obtuse angle between the asymptotes, we have the direction of the transverse axis, and we can now draw the curve, as shown in the figure.

The points determined above are marked with a cross (X) in the figure.

322. The straight lines $2x - 3y - 12 = 0$, $3x + 2y - 4 = 0$ are at right angles.

If we change the axes of co-ordinates, and make these lines our new axes of x and y respectively,

the new abscissa = the perpendicular from (x, y) upon the new axis of y, $(3x + 2y - 4 = 0)$

$$= \frac{3x + 2y - 4}{\sqrt{13}}.$$

The new ordinate = the perpendicular from (x, y) upon the new axis of x, $(2x - 3y - 12 = 0)$

$$= \frac{2x - 3y - 12}{\sqrt{13}}.$$

E.g. In the curve $\dfrac{(3x + 4y)^2}{25} - \dfrac{(4x - 3y)^2}{50} = 1,$

if we make $3x + 4y = 0$ the new axis of y,

and $4x - 3y = 0$,,   ,, x,

$$\text{the new abscissa} = \frac{3x + 4y}{5},$$

and   ,,   ordinate $= \dfrac{4x - 3y}{5}.$

$\therefore$ the transformed equation is

$$\frac{(5x)^2}{25} - \frac{(5y)^2}{50} = 1,$$

$$\text{or} \quad x^2 - \frac{y^2}{2} = 1.$$

323. Hints for curve tracing.

First examine the terms of the second degree, and thus determine the class of conic which the equation represents.

If there is no term involving the product xy, take the x terms together and the y terms together, and complete squares.

If the equation represents an ellipse or hyperbola, find the centre and transfer the origin to that point, before turning the axes through any angle.

If the coefficients of x^2 and y^2 are equal the term involving the product xy will disappear if the axes are turned through an angle of 45°, for in that case

$$\tan 2\theta = \frac{2h}{a-b} = \infty, \text{ and } 2\theta = \frac{\pi}{2}, \text{ Art. 316.}$$

It is often useful to find where the curve cuts the original, as well as the new axes of co-ordinates, and also where the curve cuts the lines $x \pm y = 0$.

In the case of a hyperbola, having transferred the origin to the centre, find and draw the asymptotes.

In the case of a parabola:

(1) The terms of the second degree equated to zero represent a straight line (or two coincident straight lines) parallel to the axis.

(2) The terms of the first degree together with the constant term represent a tangent to the curve.

A good freehand drawing of a conic can generally be made without finding the co-ordinates of many points on the curve. For example, in an ellipse, when the positions and lengths of the axes and two or three other points on the curve are determined, the curve can be drawn freehand.

If we make the straight line $ax + by + c = 0$ our new axis of x, the new ordinate y, = the perpendicular on $ax + by + c = 0$, which is equal to $\dfrac{ax+by+c}{\sqrt{a^2+b^2}}$; we should therefore write $y\sqrt{a^2+b^2}$ for $ax + by + c$.

A Slide Rule can sometimes be used with advantage.

Examples XIV. b.

Find the centres of the following conics:

1. $2x^2 - 5xy - 3y^2 - x - 4y + 6 = 0$.   2. $4x^2 - 3y^2 - 8x - 6y = 0$.

3. $7x^2 - 9xy + 4y^2 - 32x + 25y + 47 = 0$.

4. Find the directions of the axes of the conic,
$$5x^2 + 3xy + y^2 - 7x + 8y = 0.$$

5. Find the lengths of the axes of the conic,
$$x^2 + xy + y^2 = 8.$$

6. Find (i) the centre (ii) the equations of the axes of the conic,
$$x^2 - xy + y^2 + 4x - 5y - 2 = 0.$$

Trace the loci given by the following equations :

7. $(x-y)^2+(y-1)^2=0$.

8. $4(x^2+y^2-1)=x^2+y^2-4x$.   Also state its relation to the loci represented by $x^2+y^2-1=0$, and $x^2+y^2-4x=0$.

9. $x^2+2x-4y+9=0$.   **10.** $2x^2+y^2=3x$.   **11.** $3(x^2+y^2)+2xy=8$.

12. $4xy-3x^2-4y=0$.   **13.** $(3x-y)^2=6x-2y$.

14. $(y-b)(y-b')+(x-a)(x-a')=0$.   **15.** $(y-x)^2=y+x+1$.

16. $x^2+y^2-1=\pm2(x-1)(y-1)$.   Also state its relation to the straight lines $x-1=0$, $y-1=0$, and the circle $x^2+y^2-1=0$.

17. $2x^2+xy-y^2=a^2$.

18. $x^2+6xy+9y^2+4x+12y-5=0$.

19. $(x+y)^2=x-y$.   Find the equation of its directrix, and the coordinates of its focus.

20. Trace the curve given by the equations $x=2t^2$, $y=4t$, where t is a variable parameter.

21. Trace the curve given by the equations $y=4(t^2-2)$, $x=2t$, where t is a variable parameter.

22. Prove that the curve given by the equations $x=at+bt^2$, $y=bt+at^2$, where t is a variable parameter, is a parabola whose axis is parallel to $ax-by=0$.

Trace the loci given by the following equations :

23. $2x^2-3xy+y^2=0$.   **24.** $4(x^2+y^2-1)=(x+2y-1)^2$.

25. $x^2+y^2+xy=a^2$, oblique axes inclined at an angle of $60°$.

26. $x^2+y^2=a^2$, oblique axes inclined at an angle 2α.

27. $5x^2-2xy+5y^2-8x-8y-4=0$.   **28.** $(x+a)^2=c^2+(y+b)^2$,

29. $x=a\left(\mu+\dfrac{1}{\mu}\right)$, $y=b\left(\mu-\dfrac{1}{\mu}\right)$, where μ is variable.

30. $2x^2-3xy-2y^2+3x-y+1=0$.

31. $4x^2-9y^2+36=0$, when the axes are inclined at an angle 2α.

32. $7x^2+50xy+7y^2=288$.   **33.** $\sqrt{x}+\sqrt{y}=\sqrt{a}$.

34. $\dfrac{(3x-4y+12)^2}{4}+\dfrac{(4x+3y-12)^2}{9}=25$.   **35.** $xy=x+y$.

36. $(3x-4y+12)^2+4(4x+3y-12)^2=100$.   **37.** $\dfrac{1}{x}+\dfrac{1}{y}=2$.

38. $\sqrt{\dfrac{x}{2}}+\sqrt{\dfrac{y}{3}}=1$.   **39.** $x^2+y^2=(x\cos a+y\sin a-p)^2$.

40. $(4x+3y+15)^2=5(3x-4y)$.   **41.** $x^2-y^2-3xy+x=0$.

42. $\dfrac{a}{x}+\dfrac{b}{y}=1$.   **43.** $x^2+2xy+4y^2-2x-14y-3=0$.

44. $(x-y)^2=8(y-3)$.   **45.** $x^2-4y^2+x+4y=0$.

CHAPTER XV.

324. In this chapter it will be shown that if a plane section of a right cone be taken, the curves obtained are **Conic Sections** as defined earlier in this volume.

A **right circular cone**, or **right cone** is defined as the solid generated by the revolution of a right-angled triangle about one of its sides containing the right angle.

A **complete** or **double cone** is obtained by producing the hypotenuse of the right-angled triangle both ways.

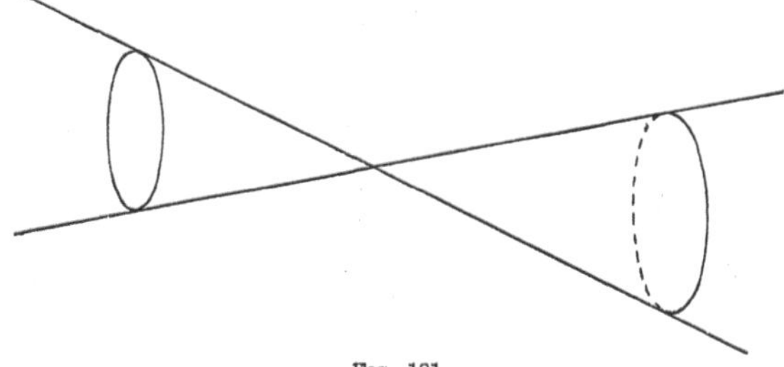

Fig. 181.

325. Let any plane XMPN cut the right cone OKK', giving the curve AP as shown in the figures, OKK' being the plane section of the cone through its axis, and at right angles to the plane XP.

Let EQE'S be the sphere inscribed in the cone and touching the plane XP at S. Let the plane of the circle of contact of this sphere, EQE', cut the plane XP in the straight line XM.

Let KPK' be the section of the cone at right angles to the axis of the cone and through any point P on the curve AP.

KPK' is a circle.

Let PN be the line of intersection of the planes XP and KPK'.

PN and XM are perpendicular to the plane OKK';

∴ if PM is drawn parallel to NX, PNXM is a rectangle.

Let OP touch the sphere at Q.

OP = OK, and OQ = OE (tangents to a sphere).

$$\therefore \ \text{QP} = \text{EK}.$$

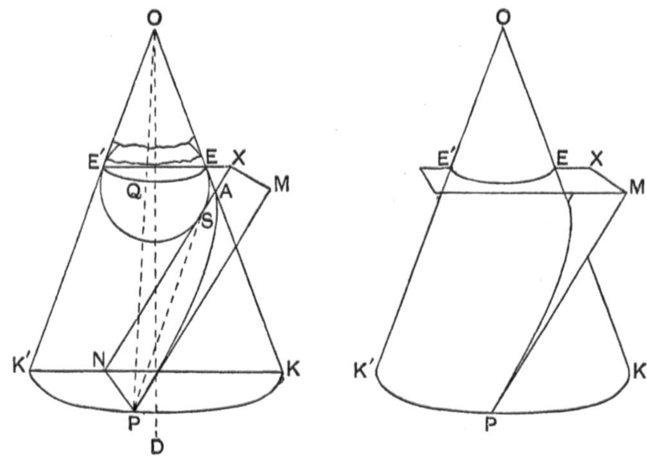

Fɪɢ. 182.

$$\frac{\text{SP}}{\text{PM}} = \frac{\text{PQ}}{\text{PM}},\ \text{for PS, PQ are tangents to the sphere,}$$

$$= \frac{\text{PQ}}{\text{NX}},\ \text{for NPMX is a rectangle,}$$

$$= \frac{\text{EK}}{\text{NX}},\ \text{for PQ} = \text{EK (proved above)},$$

$$= \frac{\text{EA}}{\text{AX}},\ \text{for EX is parallel to NK,}$$

$$= \frac{\text{SA}}{\text{AX}},\ \text{for AE, AS are tangents to the sphere.}$$

$\therefore$ the locus of P is a curve such that the distance of P from the fixed point S is in a constant ratio to the perpendicular distance of P from the fixed straight line XM, *i.e.* the curve AP is a conic section, as previously defined.

A Parabola.

If SA = AX, EA = AX ;

$\therefore$ AN = AK, from the similar $\triangle^s$ EAX, KAN ;

$\therefore$ $\angle$ ANK $= \angle$ AKK$'$ $= \angle$ AK$'$K ;

$\therefore$ XN is parallel to OK$'$;

$\therefore$ the plane XMP is parallel to OK$'$.

Thus we see that the plane section of a cone is a parabola when the plane of section is parallel to a generating line.

An Ellipse.

If SA $<$ AX,   EA $<$ AX,

$\therefore$ AK $<$ AN ;

$\therefore$ $\angle$ ANK $< \angle$ AKN,

i.e. $< \angle$ OK$'$N ;

$\therefore$ the plane XMP will cut OK$'$ produced.

Thus we see that the plane section of a cone is an ellipse when it cuts the two generating lines OK, OK$'$ on the same side of the vertex O.

A Hyperbola.

In the same way if SA $>$ AX, $\angle$ ANK $> \angle$ OK$'$N.

$\therefore$ the plane XMP will meet K$'$O produced and we have the second branch of the curve on the part of the complete cone formed by producing the generating lines KO, PO, K$'$O, etc.

Therefore the plane section of a cone is a hyperbola when it cuts both parts of the complete cone.

MISCELLANEOUS REVISION QUESTIONS.

[These may be taken orally, or the answers may be written down without any working.]

1. Interpret the constants in the equations $\dfrac{x-x_1}{\cos\theta}=\dfrac{y-y_1}{\sin\theta}=r$.

2. What is the equation of any straight line parallel to $Ax+By+C=0$?

3.   ,,   ,,   ,,   perpendicular   ,,   ,,

4. What is the equation of any straight line through the intersection of $ax+by+c=0$, and $lx+my+n=0$?

5. Give the equation of the straight line passing through the origin and also through the intersection of $3x-4y-5=0$, $4x-7y+5=0$.

6. What is the slope of the line joining the points $(3,\,4)$, $(1,\,-2)$?

7. Interpret a in the equation $x\sin a+y\cos a=p$.

8. Give (without simplification) the equations of the bisectors of the angles between the straight lines $x+y=5$, $x-2y=7$.

9. What is the tangent of the angle between the lines

$$x=0,\quad y=mx+c\,?$$

10. What is the condition that the straight lines $y=2x+3$, $y=4x-7$ may be at right angles when the axes are oblique?

What are the conditions that an equation of the second degree in x and y may represent :

11. Two straight lines.

12. Two straight lines passing through the origin.

13. Two parallel straight lines.   **14.** Two perpendicular straight lines.

15. Two coincident straight lines.   **16.** A circle.

17. A parabola.   **18.** An ellipse.   **19.** A hyperbola.

20. Give an equation of the second degree in x and y which represents a point.

Give the equation of the two straight lines passing through the origin and through the points of intersection of :

21. $x^2+y^2=a^2$ and $lx+my=n$.

22. $ax^2+2hxy+by^2+2gx+2fy+c=0$, and $lx+my=1$.

23. $ax^2+2hxy+by^2+2gx+2fy+c=0$, and $lx+my+n=0$.

24. $y^2=4ax$ and $lx+my+n=0$.   **25.** $\dfrac{x^2}{a^2}+\dfrac{y^2}{b^2}=1$ and $x^2+y^2=r^2$.

Give the equation of the tangent at the point $(x_1,\,y_1)$ in each of the following cases :

26. $x^2=ay$.   **27.** $4x^2+6y^2=1$.

28. $4x^2+7y^2=1$.   **29.** $x^2+4xy-y^2=32$.

30. $2x^2-3xy+5y^2-4x+5y+7=0$.

31. Give the equation of any parabola whose vertex is at the point (h, k) and (i) whose axis is parallel to $\mathbf{O}x$, (ii) whose axis is parallel to $\mathbf{O}y$.

32. Give the equation of any ellipse whose centre is at the point (h, k) and whose axes are parallel to the axes of co-ordinates.

33. What is the equation of a rectangular hyperbola whose centre is at (h, k) and (i) whose transverse axis is parallel to $\mathbf{O}x$, (ii) whose transverse axis is parallel to $\mathbf{O}y$?

If $\mathbf{S}=0$, and $\mathbf{S}'=0$ represent the equations of two circles :

34. What does $\mathbf{S}-\lambda\mathbf{S}'=0$ represent ?

35.   ,, $\mathbf{S}-\mathbf{S}'=0$,,

36. The straight line $ax+by-c+k(ax-by-c)=0$ passes through a fixed point for all values of k.   What are the co-ordinates of the point ?

Give the co-ordinates of any point on the curve in each of the following cases, such co-ordinates to involve only one variable :

37. $x^2+y^2=a^2$.

38. $(x-d)^2+(y-e)^2=r^2$.

39. $\dfrac{(x-a)^2}{a^2}+\dfrac{(y-b)^2}{b^2}=1$.

40. $\dfrac{(x-a)^2}{a^2}-\dfrac{(y-b)^2}{b^2}=1$.

41. $\dfrac{(x-h)^2}{a^2}+\dfrac{(y-k)^2}{b^2}=1$.

42. $\dfrac{(x+h)^2}{a^2}-\dfrac{(y+k)^2}{b^2}=1$.

What class of conic is represented in each of the following cases (give reasons for your answers) :

43. $(ax+by+c)^2+2gx+2fy+k=0$.

44. $4(x-3y)^2-(y+2x)^2=a^2$.

45. $8x^2+2xy+y^2=32$.

46. $3x^2-xy+2y^2+5x-7y+11=0$.

47. $x^2+y^2=a^2$ (oblique axes).

48. $x^2-5xy-y^2=a^2$.

49. $6y^2-xy-x^2-4x+9y+14=0$.

50. $y^2+hxy-x^2+ax+by+c=0$.

CHAPTER XVI.

REVISION PAPERS.

Revision Paper XVI. a.

1. On the axis of x is taken a point **A**, and on the axis of y a point **B**. **C** is a point in the line **AB** such that $AC = 2BC$. If the sum of the distances of **A** and **B** from the origin is a constant quantity h, find the locus of **C**. (Rect. axes.)

2. Write down the equation of a circle of radius r whose centre **A** is on the axis of x at a distance $6r$ from the origin **O**.

Write down also the equation to the tangent to this circle at the point (x', y'), and determine the abscissa of the point in which the tangent cuts Ox.

Take the circle to represent the path of one end of a connecting rod **BC**, attached to a crank **AC**, while the end **B** travels in the line **OA**; and calculate the length of **BC** if $r = 1$ foot and the abscissa of **C** is 5·7 feet when **BC** and **AC** are at right angles.

3. A circle is drawn touching a parabola at its vertex and cutting it again at **P**. From **G**, the foot of the normal at **P**, a line **GQR** is drawn at right angles to the axis cutting the circle and parabola in **Q** and **R** respectively. Show that $RG^2 = 2QG^2$.

4. An ellipse has a given focus and touches three given straight lines: find the centre of the curve and the points of contact of the given tangents.

5. Prove that the locus of the middle points of all chords of the ellipse $b^2x^2 + a^2y^2 = a^2b^2$ drawn at a constant distance c from the centre is

$$\left(\frac{x^2}{a^2} + \frac{y^2}{b^2} \right)^2 = c^2 \left(\frac{x^2}{a^4} + \frac{y^2}{b^4} \right).$$

6. A series of circles is drawn touching the axis of x at the origin. Find and draw the locus of the pole of the straight line $x + y = 1$ with respect to these circles.

7. Draw the curve given by the equations $x - a = a \sec \theta$, $y - b = 2a \tan \theta$, where θ may have any value. Give a geometrical interpretation to the angle θ.

Revision Paper XVI. b.

1. The finite straight lines **OA**, **OB**, including any angle, are bisected in **C**, **D** respectively, and points **P**, **Q** are taken on them such that $OA = n \cdot PA$ and $OB = n \cdot OQ$. Write down the equations of **CD** and **PQ** referred to **OA** and **OB** as axes, and show that **CD** bisects **PQ**.

2. Find the general equation of a circle whose centre lies in the axis of x. If the abscissae of the centres of two such non-intersecting circles be a, $-a'$, and their radii r and r', find the co-ordinates of points on the axis of x at which the circles subtend equal angles.

Find the equation of the radical axis of the two circles.

3. Find the equation of the normal at the point $(am^2, 2am)$ of the parabola $y^2 = 4ax$, and determine the ordinate of the point where it meets the curve again.

The normal at one extremity R of the latus rectum RSR′ meets the curve again at Q: prove that the diameter through the intersection of the tangents at Q and R passes through R′.

4. The ordinate and tangent at any point P on an ellipse meet the transverse axis in N and T. Show that any circle through N and T cuts the auxiliary circle orthogonally.

5. Show that the length of the normal (terminated by the major axis) at a point on the ellipse $\dfrac{x^2}{a^2} + \dfrac{y^2}{b^2} = 1$ is $\dfrac{b}{a}\sqrt{rr'}$, where r and r' are the focal distances of the point.

6. Draw the curve represented by $3x^2 + 2xy + 3y^2 = 16$.

7. ABC is a triangle; AB, AC are fixed in direction. Find the locus of the point which divides BC in the ratio $m:n$, where the perimeter of the triangle ABC is constant.

Revision Paper XVI. c.

1. Find the equation to a straight line through a fixed point (h, k), and making an angle of $\dfrac{\pi}{3}$ with the axis of x.

If the straight line rotate in a counter clock-wise direction about the fixed point through an angle of one minute, show that the intercept on the axis of y is diminished by a length equal to $\dfrac{\pi h}{2700}$ approximately.

2. Form the equation of a circle of radius a touching both the axes of co-ordinates, and lying in the quadrant in which both co-ordinates are positive.

Find the angles which a chord PQ passing through the origin O must make with the axes in order that OQ may be bisected at P.

3. Prove that if two parabolas have the same focus and axis, the locus of the point of intersection of two tangents at right angles, one to each of the parabolas, is a straight line.

4. A pair of tangents to the circle $x^2 + y^2 = a^2$ make angles θ_1, θ_2 with the axis of x such that $\tan\theta_1\tan\theta_2 + 3 = 0$. Find the equation of, and draw the locus of their point of intersection.

5. Having given the foci and the length of the major axis of an ellipse, show how to find the points on it at which the line joining the foci subtends a given angle.

If the eccentricity of the ellipse is $\frac{1}{2}$, within what limits must the angle lie?

6. The axis CY revolves round the point C, while the axis CX is fixed; hyperbolas are described having CX and CY for asymptotes, and their foci at a given distance from C: prove that the extremities of their transverse axes lie on a circle.

7. Interpret the connection between the three curves represented by $(x^2 + y^2 - 4) + (y^2 - 2x) = 0$, $x^2 + y^2 - 4 = 0$, $y^2 - 2x = 0$, and draw them.

Revision Paper XVI. d.

1. Form the equation of the two straight lines passing through the point of intersection of the lines $3x + 4y + 7 = 0$, $4x - 3y + 1 = 0$ and making angles of 60° with the former line.

2. The equation $x^2 + y^2 + 2\lambda x + c = 0$ represents a series of circles, one circle for each value of λ. Prove that, if c is negative, each of these circles passes through two fixed points, the same for all the circles.

If c is positive find the centres of the particular circles of the system whose radii are zero.

Determine the equation of a circle which cuts every circle of the system at right angles.

3. Prove that the parabolas $y^2 = 4ax$, $x^2 = 4by$ intersect at an angle of 30° if $2(a^{\frac{2}{3}} + b^{\frac{2}{3}}) = 3\sqrt{3}a^{\frac{1}{3}}b^{\frac{1}{3}}$.

4. Chords AP, AQ parallel to two conjugate diameters of the ellipse $\dfrac{x^2}{a^2} + \dfrac{y^2}{b^2} = 1$ are drawn through the vertex $(a, 0)$. Prove that PQ makes with the axis of x an angle $\tan^{-1}\dfrac{2b^2\tan\theta}{b^2 - a^2\tan^2\theta}$, θ being the angle which one of the diameters makes with the axis of x.

5. Find the lengths of the perpendiculars from the foci of an ellipse on the tangent at the point whose eccentric angle is ϕ.

Show that they may be obtained by solving the equation

$$u^2 - \frac{2abu}{b'} + b^2 = 0,$$

where a and b are the principal semi-axes, and b' the semi-diameter conjugate to the central vector of the point.

6. Find the equation of that diameter of the hyperbola

$$5x^2 + 3xy - y^2 = 8a^2$$

which is conjugate to the axis of x, and prove that the difference of the squares of these two conjugate diameters is equal to $\dfrac{512a^2}{29}$.

7. If t be a variable quantity, prove that the locus of the point (x, y) when $x = at - bt^2$, and $y = bt - at^2$ is a parabola,

Revision Paper XVI. e.

1. Show that the equation

$$ax^2 + 2hxy + by^2 + 2gx + 2fy + \frac{af^2 + bg^2 - 2fgh}{ab - h^2} = 0$$

represents two straight lines.

If these are inclined to the axis of x at angles θ_1, θ_2, express $\tan(\theta_2 - \theta_1)$, $\tan(\theta_2 + \theta_1)$ in terms of a, h, b.

2. Find the general equation (in rectangular co-ordinates) of a circle which is drawn through two fixed points **A** and **B**.

Show that the polars of a fixed point **C** relatively to all such circles will pass through another fixed point.

3. Prove that if any two chords **PQ**, **PR** at right angles to each other are drawn through the point $P(x', y')$ of the parabola $y^2 = 4ax$, **QR** meets the normal at **P** in a point on the line $y + y' = 0$.

4. Prove that if $Aa^2 + Bb^2 = 2$, the conic $Ax^2 + 2Hxy + By^2 = 1$ will pass through the extremities of a pair of conjugate diameters of the ellipse

$$\frac{x^2}{a^2} + \frac{y^2}{b^2} = 1.$$

5. Prove that the equation of the locus of a point dividing in the ratio $2 : 1$ the chords of the ellipse $\dfrac{x^2}{a^2} + \dfrac{y^2}{b^2} = 1$ which make an angle θ with the major axis is

$$\left(\frac{\cos^2\theta}{a^2} + \frac{\sin^2\theta}{b^2}\right)\left(\frac{x^2}{a^2} + \frac{y^2}{b^2} - 1\right) + 8\left(\frac{x\cos\theta}{a^2} + \frac{y\sin\theta}{b^2}\right)^2 = 0.$$

6. A series of circles touch a given straight line at a given point. Prove that the locus of the pole of any given straight line with respect to the circles is a hyperbola.

7. Find (1) the centre, (2) the equations of the asymptotes of the curve

$$y(x - 10) = x^2 - 7x + 6.$$

Trace the curve.

Revision Paper XVI. f.

1. Reduce the equation $35x^2 - 74xy + 35y^2 + x + 13y = 0$ to the forms $ax^2 + 2hxy + by^2 = c$ and $(ax + by + c)(a'x + b'y + c') = C$.

Find the co-ordinates of the centre of the conic determined by the equation.

2. Find the general equation of the circles which cut the circle $x^2 + y^2 - 2x + 2y - 2 = 0$ at right angles at the point $(1, 1)$.

3. Find the co-ordinates of the point at which $y = mx - 2am - am^3$ is normal to the parabola $y^2 = 4ax$. If y_1 is the ordinate of this point and y_2 that of the middle point of the normal chord through the point, prove that $y_1 y_2 + 4a^2 = 0$.

4. If in an ellipse n_1, n_2 are the lengths intercepted by the major axis of two perpendicular normals, show that

$$n_1^{-2} + n_2^{-2} = (a^2 + b^2)b^{-4}.$$

B.A.G. X

5. Find the angle between two lines which intersect in a focus of the conic $b^2x^2 + a^2y^2 - a^2b^2 = 0$, each line passing through the pole of the other.

6. A variable circle always passes through two fixed points A, B. Find the locus of a point on it where the tangent is perpendicular to AB.

7. Find the vertex, the latus rectum, and the equation of the axis of the parabola $(x - y)^2 = 8(y - 3)$.
Draw the curve.

Revision Paper XVI. g.

1. Find the angle between the straight lines represented by
$$x^2 + y^2 = 2xy \operatorname{cosec} a,$$
the axes being rectangular.
Find the area of the triangle formed by these lines and the line
$$lx + my + n = 0.$$

2. Find the angles between the radical axes of the circles
$$x^2 + y^2 - 4ax - 4ay + 4a^2 = 0, \quad x^2 + y^2 - 4ax + 4ay + 4a^2 = 0,$$
and
$$x^2 + y^2 - 8ax + 4a^2 = 0.$$

3. Show that the parabolas whose equations in rectangular axes are $y^2 - ax = 0$, $x^2 - by = 0$ have only one common chord which cuts the curves in real points.
If P and P' be the points on $y^2 - ax = 0$, and $x^2 - by = 0$ the tangents at which are parallel to this common chord, and S, S' be the foci of the parabolas, then SP is perpendicular to $S'P'$.

4. Prove that the equation $\dfrac{x^2}{a^2 \cos^2 a} + \dfrac{y^2}{b^2 \sin^2 a} = 1$ represents, for different values of a, a series of ellipses all inscribed in a certain rhombus.

5. If S and S' are the foci, SL the semi-latus rectum of an ellipse, and LS' produced cuts the ellipse at Q, show that the length of the ordinate at Q is $\dfrac{(1 - e^2)^2 a}{1 + 3e^2}$, where $2a$ is the length of the major axis, and e is the eccentricity of the ellipse.

6. The ordinate NP of any point P on an ellipse is produced to Q so that NQ is equal to the subtangent at P: prove that the locus of Q is a hyperbola, and draw the curve.

7. Find the equation of the axes of the central conic $ax^2 + 2hxy + ay^2 = c$.
Also prove that the lengths of the semi-axes are $\sqrt{\dfrac{c}{a + h}}$ and $\sqrt{\dfrac{c}{a - h}}$.

Revision Paper XVI. h.

1. If the straight lines $ax + by + p = 0$ and $x \cos a + y \sin a - p = 0$ enclose an angle $\dfrac{\pi}{4}$ between them, and meet the straight line $x \sin a - y \cos a = 0$ in the same point, prove that $a^2 + b^2 = 2$.

2. Find the equations of the two circles which pass through the two points $(0, a)$, $(0, -a)$ and touch the line $y = mx + c$.

If the two circles intersect at right angles, prove that $c^2 = a^2(2 + m^2)$.

3. PV is the diameter through the point P on a parabola, and QV any one of its ordinates. QV is bisected in M, and PV in N. Prove that for different positions of Q, PM and QN will always intersect on another parabola touching the given parabola.

4. CP, CD are conjugate semi-diameters of an ellipse. If on the tangent at P a point Q be taken such that PQ is equal in length to CD, show that the locus of Q is an ellipse.

5. In the ellipse whose semi-axes are a and b, prove that the acute angle between two conjugate diameters cannot be less than $\sin^{-1} \dfrac{2ab}{a^2 + b^2}$.

6. Given one asymptote of a hyperbola, one vertex, and the eccentricity, show how to find the other asymptote.

7. Find the centre and the lengths of the semi-axes of the conic
$$6x^2 + 5xy - 6y^2 - 4x + 7y + 11 = 0.$$

Revision Paper XVI. k.

1. The ends of one diagonal of a rhombus are on the co-ordinate axes Ox, Oy (rect.) at distances $2a$, a respectively from O, and the equation of one side is $4(y - a) = 3x$. Find the equations of the diagonals and of the other sides.

2. Find the equation of the chord of the circle $x^2 + y^2 = c^2$, which passes through the points P, Q where $\angle POX = a$, $\angle QOX = \beta$, OX being the axis of x.

If XOX' is a diameter of the circle, and $\angle POQ = 2\gamma$ (a constant) prove that the locus of the intersection of XP, X'Q is the circle
$$x^2 + y^2 + 2cy \tan \gamma = c^2. \quad (a > \beta).$$

3. Prove that the projection of any normal chord of a parabola on the axis is equal in length to the focal chord parallel to it.

4. A chord of an ellipse is drawn through the points whose eccentric angles are θ and ϕ. If P is the point whose eccentric angle is $\frac{1}{2}(\theta + \phi)$, and the line joining P to either focus meets the chord in Q, prove that
$$QP = a\left(1 - \cos\frac{\theta - \phi}{2}\right).$$

5. An ellipse is inscribed in a triangle, so that the line joining the points of contact of two sides is a focal chord. Show that the third side subtends a right angle at this focus.

6. Find the equation of the axes of the ellipse whose equation is
$$3x^2 + 7xy + 12y^2 = 144.$$

7. Find (1) the centre, (2) the equations of the asymptotes of the hyperbola $\dfrac{(3x - 4y - 12)^2}{100} - \dfrac{(4x + 3y - 12)^2}{225} = 1$.

Draw the curve.

Revision Paper XVI. 1.

1. Points P and Q are taken in the sides AB, AC respectively of a triangle ABC, so that PQ is parallel to BC. Find the locus of the intersection of a straight line drawn through P parallel to AC with another through Q perpendicular to BC.

2. Through a fixed point C any straight line is drawn cutting a fixed circle in two points P, Q, and a point R is taken on this line such that CR is the harmonic mean between CP, CQ. What is the locus of R ?

3. Tangents are drawn from the point $(a, 2\sqrt{2}a)$ to the parabola $y^2 = 4ax$; prove that the ratio of their lengths is $\sqrt{2} + 1 : 1$.

4. Two diameters LQM, L'Q'M' meet the parabola $y^2 = 4ax$ in Q, Q'; the tangent at P in L, L'; and the normal at P in M, M'. If LQ = QM, and $L'Q' = \frac{1}{2}Q'M'$, prove that the distance between the diameters is $\dfrac{2a(1 + m^2)}{3m}$ where $(am^2, -2am)$ are the co-ordinates of P.

5. If CP and CD be two conjugate semi-diameters of an ellipse, θ the angle between SP and CD, and ϕ the angle between SD and CP, then

$$\cot^2\theta + \cot^2\phi$$

will be constant.

6. Find the locus of the middle points of all chords of a hyperbola drawn through a fixed point.

Draw a rough sketch of the locus.

7. Find the equations of the asymptotes and axes of the conic given by the equation $48x^2 - 16xy - 15y^2 + 112x - 14y + 56 = 0$.

Revision Paper XVI. m.

1. Prove that the area of the triangle cut off by $y = x + c$ from the straight lines $ax^2 + 2hxy + by^2 = 0$ is $\dfrac{c^2\sqrt{h^2 - ab}}{a + b + 2h}$.

2. A circle passes through two points on the initial line whose distances from the pole are a, $\dfrac{c^2}{a}$; and also through two points on a line through the pole perpendicular to the initial line, whose distances from the pole are b, $\dfrac{c^2}{b}$: find its polar equation.

3. In the parabola $y^2 = 4ax$, the tangent at the point P, whose abscissa is equal to the latus rectum meets the axis in T, and the normal at P cuts the curve again in Q. Prove that $PT : PQ :: 4 : 5$.

4. Prove that all chords of an ellipse which subtend a right angle at a given point P on the ellipse cut the normal at P in a fixed point.

5. Two conjugate semi-diameters of an ellipse are inclined at angles θ and ϕ to the major axis: show that their lengths a', b', are connected by the relation $a'^2 \sin 2\theta + b'^2 \sin 2\phi = 0$.

6. Find the equation of, and draw the locus of the pole of a tangent to the ellipse $\dfrac{x^2}{a^2}+\dfrac{y^2}{b^2}=1$ with respect to the hyperbola $xy=c^2$.

7. Trace the curve $3xy-y^2=9x$.

Revision Paper XVI. n.

1. The product of the perpendiculars from the point $(a,\ b)$ on the lines represented by $ax^2+2hxy+by^2=0$ is

$$\frac{a^3+b^3+2abh}{\sqrt{(a-b)^2+4h^2}}.$$

2. A straight line passes through a fixed point and is divided by that point in a given ratio. If the line subtend a constant angle at another fixed point, prove that the loci of its ends are circles.

3. Through the vertex of a parabola any two straight lines are drawn at right angles to one another, and meet the curve again in P and Q. Show that the tangents to the parabola at P and Q intersect on a fixed straight line perpendicular to the axis.

4. Find the equation of the chord of the conic $ax^2+by^2=1$ the middle point of which is the point $(f,\ g)$.

If the point $(f,\ g)$ moves along the straight line $\mathsf{A}x+\mathsf{B}y=1$, prove that the chord always touches the parabola

$$(a\mathsf{B}x-b\mathsf{A}y)^2+4ab(\mathsf{A}x+\mathsf{B}y-1)=0.$$

5. Find the equation of the chord joining two points on the ellipse whose eccentric angles are $\theta,\ \phi$; and show that if $\sec\theta+\sec\phi=2$, the chord will touch the ellipse

$$\frac{4x^2}{a^2}+\frac{y^2}{b^2}=\frac{4x}{a}.$$

6. Prove that the curve given by the equations $\dfrac{y}{rb}=\dfrac{x}{a+na}$, $\dfrac{y}{nb}=\dfrac{x-a}{ra}$, r being a variable quantity, is an ellipse or hyperbola, whose centre is at the point $\left(\dfrac{a}{2},\ 0\right)$, according as n is less or greater than -1.

7. Trace the conic whose equation is

$$x^2+2xy+4y^2-2x-14y-3=0.$$

ANSWERS TO THE EXAMPLES.

I. a. (p. 6).

1. $P_1(5, 4)$, $P_2(9, 7)$, $P_3(-4, 4)$, $P_4(-8, 8)$, $P_5(-11, -4)$, $P_6(-6, -6)$, $P_7(5, -4)$, $P_8(11, -6)$.

3. $(4, 3)$.   4. $(0, 0)$.   6. $(-1, 2)$, $(1, -1)$, $(-2, -1)$.

7. 5.   8. 13.   9. $\sqrt{a^2+b^2}$.   10. $2\sqrt{(a^2+b^2)}$.   11. $2b$.

12. $a-b$.   13. $a\left(\dfrac{1}{m_1}-\dfrac{1}{m_2}\right)\sqrt{\left(\dfrac{1}{m_1}+\dfrac{1}{m_2}\right)^2+4}$.   14. $(1-\cos\theta)\sqrt{a^2+b^2}$.

15. $4\cdot47$.   16. $9\cdot22$.   17. $(4, 4)$, $(5, 6)$.   18. $(4, 3)$.

I. b. (p. 11).

1. 9.   2. 12.   3. 8.   4. 6.

5. 10.   6. x_1y_1.   7. $\tfrac{1}{2}(x_1y_2+x_2y_1)$.   8. $\dfrac{ch}{2}$.

9. x_1y_1.   10. 14.   11. 4.   12. $\tfrac{1}{2}(h-k)(h+k-2a)$.

13. 16.   14. 23.   15. 28.   16. 32.

I. c. (p. 13).

1. $x=6$.   2. $y=4$.   3. $x+3=0$.   4. $y+5=0$.

5. $2x+1=0$.   6. $5x+3y=10$.   7. $y=3x$.   8. $x^2+y^2=25$.

9. $2x^2+2y^2-9x+18y+45=0$.   10. $x^2+4x-10y+29=0$.

11. $x^2+y^2+4x-8y-5=0$.

12. The locus is a straight line through the origin. $x=2y$.

13. The locus is a straight line. $y=x+1$.

15. $y=x+3$.   16. $x+y=7$.

II. a. (p. 25).

1. $y=8$.   2. $y+5=0$.   3. $x=4$.   4. $x+3=0$.   5. $y=4x$.

6. $\dfrac{x}{3}=\dfrac{y}{2}$.   7. $\dfrac{x}{4}+\dfrac{y}{5}=0$.   8. $x+y=0$.   9. $y=-3x$.

10. $y = 4x + 3$.   **11.** $y = -3x - 2$.   **12.** $x + y = 3$.   **13.** $\dfrac{x}{3} - \dfrac{y}{4} = 1$.

14. (i) 2.   (ii) 3.   (iii) $-\dfrac{1}{5}$.   (iv) $-\dfrac{3}{4}$.   (v) $\dfrac{4}{5}$.   (vi) $-\dfrac{A}{B}$.   (vii) $-\dfrac{b}{a}$.
     (viii) $\cot a$.   (ix) $-\tan a$.

15. (i) -5.   (ii) 2.   (iii) $-1\frac{1}{2}$.   (iv) -3.   (v) $-\dfrac{C}{B}$.   (vi) $p \operatorname{cosec} a$.

16. (i) 4.   (ii) 3.   (iii) -5.   (iv) 6.   (v) $-\dfrac{C}{A}$.   (vi) $p \sec a$.

17. $x_1{}^2 + y_1{}^2 = a^2$.   **18.** $b = c$.   **19.** $a = \pm b$.

20. $c = 0$.   **21.** At the point (a, b).   **22.** $(2, 0)$.

23. The lines have equal slopes, and are parallel.

II. b.   (p. 26).

1. $3x - 4y = 0$.   **2.** $x - y = 3$.   **3.** $x - y = 4$.   **4.** $x + y = 4\sqrt{2}$.

5. $x - y = 10$.   **6.** $\frac{1}{2}, \frac{1}{3}$.   **7.** $1\frac{2}{3}, 1\frac{1}{4}$.   **8.** $14, -21$.

9. $\dfrac{c}{a}, \dfrac{c}{b}$.   **20.** $(3, 1)$.   **21.** $(3, 1)$.

22. $\left(4, 1\frac{1}{4}\right)$.   **23.** $(-3, -2)$.   **24.** $(-5, 4)$.

25.   (i) The line is parallel to the axis of x.
    (ii)   ,,   ,,   ,, y.
    (iii)   ,,   passes through the origin.
    (iv)   ,,   is the axis of y.
    (v)   ,,   ,,   ,, x.
    (vi)   ,,   ,,   ,, x.

26. $x + y = 0$.   **27.** $\dfrac{x}{x_1 + x_2} = \dfrac{y}{y_1 + y_2}$.   **28.** $-\frac{1}{2}$.

II. c.   (p. 40).

6. $3x - 5y + k = 0$.

7. (1) $5x + 3y = k$;   (2) $2x - 3y = k$;   (3) $my + x = k$;
    (4) $Bx - Ay = k$;   (5) $Bx + Ay = k$;   (6) $ax + by = k$.

8. $-\frac{1}{2}, 2$.   **9.** (1) $\frac{4}{3}$.   (2) $\dfrac{k}{h}$.   (3) 1.   (4) $-\frac{8}{5}$.   (5) 1.

10. (1) $-\dfrac{1}{b}$.   (2) $\dfrac{1}{b}$.   (3) b.   (4) $\frac{4}{3}$.   (5) $-\frac{2}{3}$.   (6) $\frac{4}{5}$.

17. Line (3) passes through the intersection of (1) and (2).

18. $3y + 4x = 0$.   **19.** $2(y + 6) - 3(x + 10) = 0$.

II. d.   (p. 41).

1. $3y - 5x = 0.$ **2.** $y = 3x - 11.$ **3.** $3x + 7y = 4.$ **4.** $y - x = 1.$

5. $bx - ay = 0.$ **6.** $5x + 4y = 0.$ **7.** $y = 0.$ **8.** $x = 6.$

9. $12x + 5y = 1.$ **10.** $x = 0.$ **11.** $30°.$ **12.** $90°.$

13. $\frac{1}{8}.$ **14.** $\tan^{-1}\left(\frac{27}{11}\right) = 67° \, 50'.$ **15.** $90°.$

16. They are equally inclined to Ox but in opposite directions.

17. $2y + x + 4 = 0.$ **18.** $4y + 3x = 0.$ **19.** $y - x + 7 = 0, \; y + x + 1 = 0.$

21. $4y + x = 3.$ **22.** $x + y = 13.$ **23.** $3x + 2y = 12.$ **24.** $y - x = 3.$

25. $\dfrac{a}{m} = -b = \dfrac{c}{k}.$ **26.** $\sqrt{3}(y + 4) = \pm(x + 2).$ **27.** $6y + 5x = 9.$

28. $2\frac{4}{5}.$ **29.** $\frac{7}{13}$ **30.** $0.$ **31.** $\sqrt{a^2 + b^2}.$

32. $1\frac{3}{5}.$ **33.** $9.$ **34.** $\dfrac{8}{\sqrt{17}}.$ **35.** $\frac{9}{5}.$

36. $\dfrac{2n}{\sqrt{l^2 + m^2}}.$ **37.** $\dfrac{14ab}{\sqrt{a^2 + b^2}}.$ **38.** $23y - 448x - 150 = 0.$

39. $13x - 19y + 14 = 0.$ **40.** $5x - 10y + 6 = 0.$ **41.** $25x - 13y = 0.$

42. $7x - 7y + 12 = 0, \; 2x + 2y - 1 = 0.$ **43.** $24.$ **44.** $\dfrac{c^2}{2m}.$

45. $k^2.$ **46.** $30.$ **47.** $17\frac{1}{2}.$ **50.** $p = \dfrac{C}{\sqrt{A^2 + B^2}}.$

51. $x - 7y + 2 = 0, \; 7x + y - 18 = 0.$ **52.** $4x + 2y - 1 = 0, \; 2x - 4y + 9 = 0.$

53. $x = 0, \; y = 0.$ **54.** $(a + b)x - (a + b)y + 2c = 0, \; x + y = 0.$

55. $y - b = \dfrac{m + m'}{1 - mm'}(x - a), \; y - b = \dfrac{mm' - 1}{m + m'}(x - a).$

56. $y = x \tan 2\theta, \; y + x \cot 2\theta = 0.$ **58.** $(-2, \, 5).$ **60.** $\left(-\frac{5}{3}, \; -\frac{5}{3}\right).$

II. e.   (p. 44).

1. $x = 1.$ **2.** $x = 2.$

3. A straight line through the intersection of the given lines, and making an angle $\tan^{-1} 2$ with the first.

4. $y = x \pm 1.$ **5.** $7x + y + 2 = 0.$

6. The straight line $Ax + By - C + D\sqrt{A^2 + B^2} = 0.$

8. The straight line $x + y = 6$ if the perpendicular lines are taken as axes.

9. The straight line $2Ax + 2By + C = 0.$ **10.** The straight line $x + y = 4.$

11. The straight line $x - y = 5$ if OA, OB are taken as axes of co-ordinates.

12. The straight line $3Ax + 3By + C = 0.$

15. $x - y + 12 = 0, \; 7x + 7y - 36 = 0.$

II. f. (p. 46).

1. $(16, 8)$.   **2.** (1) $(2, 0)$, $(3, 0)$; (2) $(0, 1)$, $(0, 3)$.   **3.** $y^2 = 4x$.   **4.** Yes.

5. $5x - 13y = 0$. The given line passes through the point $(-5, 2)$.

6. $(1, 2)$, $(1, -2)$.   **7.** $x^2 + y^2 = 13$.   **8.** $a \cos\left(\dfrac{a - \beta}{2}\right)$.

10. $y = 4$, $10y - 3x = 35$, $8y - 3x = 27$.   **12.** $\dfrac{3\sqrt{10}}{2}$.

13. $y - b = (x - a)\tan 3\theta$, $y - b = -(x - a)\cot 3\theta$.

14. $3x + y = 0$, $5y - 9x + 12 = 0$.   **16.** $2xy - kx - hy = 0$.

17. $ax - by = k$, $ax - by = a^2 - b^2$.   **18.** $y + 4x = 8$.

19. $\dfrac{x}{16} + \dfrac{y}{8} = 1$.   **20.** $(0, 2)$.   **21.** $\left(0, 3\tfrac{1}{2}\right)$.   **22.** $\left(6\tfrac{1}{2}, 6\tfrac{1}{2}\right)$.

23. $\left(\tfrac{1}{6}, \tfrac{1}{2}\right)$.   **24.** $\left(-\tfrac{71}{12}, \tfrac{29}{12}\right)$, $\left(-\tfrac{41}{2}, \tfrac{69}{2}\right)$.   **25.** $3x + 4y = 9$.

26. $12x - 5y + 19 = 0$, $12x - 5y - 33 = 0$.   **27.** $x + 2y = 9$.

28. $ax - by = 0$.   **29.** $\left(1, 2\tfrac{2}{5}\right)$, $\left(-3, 3\tfrac{1}{5}\right)$.   **30.** $\left(0, \dfrac{a}{m}\right)$.

31. $(a - a')\sin a$.   **32.** $y = \pm c$, $x = a \pm c$.

33. $(a - b)(x \cos a + y \sin a - p_1) = (p_1 - p_2)(x - a)$,
$(a - b)(x \cos a + y \sin a - p_2) = (p_1 - p_2)(x - b)$.

36. $\dfrac{x}{x_1} = \dfrac{y}{y_1}$.   **37.** $\left(4, 2\tfrac{2}{3}\right)$.   **38.** $23x + 23y = 84$.

39. $y(ab' - a'b) + ac' - a'c = 0$.   **40.** $x(ab' - a'b) + cb' - c'b = 0$.

III. a. (p. 53).

1. The two straight lines $x = a$, $x = b$.   **2.** The point (a, b).

3. The two straight lines $y = mx + c$, $y = m'x + c'$.

4. The point of intersection of the two straight lines
$$y = mx + c, \quad y = m'x + c'.$$

5. The two straight lines $x - y = 0$, $x + y = 0$.

6.   ,,   ,, $y = 0$, $y - x = 0$.

7.   ,,   ,, $x + y - a = 0$, $x + y + a = 0$.

8.   ,,   ,, $y - 3x = 0$, $y + 2x = 0$.

9. The point $(a, 0)$.   **10.** The two straight lines $x = c$, $y = c$.

11. $45°$.   **12.** $90°$.   **14.** $2\sqrt{10}$.   **15.** $45°$.

16. $\tan^{-1}\dfrac{\sqrt{b^2 - 4ac}}{a + c}$.   **18.** $a^2 + bc = 0$.

III. b.   (p. 58).

1. $y=\dfrac{1-x}{2}$ or $1-2x$.   **2.** $x=y-1$ or $2y-3$.   **3.** $y=\dfrac{x}{2}$ or $\dfrac{7x}{3}$.

4. -15.   **5.** 15.   **6.** 3.   **7.** 4.   **8.** -1 or $\frac{13}{2}$.

9. $\tan^{-1}\frac{1}{3}=18° \, 26'$.   **10.** $90°$.   **11.** $5x-7y+1=0,\ 7x+5y-1=0$.

12. $y^2-x^2+6x-9=0$.   **13.** $\dfrac{10a}{\sqrt{2}}$.   **14.** $a+b=0$.

15. (1) $q^2=4pr$; (2) $p+r=0$; (3) $q^2<4pr$.   **16.** $\tan^{-1}\dfrac{2\sqrt{h^2-ab}}{a+b}$.

18. $y-2x+3=0$ and $y-2x=0$.   **20.** The point $(0,\,4)$.

21. $90°$.   **23.** $(1,\,2)$.   **24.** $10x^2-2xy-13y^2=0$.

IV. a.   (p. 67).

1. $\sqrt{171}$.   **2.** 4.   **3.** 4.   **4.** $(2,\,3)$.   **5.** $(0,\,5)$.

6. $(2,\,4)$.   **7.** $\tan^{-1}\dfrac{\sqrt{3}}{2}=40° \, 54'$.   **8.** $45°$.   **9.** $30°$.

10. $\tan^{-1}\dfrac{3\sqrt{3}}{5}=46° \, 6'$.   **11.** $3x-2y=0$.   **12.** $5x+4y=0$.

13. $x+y=4$.   **14.** $3x-4y=12$.   **15.** $x=y$.

16. $y=x+2$.   **17.** $(y-y_1)\cos\omega+(x-x_1)=0$.

18. $y-y_1+\cos\omega(x-x_1)=0$.   **21.** $m+m'+2mm'\cos\omega=0$.

23. $-\frac{2}{5}$.   **26.** $y=0$.   **27.** $5y-x+17=0$.   **28.** $90°$.

IV. b.   (p. 72).

7. $r\cos(\theta-a)=p$.   **8.** $r=a$.   **9.** $r^2\cos2\theta=a^2$.

10. $r\sin^2\theta=4a\cos\theta$.   **11.** $r^4\cos2\theta=a^4$.   **12.** $x=a$.

13. $x\cos a-y\sin a=p$.   **14.** $Ax+By+C=0$.   **15.** $x^2+y^2=ax$.

16. $(x^2+y^2)^2=a^2(x^2-y^2)$.   **17.** $x^2+y^2=ax\cos a+ay\sin a$.

18. $r\cos\theta=a$.   **19.** $r\sin\theta=a$.   **20.** $\theta=a$.

21. $r\cos\theta=\pm a,\ r\cos(\theta-60°)=\pm a,\ r\cos(\theta+60°)=\pm a$.

22. $(2\sqrt{2},\,90°)$.   **29.** $p=r\cos(\theta+a)$.   **30.** $\theta=a$.

V. a.   (p. 76).

1. $2a$.   **2.** $4a$.   **3.** $(a,\,-a),\ 2a$.   **4.** $(-1,\,1),\ \sqrt{2}$.

5. $(2,\,0),\ 2$.   **6.** $(0,\,-3),\ 3$.   **7.** $(-g,\,-f),\ c$.   **8.** $(\frac{3}{4},\,-1),\ \frac{5}{4}$.

9. $\left(\dfrac{5}{2},\,\dfrac{7}{2}\right),\ \dfrac{\sqrt{74}}{2}$.   **10.** $(\frac{3}{2},\,-3),\ 2$.   **11.** $x^2+y^2-2ax+2ay+a^2=0$.

12. $x^2 + y^2 - 2ax - 2by = 0.$ **13.** $x^2 + y^2 - 6x - 6y + 9 = 0.$
14. $x^2 + y^2 - ax - ay = 0.$ **15.** $x^2 + y^2 - ax - by = 0.$
16. $x^2 + y^2 - 2x - 2y + 1 = 0,\ x^2 + y^2 - 10x - 10y + 25 = 0.$
17. $x^2 + y^2 + 8y = 0.$ **18.** $x^2 + y^2 - xx_1 - yy_1 = 0.$

V. b. (p. 83).

1. $3x + 4y = 25.$ **2.** $2y - x = 5.$ **3.** $y = \sqrt{3}x \pm 4.$
4. $y = x \pm 4\sqrt{2}.$ **5.** $y = 3x \pm 8\sqrt{10}.$ **6.** 2.   **7.** $2\frac{1}{2}.$
8. $Ax + By \pm a\sqrt{A^2 + B^2} = 0.$ **9.** $Bx - Ay \pm a\sqrt{A^2 + B^2} = 0.$
10. $x + y = a\sqrt{2}.$ **11.** $a + \beta m = 0.$ **12.** $(5,\ 7),\ (-1,\ -1).$
13. $4\sqrt{3}.$ **14.** $\left(-\dfrac{a}{\sqrt{2}},\ \dfrac{a}{\sqrt{2}} \right).$ **15.** $(\frac{1}{6},\ \frac{1}{2}).$
16. $x^2 + y^2 - 3x - 4y + 4 = 0.$ **17.** $(\frac{44}{13},\ -\frac{14}{13}).$
18. $(\frac{2}{5},\ \frac{1}{5}),\ (3\frac{3}{5},\ -2\frac{1}{5}),\ 4.$ **19.** $\left(\dfrac{2\sqrt{3}}{7},\ \dfrac{4}{7} \right).$

V. c. (p. 92).

1. $\left(-\dfrac{a^2 m}{c},\ \dfrac{a^2}{c} \right).$ **2.** $\left(\dfrac{a^2 x_1}{b^2},\ \dfrac{a^2 y_1}{b^2} \right).$ **3.** $\left(0,\ \dfrac{a}{2} \right).$
4. $(2a,\ 0).$ **6.** $k = 80,$ or $-20.$ **7.** $3x + 5y = 0.$
8. The straight line touches the circle at the point $(-a,\ -b).$
9. 1, 2.   **10.** $x^2 + y^2 - 18x - 8y - 3 = 0\ ;\ 9x + 4y + 3 = 0.$
11. $a(y - k) = \pm x\sqrt{k^2 - a^2}.$ **12.** $C = \pm a\sqrt{A^2 + B^2}.$
13. $x^2(1 + m^2) + 2mcx + c^2 - a^2 = 0.$ (1) $c^2 < a^2(1 + m^2).$ (2) $c^2 > a^2(1 + m^2).$
14. $x^2(x_1^2 + y_1^2) - 2a^2 x x_1 + a^2(a^2 - y_1^2) = 0.$
     (1) $x_1^2 + y_1^2 < a^2.$ (2) $x_1^2 + y_1^2 = a^2.$ (3) $x_1^2 + y_1^2 > a^2.$
16. $(Ah + Bk + C)^2 = r^2(A^2 + B^2).$ **18.** $xx_1 + yy_1 = x_1^2 + y_1^2.$
19. $\sqrt{4c^2 - 2(a - b)^2}.$ **20.** $x_1^2 + y_1^2 = 2a^2.$ **22.** $2\tan^{-1}\dfrac{r}{s}.$
23. (1) $g^2 = c.$ (2) $f^2 = c.$ (3) $f = 0.$ (4) $g = 0.$
24. The equation represents the common chord of the circles.
25. A circle passing through the common points of the given circles, and
     having its centre at the origin.
26. The straight line is the common chord of the two circles.
27. $x^2(1 + 2ag) + y^2(1 + 2bf) + 2xy(af + bg) = 0.$ **28.** $21x^2 - 70xy + 37y^2 = 0.$
29. $y = 6,\ 7y - 24x + 150 = 0.$ **30.** $(6,\ 8).$
31. $(x - 4)^2 + (y - 3)^2 = 4.$ **32.** $y = 0,\ 4x - 3y = 0.$
35. $(1 + m^2)(x^2 + y^2) - ax - amy = 0.$ **38.** $y = x \pm 3\sqrt{2}.$

39. $x^2 + y^2 - 4ax - 2y\left(\dfrac{c^2 + 2a^2}{c}\right) + 4a^2 = 0.$ **40.** $y - k = -\cot a\,(x - h).$

41. $x^2 + y^2 - 2by = a^2.$ $ax + \beta y - b\,(y + \beta) = a^2.$ $\left(-\beta,\ \dfrac{a^2 + \beta^2}{a}\right)$ is the fixed point.

V. d. (p. 97).

1. A circle on the line joining the fixed point and the centre of the given circle as diameter.

2. A circle whose centre is at the point of intersection of the given lines.

3. The circle $(x - 2)^2 + (y - 3)^2 = 25.$

5. The circle $3x^2 + 3y^2 - 2gx - 2fy - c = 0.$

6. The circle $(x - a)^2 + (y - \beta)^2 = c^2.$ **7.** The circle $(x - a)^2 + (y - a)^2 = a^2.$

11. The circle $x^2 + y^2 = a^2.$

12. The circle $(x - x_1)(x - x_2) + (y - y_1)(y - y_2) = 0.$

15. The circle $x^2 + y^2 = a^2 + b^2.$ **16.** The circle $x^2 + y^2 = a^2.$

17. The circle $x^2 + y^2 = a^2 + b^2.$

VI. a. (p. 100).

1. (4, 6), (6, 8), (8, 10).   **2.** 10 sq. ft.

3. $-2,\ 2,\ y - x = 2.$ **4.** $25°.$ **6.** Radii $2a,\ \dfrac{5a}{2}.$

VI. b. (p. 100).

1. $\left(\frac{10}{3},\ \frac{17}{3}\right),\ \left(\frac{17}{3},\ \frac{13}{3}\right).$ **2.** $x - y = 1.$

3. If the line makes an angle a with $\mathsf{O}x,$

   (i) $\mathsf{B} = -\cot a,\ \mathsf{C} =$ the intercept of the line on $\mathsf{O}x.$

   (ii) $\mathsf{A} = -\tan a,\ \mathsf{C} =$,,   ,,   ,, $\mathsf{O}y.$

   (iii) $\dfrac{1}{\mathsf{A}},\ \dfrac{1}{\mathsf{B}}$ are the intercepts on the axes.

   (iv) C is the perpendicular on the line from the origin ; $\mathsf{A} = \cos a,$ $\mathsf{B} = \sin a,$ where a is the inclination of this perpendicular to $\mathsf{O}x.$

4. $12x + 5y = 0,\ 5x - 12y = 0.$ **5.** $x^2 + y^2 - 2ax = 0,\ \left(\dfrac{2a}{5},\ \dfrac{4a}{5}\right).$

6. $\left(\frac{1}{2},\ -\frac{1}{2}\right),\ x + y = 0.$

VI. c. (p. 101).

1. $(-1, 2),\ (0, -2),\ -\frac{1}{2}.$ **2.** $\left(\frac{4}{3}, 1\right).$

3. $3x - 2y + 5 = 0,\ 19x + 9y = 0.$

5. $3y - x = 0,\ y - 3x = 0,\ \tan^{-1}\frac{4}{3} = 53°\ 8',\ x^2 + y^2 + 10x + 10y = 0.$

6. $2,\ \sqrt{21}.$ The point is outside.

VI. d. (p. 101).

1. $1\cdot2$ inches.   2. $-\dfrac{x}{2\cdot6}+\dfrac{y}{4\cdot3}=1,$ $-\dfrac{x}{2\cdot6}-\dfrac{y}{1\cdot5}=1,$ $\dfrac{x}{0\cdot9}-\dfrac{y}{1\cdot5}=1.$

3. $c\sin\mathsf{A}y+(c\cos\mathsf{A}-b)x=0$, when A is the origin, AB the axis of x, and the line is parallel to BC.

4. $8\cdot1$.   6. $2(x+y)=a(\sqrt{2}+1),\ \ 2(x+y)+a(\sqrt{2}-1)=0$

VI. e. (p. 102).

1. $6\frac{1}{2}$ sq. in.   2. $2,\ (1\cdot2,\ 1\cdot6)$.

4. $(f+g)^2$ is not less than $2c$.   6. $2ax-2(b-c)y=a^2+bc-b^2$.

VI. f. (p. 102).

1. $3x=4y$.   2. $\dfrac{aa'+bb'}{\sqrt{(a^2+b^2)(a'^2+b'^2)}}.$ 3. $\dfrac{y'-mx'-b}{\sqrt{1+m^2}}.$

4. $(-3\cdot4,\ 2\cdot2)$ 6. $3y=x,\ \ y=3x$.

VI. g. (p. 103).

2. 3.   3. $11y-3x+20=0,\ \ 11x+3y-30=0$.

4. $y=1-2x,\ \ 2y=1-x.$ $\cdot8$ and $\cdot45,\ \ \cdot6$ and $\cdot4,\ \ 0$ and $\cdot25$.

6. A circle of radius ma, passing through A, and having its centre in the line (produced) joining A to the centre of the given circle.

VI. h. (p. 103).

1. $(a-l)x+(b-m)y=0$.   2. $12x-5y=13,\ \ 12x-5y=39$.

3. $[(m+n)k-2mn]x+(m+n-2k)y=0$.   4. $\tan^{-1}\dfrac{\sqrt5}{2}=48°\ 11'$.

5. $12y+5x-5a=\pm13a$.   6. $x=0,\ \ \sqrt3y=\pm(x+3)$.

VII. a. (p. 104).

2. $\cos^{-1}\left[\dfrac{(a-a')^2+(\beta-\beta')^2-r^2-r'^2}{2rr'}\right].$

5. $x^2+y^2+6x-3y=0$.   6. $x^2+y^2-2hx-2ky+a^2=0$.

7. $x^2+y^2-ax-by+r^2=0$ 8. $x^2+y^2-2y-2\lambda(x-1)=0$.

9. $c^2=ab$.   10. $3x^2+3y^2-8x+29y=0$.

11. $8x-12y+5=0$.   12. $x^2+y^2-12x-5=0$.

VII. b. (p. 109).

1. $2x+3y=6.$ **3.** $y+6=0.$ **4.** The axis of y.

5. $x^2+y^2+\lambda(x-y)-c^2=0.$ **7.** $\left(\frac{7}{6},\ \frac{11}{6}\right).$

8. $7x^2+7y^2+6x-8=0.$ **9.** $x^2+y^2-4x+2y-12=0.$

12. $x+y=0.$ **13.** $x^2+y^2+\lambda(2x-y)-c^2=0.$

14. A system of circles having $x-2y-1=0$ for a common radical axis and passing through the intersection of this line and
$$(x-2)^2+(y-3)^2=4.$$

15. $x^2+y^2+\lambda(x+y)=25.$ **16.** $c^2+\lambda\lambda'=0.$

17. $x^2+y^2-8x-12y+12=0.$ Centre $(4,\ 6)$.   **18.** $x^2+y^2-4y=0.$

19. $(4,\ -2),\ (-4,\ 2).$ **20.** $(x\pm3)^2+(y\mp4)^2=49.$

VII. c. (p. 113).

2. $x^2+y^2+2xy\cos\omega-hx-ky=0.$

3. $\left[\dfrac{a^2(l-m\cos\omega)}{n\sin^2\omega},\ \dfrac{a^2(m-l\cos\omega)}{n\sin^2\omega}\right].$

4. $\cos^{-1}\frac{16}{25}=50°\ 12'.$ **5.** $\left(\dfrac{a}{1+\cos\omega},\ \dfrac{a}{1+\cos\omega}\right).$

6. $(g+fm)^2=c(1+2m\cos\omega+m^2).$ **7.** $150°,\ \sqrt{2}a,\ (2a,\ \sqrt{3}.\,a).$

8. $x^2+y^2+xy-6x-3y=0.$ **9.** $(a,\ 0).$

10. $x+2y=\pm2a.$ **11.** $4x+5y=0.$

12. $gx+fy=0,\ y(f\cos\omega-g)+x(f-g\cos\omega)=0.$

VII. d. (p. 118).

1. A circle of radius a, whose centre is at the point $\left(a,\ \dfrac{\pi}{2}\right).$

2. A circle of radius a, whose centre is at the point $\left(a,\ \dfrac{\pi}{2}-a\right).$

3. A circle of radius $\sqrt{2}$, whose centre is at the point $\left(\sqrt{2},\ \dfrac{\pi}{4}\right).$

4. A circle of radius a, whose centre is at the point $\left(a,\ \dfrac{\pi}{2}+a\right).$

5. $a^2q^2=1-2ap.$ **6.** $(\sqrt{5},\ \tan^{-1}2).$ **7.** A circle.

9. $\left(\frac{1}{2}\sqrt{a^2+b^2},\ \tan^{-1}\dfrac{b}{a}\right).$ **10.** $r^2\pm2r\cos\theta\sqrt{ab}-r\sin\theta(a+b)+ab=0.$

11. $3=2r\cos\left(\theta-\dfrac{\pi}{3}\right).$ **12.** $a\sqrt{3}=2r\cos\left(\theta+\dfrac{\pi}{6}\right).$ **13.** $\theta=\dfrac{\pi}{2}+a.$

14. $c\pm a=r\cos\theta.$ **15.** A circle passing through the point O.

16. A circle passing through the point O.

VIII. a. (p. 125).

4. $x^2 = 4a(y - a)$.   5. $18\frac{1}{4}$.

6. A parabola, whose focus is at the fixed point, and directrix the fixed line.

7. $(x + y)^2 - 22x + 26y + 25 = 0$.   8. $y^2 = 4ax$.

9. $y^2 = -4ax$.   10. $x^2 = 6by + 3b^2$.

11. Vertex $(2, 0)$, focus $(3, 0)$, axis $y = 0$, directrix $x = 1$.

12. ,, $(0, 3)$, ,, $(0, 4)$, ,, $x = 0$, ,, $y = 2$.

13. ,, $(-3, 3)$, ,, $\left(-\frac{9}{4}, 3\right)$, ,, $y = 3$, ,, $4x + 15 = 0$.

14. ,, $(-a, b)$, ,, $(0, b)$, ,, $y = b$, ,, $x + 2a = 0$.

15. ,, $\left(-2, -\frac{2}{3}\right)$, ,, $\left(-2, \frac{5}{6}\right)$, ,, $x + 2 = 0$, ,, $6y + 13 = 0$.

VIII. b. (p. 137).

1. $y = x + 2,\ x + y = 6$.   3. $(m + n)y = 2mnx + 2a$.   4. $(5, 10)$.

5. $y = \pm(x + a)$.   6. $12y = 8x + 27$.

7. It is a tangent at the point $(2, -3)$.   8. $\left(\frac{3}{2}, 1\right)$.

9. $xx_1 = 2b(y + y_1),\ 2b(x - x_1) + x_1(y - y_1) = 0$.   10. $ax + by = 0,\ bx - ay = 0$.

11. $2hy = [k \pm \sqrt{k^2 + 4ah}]x$.   13. $y + 2x = 7$.   15. $\dfrac{312\sqrt{6}}{25}$.

16. $2y = 5x + 1,\ 10y = x + 125$.   17. $3y + 4 = 0$.   18. $\dfrac{-4a(2a + x_1)}{y_1}$.

22. The straight line is a tangent at the point $(1, 1)$.   23. $177\frac{7}{9}$.

24. $\tan^{-1}\frac{3}{5} = 30° 58'$.   25. The straight line is a common chord.

VIII. c. (p. 143).

1. A parabola, co-axial with the given parabola, and having the same vertex.

2. $y^2 = 2ax$.

3. A co-axial parabola, having its vertex at the focus of the given curve.

5. A parabola whose focus is at the centre of the given circle and directrix parallel to the given line.

6. $y = m(x + a) + \dfrac{a}{m}$.   8. The parabola $y^2 = 2a(x - h)$.

9. If $y^2 = 4ax$ is the equation of the given parabola, the locus of **Q** is the parabola $y^2 = 4a(x + 2a)$.

10. A co-axial parabola, having its vertex at the focus of the given curve.

11. A straight line through the vertex.

12. A co-axial parabola, having its vertex at the focus of the given curve.

13. $\left(h - \dfrac{k^2}{8a}, \dfrac{k}{2}\right)$.   14. The parabola $x^2 = \dfrac{2a}{k}\left(y + \dfrac{ak}{2}\right)$.

15. A straight line parallel to the axis of y.

17. The tangent at the vertex.

18. (1) The straight line $y = 2k$.

(2) ,, ,, $x = \dfrac{k}{4a}$.

19. A parabola whose axis bisects the base at right angles.

20. The latus rectum.   **22.** The circle $x^2 + y^2 - ax = 0$.

VIII. d. (p. 146).

1. $m^3 y + m^2 x = a(1 + 2m^2)$.   **4.** $\left[\dfrac{a(m^2+2)^2}{m^2},\ \dfrac{2a(m^2+2)}{m}\right]$.

5. $(am^2 - 2am)$.   **6.** $y^4 + 8a^4 = 2ay^2(x - 2a)$.

8. The straight line $y + kx = 2ak + k^3$. **9.** The parabola $y^2 = a(x - a)$.

VIII. e. (p. 161).

4. $y = 2x + 1,\ y = x + 2$.   **6.** $y^2 = 4(a' - a)(x - a)$.

9. $\dfrac{4a}{\sin^2\theta},\ \dfrac{4a^2}{\sin^2\theta}$.

15. $(x - y)^2 - 2ax - 2ay + a^2 = 0$, or $\sqrt{x} + \sqrt{y} = \sqrt{a}$, when the curve touches the axes at points at a distance a from the origin.

18. $x_1 y^2 - 2y_1 xy + 4ax^2 = 0$.

19. If $y^2 = 4ax$ is the equation of the given curve, the locus is the parabola $y^2 = 2a(x - 4a)$.

30. $\dfrac{4a(1 + k)^2}{1 + 2k}$, where $k = \dfrac{\mathsf{PZ}}{\mathsf{PT}}$.

IX. a. (p. 165).

1. $(3, -1), (3, 3), (1, 1)$. Area 4. **2.** $612x + 34y - 217 = 0$.

3. (1) The circle touches the axis of y at the point $(0, a)$.

(2) ,, ,, ,, x ,, $(a\tan a, 0)$.

4. $x^2 + y^2 - 5x - 4y - 8 = 0$.   **5.** $(4x + 3y)^2 = 100(x - 1)$.

6. $2x + 7 = 0$.   **7.** $x + y = 2$.

IX. b. (p. 165).

1. $ay + bx = ab,\ \dfrac{ab}{\sqrt{a^2 + b^2}}$.   **3.** $x^2 + y^2 - 4ax - 2ay + 4a^2 = 0$.

4. $y = 0,\ y = x\tan 2a$.   **5.** $\dfrac{1}{12}$.

6. $y^2 = \dfrac{4ax}{9}$.   **7.** $(4a, 4a)$.

B.A.G.   Y

IX. c.   (p. 166).

2. Area $15\frac{1}{2}$. $\left(\frac{1}{62}, -\frac{159}{62}\right)$.   3. $11x - 7y = 0$.

4. A circle on the line joining (a, β), (a', β') as diameter.

5. $1\cdot58$, $-0\cdot96$. The other roots are imaginary.

7. If S is the given point, and **SY** the perpendicular on the given tangent, the locus is a circle on **SY** as diameter.

IX. d.   (p. 166).

1. $3x - y = 0$.   3. $5x + 12y = 169$, $5x + 12y = 143$.

4. $x^2(k^2 - 4k + 16) + 48xy + y^2(k^2 + 9k + 36) = 0$, $k = 4$ or $-\frac{13}{2}$.

5. $(b - a, \pm 2\sqrt{ab})$.   6. $y = 2$.   7. The parabola $y^2 = 16a(x + 2a)$.

IX. e.   (p. 167).

1. $4x - 3y = 26$. Area $4\frac{1}{6}$.   2. $\tan^{-1}7 = 82°$ nearly.

3. (1) $A^2 = 4C$.   (2) $C = 0$ and $A = 0$.   4. $13y^2 - 4x^2 = 0$. $\dfrac{4\sqrt{13}}{9}$.

5. $(3\cdot24, -4)$, $(-1\cdot24, -4)$.

IX. f.   (p. 167).

1. $y - x = a$, $y + 2x = 4a$, $\tan^{-1}3 = 71° \, 34'$.

2. $x^2 + 5xy + 2y^2 - 7x - 9y + 8 = 0$, $x - y = 0$, $7x + 9y = 8$.

4. $12y^2 + 8xy - 3x^2 = 0$, $\tan^{-1}\dfrac{4\sqrt{13}}{9} = 58° \, 2'$.

6. $b^{\frac{1}{3}}y - a^{\frac{1}{3}}x = 0$, $2b^{\frac{1}{3}}y = a^{\frac{1}{3}}x + 4a^{\frac{2}{3}}b^{\frac{2}{3}}$, $2a^{\frac{1}{3}}x = b^{\frac{1}{3}}y + 4a^{\frac{2}{3}}b^{\frac{2}{3}}$.

IX. g.   (p. 168).

1. $y = \dfrac{2x + 4}{3}$ or $\dfrac{-2x + 5}{3}$.   5. $xx_1 = a(y + y_1)$; $a(x - x_1) + x_1(y - y_1) = 0$.

7. $3x^2 = 4(y - 2)$.

IX. h.   (p. 168).

1. $38x + 19y - 53 = 0$.   2. $y + 6 = 0$, $y + 2 = 0$.   3. $(8, 1)$.

4. $3x + 4y = 30$, $3x - 4y = 30$.   5. $y = 2$.   6. $y = x$, $4y + x = 0$.

7. $\left(\dfrac{2a - cm}{m^2}, \dfrac{2a}{m}\right)$.

IX. k.   (p. 169).

1. $x = 2$.   2. $x = 0$, $x + 2 = 0$.   3. $\left(5\frac{1}{3}, 2\frac{2}{3}\right)$.

4. $x^2 + y^2 - 18x - 10y + 56 = 0$.   5. $(an^2, 2an)$.

7. The straight line is a diameter of the parabola.

IX. 1. (p. 169).

1. $(kx - hy)^2 = p^2(x^2 + y^2)$.

2. $y = 0$, $4y - 3x = 0$.

4. $a^2m^2 = c^2 - 2ac$.

5. $y + 2x + 1 = 0$.

6. $y - 2 = m(x + 2) + \dfrac{9}{4m}$.

7. $(3x + 4y - 12)^2 = 80(4x - 3y - 16)$.

X. a. (p. 176).

1. $\frac{3}{5}$, 6, 6·4.

2. $\frac{1}{2}$, 2, 3.

3. $\frac{1}{2}$, $2\sqrt{2}$, $3\sqrt{2}$.

4. $3x^2 + 7y^2 = 115$.

5. $9x^2 + 12y^2 = 25$.

6. $8x^2 + 9y^2 = 648$.

7. 5, 3.

8. $\dfrac{\sqrt{3}}{2}$.

9. $(x - c)^2 + y^2 = e^2x^2$.

10. $3x^2 + 4y^2 - 14x + 11 = 0$.

13. $(0, 1·5)$, $(0, -1·5)$.

14. $17x^2 - 2xy + 17y^2 - 104x - 140y + 446 = 0$.

X. b. (p. 184).

1. $y + 2x \pm \sqrt{19} = 0$.

3. The four points $\left(\pm \dfrac{3\sqrt{2}}{2}, \ \pm\sqrt{2} \right)$.

7. $Ct = \dfrac{a^2}{x_1}$.

9. $Ct = \dfrac{b^2}{y_1}$.

10. $y = 3$, $x + y = 5$.

12. $x = 12$, $y = x + 13$.

13. 20.

16. $\left(\dfrac{a^2}{\sqrt{a^2 + b^2}}, \ \dfrac{b^2}{\sqrt{a^2 + b^2}} \right)$.

17. $\dfrac{a^2}{\sqrt{a^2 + b^2}}$, $\dfrac{b^2}{\sqrt{a^2 + b^2}}$.

20. $x = 5$, $7x + 30y = 125$.

21. $\dfrac{a^2}{m^2} + \dfrac{b^2}{n^2} = 1$.

22. $gx + fy = 0$.

24. If $f^2 = bc$, the axis of y is a tangent.

26. The line is a tangent at the point (a, b).

27. $x = 3$, $2y = x + 5$.

X. c. (p. 190).

4. Centre at the point $(2, 0)$, semi-axes 2, 1.

5. ,, ,, $(-1, 1)$, ,, 2, 3.

6. ,, ,, $(0, 2)$, ,, 5, 2.

7. ,, ,, $(-2, 0)$, ,, 2, 1.

8. ,, ,, $(-1, 2)$, ,, 1, 2.

X. d. (p. 194).

3. $\left(-\dfrac{a^2m}{c}, \ \dfrac{b^2}{c} \right)$.

4. $\left(\dfrac{ax_1}{b}, \ \dfrac{by_1}{a} \right)$.

5. $\left(\dfrac{a^2}{c}, \ 0 \right)$.

6. The corresponding foci.

7. $3x - 14y = 31$.

10. $\left(\dfrac{a}{e}, \ -\dfrac{b^2}{aem} \right)$.

12. $a^2l^2 + b^2m^2 = 4$.

X. e.   (p. 202).

1. The straight line $10y + 3x = 0$.   **3.** $4x + 5y = 13$.

4. (1) $a^2y + b^2x = 0$,   (2) $a^2by - b^2mx = 0$,   (3) $ay + bx = 0$.

5. $5y + 2x = 0$.   **6.** $4y + x = \pm 4\sqrt{13}$.   **7.** $16y = 3x$.

8. $2\tan^{-1}\frac{1}{2} = 53° \, 8'$.   **12.** $\left(-y_1\sqrt{\dfrac{b}{a}}, \; x_1\sqrt{\dfrac{a}{b}}\right)$.

13. $axx_1 + h(x_1y + y_1x) + b(yy_1) = 1$.

16. $5y + 3x = k$.   **17.** An ellipse, foci C and S.

20. $\dfrac{x}{\sqrt{7}} = \pm \dfrac{y}{\sqrt{3}}$.   Length $2\sqrt{5}$.   **21.** $\left(\mp \dfrac{ay_1}{b}, \; \pm \dfrac{bx_1}{a}\right)$.

X. f.   (p. 206).

1. An ellipse whose foci are at the ends of the base.

2. The ellipse $x^2 + 4y^2 = a^2$.   **3.** The ellipse $\dfrac{2x^2}{a^2} + \dfrac{2y^2}{b^2} = 1$.

4. The ellipse $k^2x^2 + y^2 = c^2k^2$.   **5.** The ellipse $\dfrac{x^2}{a^2\sec^2 a} + \dfrac{y^2}{b^2\sec^2 a} = 1$.

6. The ellipse $\dfrac{x^2}{a^2} + \dfrac{y^2}{b^2} - \dfrac{hx}{a^2} - \dfrac{ky}{b^2} = 0$.   Its centre is at the point $\left(\dfrac{h}{2}, \dfrac{k}{2}\right)$, and it passes through the origin and the point (h, k).

8. The straight line $\dfrac{x\sin a}{a} = \dfrac{y\cos a}{b}$.   **9.** $kx^2 - 2xy = a^2k$.

10. $k(x^2 - y^2) - 2xy = k(a^2 - b^2)$.   **12.** The ellipse $4\left(x - \dfrac{a}{2}\right)^2 + y^2 = r^2$.

13. $\dfrac{a^2}{x^2} + \dfrac{b^2}{y^2} = 4$.   **16.** $4x^2 + 3y^2 - 6y - 9 = 0$.

17. The director circle $x^2 + y^2 = a^2 + b^2$.

19. An ellipse whose foci are the given focus and the centre of the given ellipse.

20. Two circles whose centres are at the foci of the ellipse, and whose radii are equal to the semi-major axis.

X. g.   (p. 225).

1. Foci distant 2·24 inches from the centre.

5. $(a\cos\theta\sec a, \; b\sin\theta\sec a)$.   **7.** $4x + y = 0$.

8. Foci $(0, 0)$ and $\left(-\dfrac{2e^2}{1 - e^2}p\cos a, \; -\dfrac{2e^2}{1 - e^2}p\sin a\right)$.

Directrices $x\cos a + y\sin a = p$, $x\cos a + y\sin a + p\dfrac{1 + e^2}{1 - e^2} = 0$.

10. $\left[\dfrac{2ae}{1 + e^2}, \; -\dfrac{a(1 - e^2)^{\frac{3}{2}}}{1 + e^2}\right]$.

12. $36x^2 + 100y^2 = 225$. The curve meets the axes at points distant $2\frac{1}{2}$, $1\frac{1}{2}$ inches from the middle point of AB.

13. (i) $\left(\dfrac{5}{2},\ 0\right)$.   (ii) $\left(\dfrac{5 \pm \sqrt{15}}{2},\ 0\right)$.   (iii) 5.   (iv) $\sqrt{\dfrac{3}{5}}$.

16. A concentric and co-axial ellipse, semi-axes $\dfrac{ab}{\sqrt{a^2 + b^2}}$, b.

20. The straight line $\dfrac{x}{l} - \dfrac{y}{m} = a^2 - b^2$.

24. Common tangents $ax \pm by = \pm \sqrt{a^4 + a^2 b^2 + b^4}$.   **25.** $\dfrac{b^2}{\sqrt{b^2 \sin^2 \alpha + a^2 \cos^2 \alpha}}$.

27. Intercepts $-y_1 \left(1 + \dfrac{a^2}{b^2}\right)$, $e^2 x_1$.

33. $2ab \sqrt{\left(\dfrac{x_1^2}{a^2} + \dfrac{y_1^2}{b^2} - 1\right)\left(\dfrac{x_1^2}{a^4} + \dfrac{y_1^2}{b^4}\right)} \Big/ \dfrac{x_1^2}{a^2} + \dfrac{y_1^2}{b^2}$.

XI. a. (p. 230).

2. $x^2 + y^2 + 2x - 2y + 1 = 0$, $x^2 + y^2 + 10x - 10y + 25 = 0$.

4. $\left(y - 2a \cot \dfrac{\theta}{2}\right)\left(y + 2a \tan \dfrac{\theta}{2}\right) + \left(x - a \cot^2 \dfrac{\theta}{2}\right)\left(x - a \tan^2 \dfrac{\theta}{2}\right) = 0$.

6. $x \pm \sqrt{2} \cdot y + 2 = 0$.

XI. b. (p. 230).

1. $8x^2 - 2xy - 15y^2 + 11y - 2 = 0$, $8x^2 - 2xy - 15y^2 = 0$.

2. $\dfrac{4a}{\sqrt{5}}$.   **3.** $\left(-\dfrac{9a}{8},\ \dfrac{3a}{4}\right)$.   **5.** $\left(\pm \dfrac{4\sqrt{5}}{5},\ \pm \dfrac{\sqrt{15}}{5}\right)$.

6. $7x^2 - 2xy + 7y^2 - 48\sqrt{2}x + 144 = 0$.   Major axis, 4.

XI. c. (p. 231).

1. $A = 1\frac{3}{8}$, $B = -2\frac{1}{8}$.   **2.** $\tan^{-1}\left(-\dfrac{\sqrt{7}}{3}\right) = 138° \, 35'$.

3. $\sqrt{35} = 5 \cdot 92$.   **4.** $y^2 = bx - ay$, $bx - ay = 0$.

5. $y = x + 5$, $x = 4$.   **6.** $\dfrac{x^2}{32} + \dfrac{y^2}{36} = 1$.

XI. d. (p. 231).

2. $x^2 + y^2 - x - 3y + 1 = 0$.   **3.** $y = \pm (x + 1)$.

4. If $y^2 = 4ax$ is the equation of the given parabola, the locus is the parabola $y^2 = a(x - 3a)$.

5. $axx_1 + byy_1 + g(x + x_1) + f(y + y_1) + c = 0$.

6. $21x + 16y = 79$, $\left(\dfrac{336}{79},\ \dfrac{112}{79}\right)$.   **7.** $16x + 33y = 0$.

XI. e. (p. 232).

1. $\left(\dfrac{c}{3},\ \dfrac{b}{3}\right).$ 2. $y\sqrt{h^2-a^2}=\pm a(x-h).$

4. The equations of the tangents are $\pm y\sqrt{2}=x-1.$

6. $\left(\dfrac{24}{17},\ \dfrac{18}{17}\right).$

XI. f. (p. 232).

1. $y-b=(x-a)\tan\dfrac{3\theta}{2},\ \ y-b=-(x-a)\cot\dfrac{3\theta}{2}.$

2. $y-x=4.$ 3. $x^2+4xy+4y^2-20x-10y+25=0.$

4. At an angle $\tan^{-1}\left(\dfrac{3}{5}\right)=30°\ 58',$ and at right angles.

5. $y+11x=0,\ \ y-x=0.$ 6. A straight line through P.

7. $y=mx\pm\sqrt{33m^2+21}.$ The locus is the circle $x^2+y^2=54.$

XI. g. (p. 233).

2. $x^2+y^2-[a+b-\sqrt{a^2+b^2}](x+y)+2[a^2+b^2+ab-(a+b)\sqrt{a^2+b^2}]=0.$

4. $\left(\dfrac{x_1^2+2ax_1+2a^2}{x_1},\ -\dfrac{4a}{y_1}\right).$ 5. $(-2,\ \pm 2\sqrt{3}).$

7. $(mx-y)\sqrt{a^2+m^2b^2}=m(a^2-b^2).$

XI. h. (p. 233).

1. 12.   2. $\left(x-\dfrac{25}{17}\right)^2+\left(y-\dfrac{70}{17}\right)^2=\dfrac{935}{289}.$

3. $\left(1,\ \dfrac{4}{7}\right).$ 4. $\dfrac{22}{3},\ -6.$ 5. $x^2+(y-b)^2=e^2y^2.$

6. $\left(\dfrac{a}{e},\ -\dfrac{b^2}{aem}\right).$ 7. $\dfrac{x^2}{4a^2}+\dfrac{y^2}{4b^2}=1.$

XII. a. (p. 237).

1. Eccentricity $=\dfrac{5}{3},$ distance between foci $=10,$ latus rectum $=10\dfrac{2}{3}.$

2.   ,, $=\sqrt{\dfrac{7}{3}}=1{\cdot}53,$,, $=2\sqrt{14}=7{\cdot}48,$,, $=\dfrac{8\sqrt{6}}{3}=6{\cdot}53.$

3.   ,, $=\sqrt{2}=1{\cdot}414,$,, $=8\sqrt{2}=11{\cdot}314,$,, $=8.$

4.   ,, $=\dfrac{13}{5}.$,, $=26,$,, $=57\dfrac{3}{5}.$

5. $\left(\pm\dfrac{15}{2},\ 0\right).$ 6. $(0,\ \pm 5).$ 7. $(x+c)^2+y^2=e^2x^2.$

10. $x^2-4xy+y^2+8x-16=0.$ 13. $x^2-y^2=a^2.$

XII. b. (p. 243).

1. $2y = 4x \pm \sqrt{65}$.   2. $7\frac{1}{9}$.   3. $\left(\pm \frac{5}{3}, \ \pm 2 \right)$.

4. $(6, \ -2)$.   5. $2y - 5 = \pm 3x$.   6. $x = 8, \ 25x - 12y = 56$.

7. $p^2 = m^2(a^2m^2 - b^2)$.   9. $\dfrac{\sqrt{840}}{5} = 5\cdot8$ nearly.

10. Q is a focus of the curve, and the straight line $x = 3\cdot2$ is the corresponding directrix.

14. $xx_1 - yy_1 = a^2$.

XII. c. (p. 250).

1. $2x - y = 0, \ \ x + 2y = 0$.   2. $\sqrt{3}(x - 2) \pm 2(y + 1) = 0$.

3. $x - y - 1 = 0, \ \ 2x + 3y - 2 = 0$.   4. $x - y = 0, \ \ 5x + 3y - 4 = 0$.

5. $x^2 + xy - 2y^2 = 4$.   6. $9y^2 - 72y - 16x^2 = 0$.

7. $2x - y + 2 = 0, \ \ 2x + y + 2 = 0, \ \ e = \sqrt{5} = 2\cdot24$.

8. $y + 2x = 1, \ \ y - 2x = 1, \ \ e = \dfrac{\sqrt{5}}{2} = 1\cdot12$.

9. $x + y - 1 = 0, \ \ x - y + 3 = 0, \ \ e = \sqrt{2} = 1\cdot414$.

10. $3x + 2y = 0, \ \ 2x - 3y = 0, \ \ e = \sqrt{2} = 1\cdot414$.

11. $3x - 2y = 0, \ \ 2x - 3y = 0, \ \ e = \dfrac{\sqrt{26}}{5} = 1\cdot02$.

12. $3x - 2y = 0, \ \ 2x - 3y = 0, \ \ e = \sqrt{26} = 5\cdot1$.

XII. d. (p. 253).

4. The centre is at the point $(0, 1)$; 2, 1 are the lengths of the semi-axes which are parallel to the axes of co-ordinates.

6. The centre is at the point $(-1, 1)$; $2\sqrt{2}$, 3 are the lengths of the semi-axes which are parallel to the axes of co-ordinates.

XII. e. (p. 258).

1. $(12, 3)$.   2. $16y - 75x = 0$.   3. $\dfrac{a^2}{l^2} - \dfrac{b^2}{m^2} = 1$.

4. $b^2 \sin\theta \sin\phi + a^2 = 0$.   7. $9y - 64x + 741 = 0$.   8. $\dfrac{5\sqrt{3}}{3}$.

9. $a^2 \tan\dfrac{\theta}{2} \tan\dfrac{\phi}{2} + b^2 = 0$.   10. $2x + 4y - 1 = 0, \ \ 2x - 4y + 3 = 0$.

13. $x + a = 0, \ \ y + b = 0$.   19. $48y - 125x + 481 = 0$.

XII. f. (p. 262).

8. $\left(n, \dfrac{1}{n} \right), \ \left(-\dfrac{1}{n^3}, n^3 \right)$.

XII. g.   (p. 265).

1. The rectangular hyperbola $x^2 - y^2 = 16$.

2. The hyperbola $a^2x^2 - b^2y^2 = (a^2 + b^2)^2$.

4. $(x^2 + y^2)^2 = a^2(x^2 - y^2)$.   **8.** The hyperbola $\dfrac{x^2}{a^2} - \dfrac{y^2}{b^2} = 1$.

10. Take **AB**, and the perpendicular from **P** on **AB** as axes of co-ordinates, co-ors. of **P** $(0, -k)$, **CD** $= a$.
The equation of the locus is $y(x - a) = ak$: a rectangular hyperbola whose centre is at the point $(a, 0)$, and whose asymptotes are $x = a$, $y = 0$.

12. The equation of the locus is $xy + ab = (bx - ay)\cot(a - \beta)$. A hyperbola whose asymptotes are the lines $\dfrac{x}{a} + \cot(a - \beta) = 0$, $\dfrac{y}{b} - \cot(a - \beta) = 0$.

14. The equation of the parabola is $2x^2 + ay = 0$.

20. The asymptotes are parallel to **AB** and **AC**, and the centre of the curve is the middle point of **AD**.

XII. h.   (p. 277).

5. The straight line $x(1 + m) + y(1 - m) + (a + bm) = 0$.

6. $Ax + By = 0$, $Bx + Cy = 0$. The lines coincide if $B^2 = AC$, and the given equation then represents two parallel straight lines.

8. $x \pm y = \sqrt{a^2 - b^2}$.   **11.** $2y - 3x = 0$, $y + x = 0$.

17. The hyperbola $4x^2 - y^2 = 4a^2$.   **21.** $y^2 = 2k(y - x)$.

23. $x^2 + y^2 + 2xy \cos 2a = a^2 - b^2$, where $\tan a = \dfrac{b}{a}$.

XII. k.   (p. 279).

1. Eccentricity $= \cdot 72$.   **4.** Eccentricity $= 1 \cdot 15$.

7. $\begin{cases} \text{Lengths of sides, } 9\cdot6,\ 15\cdot4 \text{ cm.} \\ \dfrac{4x^2}{625} + \dfrac{y^2}{121} = 1, \ \ x^2 + y^2 - 9\cdot1y = 36. \end{cases}$ **12.** $2, \frac{10}{19}$.

XIII.   (p. 291).

2. $\dfrac{l}{r} = e \cos(\theta - a)$.   **6.** $r \pm 2\sqrt{ab} \cos\theta - (a + b)\sin\theta = 0$.

7. $r = l(\sin\theta - 7\cos\theta)$.

8. A confocal and co-axial conic, whose eccentricity $= e \cos a$, and latus rectum $2l \cos a$.

9. $\tan^{-1}\left(\dfrac{1 + e \cos a}{e \sin a}\right)$.

14. If $2a$ is the given angle, the equation of the locus is
$$\frac{l \sec a}{r} = 1 + e \sec a \cos \theta.$$

18. $\theta = \dfrac{\pi}{4}, \quad r \cos\left(\theta - \dfrac{\pi}{4}\right) = \dfrac{a}{2}\left(\dfrac{1}{\sqrt{2}} \pm 1\right).$

20. (1) The two straight lines $\theta = \dfrac{\pi}{3}, \quad \theta = \dfrac{\pi}{2}.$

(2) The two straight lines $\theta = \dfrac{\pi}{3}, \quad \theta = \dfrac{\pi}{4}.$

(3) A circle, centre $\left(\dfrac{a}{\sqrt{2}}, \dfrac{\pi}{4}\right)$, radius $\dfrac{b}{\sqrt{2}}.$

When $b = 0$, the equation represents the point $\left(\dfrac{a}{\sqrt{2}}, \dfrac{\pi}{4}\right).$

21. $r = 4a \cos \theta \operatorname{cosec}^2 \theta.$

22. A straight line perpendicular to the initial line.

23. A rectangular hyperbola.   **24.** A parabola, whose focus is the pole.

25. $r^2 \sin 2\theta = a^2$, where $2a$ is the length of the transverse axis.

XIV. a.   (p. 299).

1. $x + y - 1 = 0,\ (0,\ 1).$ **2.** $x - y + 3 = 0,\ \left(-\tfrac{3}{2},\ \tfrac{3}{2}\right).$

3. $5y - 4x - 5 = 0,\ (5,\ 5).$ **4.** $2x + y + 3 = 0,\ (-2,\ 1).$

5. $y = B,\ \left(-\dfrac{B^2 + C}{2A},\ B\right),\ 2A.$

7. If OAB is drawn from the centre to cut the circle in A and the straight line in B, the middle point of AB is the vertex of the parabola.

9. The parabola $(x - 1)^2 = 4y.$ **10.** The parabola $y^2 = 12(x + 3).$

11. Vertex $(0,\ 2),$ axis along $Oy,$ latus rectum 4.

12. ,, $(4,\ -4),$ axis parallel to $Oy,$,,   4, directrix $y + 5 = 0.$

13. ,, $(2,\ -1),$,, ,, $Ox,$,,   3, focus $\left(2\tfrac{3}{4},\ -1\right).$

14. ,, $(-3,\ 2),$,, ,, $Oy,$,,   8, directrix $y = 0.$

15. Axis $x + y = 0,$ tangent at vertex $x - y = 0,$ latus rectum 8.

16. ,, $x - y = 0,$,, $x + y = 0,$,,   8.

17. ,, $x - y + 4 = 0,$,, $x + y - 1 = 0,$,,   4.

18. ,, $2x + y = 0,$,, $x - 2y - 1 = 0.$

The curve touches Oy at $(0,\ -1).$

19. Axis $4x - 3y = 0,$ tangent at vertex $3x + 4y = 0,$ latus rectum 4.

20. ,, $12x - 5y + 60 = 0,$,, $5x + 12y + 900 = 0,$,, $\tfrac{4}{13}.$

21. The curve touches the axes at the points $(3,\ 0),\ (0,\ 4).$

The focus is at $\left(\tfrac{48}{25},\ \tfrac{36}{25}\right).$ Latus rectum $\tfrac{288}{125}.$

XIV. b.   (p. 311).

1. $\left(-\frac{2}{7},\ -\frac{3}{7}\right)$.   2. $(1,\ -1)$.   3. $(1,\ -2)$.

4. $\tan^{-1}\left(-\frac{1}{3}\right)$, $\tan^{-1}3$ with Ox.   5. $\dfrac{8\sqrt{3}}{3}$, 8.

6. $(-1,\ 2)$, $y-x-3=0$, $y+x-1=0$.   7. The point $(1,\ 1)$.

8. A circle, centre $\left(-\frac{2}{3},\ 0\right)$, radius $\frac{4}{3}$.

9. A parabola, vertex $(1,\ 2)$, latus rectum 4, axis parallel to Oy. The curve lies on the positive side of Ox.

10. An ellipse through the origin, centre $\left(-\frac{3}{4},\ 0\right)$ and touching the axis of x at the origin.

11. An ellipse whose centre is at the origin. Its axes bisect the angles between the axes of co-ordinates and their lengths are 4 and $2\sqrt{2}$.

12. A hyperbola, centre $\left(1,-\frac{3}{2}\right)$, asymptotes $x=1$ and $4y-3x=3$. The curve passes through the origin and through the point $(2,\ 3)$.

13. The two straight lines $3x-y=0$, $3x-y=2$.

14. The circle on the line joining $(a,\ b)$, $(a',\ b')$ as diameter.

15. A parabola, axis $y-x=0$, tangent at the vertex $y+x+1=0$, latus rectum $\dfrac{\sqrt{2}}{2}$, $\left(-\frac{1}{2},\ -\frac{1}{2}\right)$ is a point on the curve.

16. A parabola whose axis bisects the angle xOy, latus rectum $\sqrt{2}$, vertex $\left(\frac{3}{4},\ \frac{3}{4}\right)$. The origin lies within the curve.

17. A hyperbola, centre $(0,\ 0)$, asymptotes $2x-y=0$, $x+y=0$, passing through the point $\left(\dfrac{a}{\sqrt{2}},\ \dfrac{a}{\sqrt{2}}\right)$.

18. The two parallel straight lines $x+3y+5=0$, $x+3y-1=0$.

19. A parabola, axis $x+y=0$ tangent at vertex $x-y=0$, directrix $4y-4x=1$, focus $\left(\frac{1}{8},\ \frac{1}{8}\right)$.

20. The parabola $y^2=8x$.   21. The parabola $x^2=y+8$.

22. The equation of the curve is $(ax-by)^2=(a^2-b^2)(ay-bx)$.

23. The two straight lines $y=2x$, $y=x$.

24. The two straight lines $x=1$, $3x-4y+5=0$.

25. A circle ; centre at the origin ; radius a.

26. An ellipse referred to its equi-conjugate diameters as axes of co-ordinates.

27. An ellipse ; centre $(1,1)$; semi-axes $\sqrt{3}$, $\sqrt{2}$; the major axis bisects $\angle xOy$.

28. A rectangular hyperbola, centre $(-a,\ -b)$, axes parallel to Ox and Oy.

29. The hyperbola $\dfrac{x^2}{4a^2}-\dfrac{y^2}{4b^2}=1$.

30. Two straight lines at right angles, $2x+y+1=0$, $x-2y+1=0$.

31. A hyperbola referred to two conjugate diameters as axes ; lengths of these semi-diameters 3, 2 ; asymptotes $2x-3y=0$, $2x+3y=0$; the curve passes through the points $(0,\ \pm2)$.

32. A hyperbola whose transverse axis bisects $\angle xOy$; semi-axes 3, 4.

33. A parabola symmetrically situated between the axes and touching them at the points $(0, a)$, $(a, 0)$.

34. An ellipse whose axes are the straight lines
$$3x - 4y + 12 = 0, \quad 4x + 3y - 12 = 0;$$
lengths of semi-axes 2, 3.

35. A rectangular hyperbola, centre $(1, 1)$; passing through the origin; asymptotes parallel to Ox and Oy.

36. An ellipse whose axes are the straight lines
$$3x - 4y + 12 = 0, \quad 4x + 3y - 12 = 0;$$
lengths of semi-axes 2, 1.

37. A rectangular hyperbola, centre $(\tfrac{1}{2}, \tfrac{1}{2})$; asymptotes parallel to Ox and Oy; passing through the points $(1, 1)$, $(0, 0)$.

38. A parabola touching Ox at $(2, 0)$, Oy at $(0, 3)$.

39. A parabola whose focus is at the origin, and whose directrix is the straight line $x \cos a + y \sin a - p = 0$.

40. A parabola, axis $4x + 3y + 15 = 0$, tangent at vertex $3x - 4y = 0$.

41. A rectangular hyperbola, centre $\left(-\tfrac{2}{13}, \tfrac{3}{13}\right)$; passing through the origin; asymptotes parallel to $y = \dfrac{-3 \pm \sqrt{13}}{2}\, x$.

42. A rectangular hyperbola, centre (a, b); asymptotes parallel to Ox, Oy; passing through the points $(0, b)$, $(a, 0)$, $(0, 0)$.

43. An ellipse, centre $(-1, 2)$; passing through the points $(3, 0)$, $(-1, 0)$. When the centre is at the origin it passes through the points $(0, \pm 2)$, $(\pm 4, 0)$.

44. A parabola; axis $x - y + 2 = 0$; tangent at vertex $x + y - 5 = 0$; passing through $(3, 3)$; latus rectum $2\sqrt{2}$.

45. A hyperbola, centre $\left(-\tfrac{1}{2}, \tfrac{1}{2}\right)$; semi-axes $\dfrac{\sqrt{3}}{4}$, $\dfrac{\sqrt{3}}{2}$; asymptotes
$$2x + 4y - 1 = 0, \quad 2x - 4y + 3 = 0;$$
axes parallel to axes of co-ordinates.

XVI. a. (p. 318).

1. The straight line $6x + 3y = 2h$.

2. $(x - 6r)^2 + y^2 = r^2$, $\quad xx' + yy' - 6r(x + x') + 35r^2 = 0$, $\quad \dfrac{6rx' - 35r^2}{x' - 6r}$,

$$BC = \dfrac{\sqrt{91}}{3} = 3\cdot 18 \text{ feet.}$$

6. The hyperbola $y(y - x) = x$; asymptotes $y + 1 = 0$, $y - x = 1$; passing through the origin.

7. The hyperbola $\dfrac{(x - a)^2}{a^2} - \dfrac{(y - b)^2}{4a^2} = 1$.

XVI. b.   (p. 318).

1. $\dfrac{2x}{a}+\dfrac{2y}{b}=n$, $\dfrac{x}{(n-1)a}+\dfrac{y}{b}=1$, where $OA=na$, and $OB=nb$.

2. $\left(\dfrac{ar'-a'r}{r'+r},\ 0\right)$, $\left(\dfrac{ar'+a'r}{r'-r},\ 0\right)$. $2x(a'+a)+a'^2-a^2=r'^2-r^2$.

3. $y+mx=2am+am^3$. $-\dfrac{2a(m^2+2)}{m}$.

6. An ellipse whose minor axis bisects the angle xOy.   Centre at the origin.   Lengths of semi-axes $2\sqrt{2}$, 2.

7. A hyperbola whose asymptotes are parallel to AB and AC.

XVI. c.   (p. 319).

1. $y-k=\sqrt{3}(x-h)$.

2. $x^2+y^2-2a(x+y)+a^2=0$; $\dfrac{1}{2}\sin^{-1}\left(\dfrac{1}{8}\right)$, $\dfrac{\pi}{2}-\dfrac{1}{2}\sin^{-1}\left(\dfrac{1}{8}\right)$.

4. The ellipse $y^2+3x^2=4a^2$.   **5.** Between $0°$ and $60°$.

7. The first equation represents an ellipse whose centre is at the point $(1, 0)$, semi-axes $\sqrt{5}$, $\dfrac{\sqrt{10}}{2}$. It passes through the common points of the circle and parabola.

XVI. d.   (p. 320).

1. $39x^2-96xy+11y^2-18x-74y-46=0$.

2. $(0, \pm\sqrt{-c})$ are the fixed points.   The centres are the points $(\pm\sqrt{c}, 0)$. $x^2+y^2-2\beta y=c$.

6. $10x+3y=0$.

XVI. e.   (p. 321).

1. $\dfrac{2\sqrt{h^2-ab}}{a+b}$, $\dfrac{2h}{a-b}$.

2. $x^2+y^2-2by=a^2$ is the equation of the system of circles through the points $(\pm a, 0)$.

7. (1) $(10, 13)$,   (2) $x=10$, $y-x=3$.

XVI. f.   (p. 321).

1. $35x^2-74xy+35y^2=-12$. $(7x-5y-4)(5x-7y+3)=-12$. $\left(\tfrac{43}{24}, \tfrac{41}{24}\right)$.

2. $x^2+y^2-2\lambda(x-1)-2y=0$.   **3.** $(am^2, -2am)$.   **5.** $90°$.

6. A rectangular hyperbola.

7. Vertex $\left(\tfrac{3}{2}, \tfrac{7}{2}\right)$; latus rectum $2\sqrt{2}$; axis $x-y+2=0$.

XVI. g. (p. 322).

1. $\dfrac{\pi}{2} - a.$ $\dfrac{n^2 \cot a}{l^2 - 2lm \cosec a + m^2}.$ 2. $45°,\ 45°,\ 90°.$

6. The hyperbola $x(x+y)=a^2.$ 7. $x^2 - y^2 = 0.$

XVI. h. (p. 322).

2. $x^2 + y^2 - 2x(mc \pm \sqrt{(1+m^2)(c^2 - a^2)}) = a^2.$

7. $\left(\frac{1}{13},\ \frac{8}{13}\right).$ Each semi-axis is $\sqrt{2}$ in length.

XVI. k. (p. 323).

1. Diagonals $x+2y=2a,\ 4x-2y=3a.$ Sides $4y=3(x-2a),\ x=0,\ x=2a.$

2. $x \cos \dfrac{a+\beta}{2} + y \sin \dfrac{a+\beta}{2} = a \cos \dfrac{a-\beta}{2}.$ 6. $7y^2 - 18xy - 7x^2 = 0.$

7. (1) $\left(-\frac{12}{25},\ \frac{12}{175}\right).$ (2) $x - 18y - 12 = 0,\ 17x - 6y - 60 = 0.$

XVI. l. (p. 324).

1. A straight line through A.   2. A straight line.

6. A hyperbola whose axes are parallel to those of the given curve, and whose centre is at the middle point of the line joining the centre of the given hyperbola and the given point.

7. Asymptotes $4x - 3y + 5 = 0,\ 12x + 5y + 13 = 0.$ Axes $56x - 7y + 65 = 0,$ $x + 8y = 0.$

XVI. m. (p. 324).

2. $r^2 - \dfrac{r \cos \theta}{a}(a^2 + c^2) - \dfrac{r \sin \theta}{b}(b^2 + c^2) + c^2 = 0.$

6. The ellipse $a^2y^2 + b^2x^2 = 4c^4.$

7. A hyperbola, centre $(2,\ 3)$; asymptotes $y=3,\ 3x - y = 3.$

XVI. n. (p. 325).

4. $af(x-f) + bg(y-g) = 0.$ 5. $\dfrac{x}{a} \cos \dfrac{\theta+\phi}{2} + \dfrac{y}{b} \sin \dfrac{\theta+\phi}{2} = \cos \dfrac{\theta-\phi}{2}.$

7. An ellipse, centre $(-1,\ 2)$; central equation $x^2 + 2xy + 4y^2 = 16.$ Points on curve $(3,\ 0),\ (-1,\ 0).$